Praise for *Cott*

"With *Cotton*, Stephen Yafa . . . employ[s] a playwright's sense of drama to show how cotton was woven into the American experience. With wit and intelligence, Yafa demonstrates how a good deal of history can be learned by following a single thread. . . . His ambitious narrative energetically summarizes the approximately 5,000 years from cotton's domestication to the latest genetically modified boll."

—*The Washington Post*

"Exceptional details abound in *Cotton*, Stephen Yafa's first-rate history of the humble fiber. . . . Sprawling and fascinating."

—*San Francisco Chronicle*

"After reading *Cotton* you may never look at your socks quite the same way again. Yes, cotton has a story to tell, and Yafa spins a good yarn. That scrawny plant has caused more death, destruction and bad behavior, and inspired more flights of human ingenuity and innovation than perhaps anything else that grows under the sun. *Cotton* is filled with colorful characters and powerful motives; the very best and worst of human character is revealed through a simple plant. Yafa succeeds in weaving a great deal of insight and research into a rich and thought-provoking tapestry, one that ultimately reveals our special relationship with cotton."

—*Milwaukee Journal Sentinel*

"An extraordinary book . . . This is not just good history, it is good reading. Yafa is a fine write with a keen eye for the telling details."

—*Louisville Courier-Journal*

"A readable, informative and often entertaining book about a complex and perhaps surprisingly important subject."

—*San Jose Mercury News*

"A history that touches on economics, race, science, fashion, and popular culture. . . . Yafa has a knack for unearthing unusual material that breathes life into his work." —*The Baltimore Sun*

ABOUT THE AUTHOR

Stephen Yafa, a novelist, playwright, video producer, and award-winning screenwriter, has written for *Playboy, Details, American Heritage, Rolling Stone,* and frequently writes about wine for the *San Francisco Chronicle*. He lives in Mill Valley, California. For more information, please visit his Web site, www.stephenyafa.com.

COTTON

THE BIOGRAPHY OF A REVOLUTIONARY FIBER

Stephen Yafa

Previously published as *Big Cotton*

PENGUIN BOOKS

PENGUIN BOOKS
Published by the Penguin Group
Penguin Group (USA) Inc., 375 Hudson Street, New York, New York 10014, U.S.A.
Penguin Group (Canada), 90 Eglinton Avenue East, Suite 700, Toronto,
Ontario, Canada M4P 2Y3 (a division of Pearson Penguin Canada Inc.)
Penguin Books Ltd, 80 Strand, London WC2R 0RL, England
Penguin Ireland, 25 St Stephen's Green, Dublin 2, Ireland (a division of Penguin Books Ltd)
Penguin Group (Australia), 250 Camberwell Road, Camberwell,
Victoria 3124, Australia (a division of Pearson Australia Group Pty Ltd)
Penguin Books India Pvt Ltd, 11 Community Centre, Panchsheel Park,
New Delhi – 110 017, India
Penguin Group (NZ), cnr Airborne and Rosedale Roads, Albany,
Auckland 1310, New Zealand (a division of Pearson New Zealand Ltd)
Penguin Books (South Africa) (Pty) Ltd, 24 Sturdee Avenue,
Rosebank, Johannesburg 2196, South Africa

Penguin Books Ltd, Registered Offices:
80 Strand, London WC2R 0RL, England

First published in the United States of America by Viking Penguin,
a member of Penguin Group (USA) Inc. 2005
Published in Penguin Books 2006

3 5 7 9 10 8 6 4 2

THE LIBRARY OF CONGRESS HAS CATALOGED THE HARDCOVER EDITION AS FOLLOWS:
Yafa, Stephen H., ——
Big cotton : how a humble fiber created fortunes, wrecked civilizations,
and put America on the map / Stephen Yafa.
p. cm.
Includes index.
ISBN 0-670-03367-7 (hc.)
ISBN 0 14 30.3722 6 (pbk.)
1. Cotton textile industry—United States—History.
2. Cotton manufacture—United States—History. I. Title.
HD9875.Y34 2005
338.4'767721'0973—dc22 2004057185

Printed in the United States of America • Designed by Nancy Resnick

To Bonnie

A sorry farmer on a sorry farm is a sorry spectacle. A good farmer on poor land and a poor farmer on good land are purty well balanced, and can scratch along if the seasons hit; but I reckon a smart and diligent man with good hands to back him is about as secure against the shiftin' perils of this life as anybody can be.

—Bill Arp, *The Uncivil War to Date,* 1903

Preface

The town of Lowell, Massachusetts, was as worn and shabby as an old sweater by the 1950s, when I was growing up in it. The stone and brick façades of its downtown buildings loomed over me behind a thin gray skin of grime and soot, still respectable but tattered by time, weather, and neglect. There were sunny parks and brooks and birch forests, too, and a high school football team with a terrific coach, Ray Riddick, that rarely lost. The city itself, however, was in a prolonged state of decline—officially a "depressed area," as were other cities along the Merrimack River north of Boston—particularly Lawrence and Haverhill—that once flourished as mill towns. About 100,000 people lived in and around Lowell, many of them Irish Catholics whose families migrated to the mills after the potato famine in Ireland a century earlier, followed by impoverished French Canadian farmers from Quebec a few decades later along with Greeks and numerous other nationalities. They all had two things in common: no money and a desperate need for work.

The main streets of Lowell, curving and twisting as in a maze, seemed inevitably to end up at the river, bridging over miles of interconnected canals that were no longer in use. At the end of these streets, you came to rows of massive rectangular brick buildings with tiny windows, mostly abandoned. They were intimidating and forbidding fortresses, four stories high, stolid, slowly decaying but, in their own way, formidable. These were the textile mills of Lowell, built during the first forty years of the nineteenth century, heralding the birth of the industrial revolution in America. Andrew Jackson, among several United States presidents, made a special trip to Lowell to view the phenomenon of these water-powered factories in person and to pay homage to them. A few were still in operation, even if marginally. By the time I was growing up they had become a source of both pride and embarrassment.

As schoolchildren we learned that, once upon a time, Lowell was a powerful center of Yankee ingenuity and industry. In these lumbering, fatigued buildings American textile manufacturing was born, freeing us for the first time from our dependence on British production facilities. In the way that money in the pocket defines self-reliance more meaningfully than rhetoric alone, our young nation was finally striking out on its own. Raw cotton from the South, spun and woven into fabric in our city, clothed the entire country, including its slaves, in the form of coarse, cheap cloth called osanburg. We learned, too, that Lowell's factory girls, a group of young women fresh off the farm, distinguished themselves nationally as early mill workers with a bent for poetry, essays, and related cultural accomplishments—gathered in a monthly magazine, the *Lowell Offering*, that won praise from Charles Dickens among many others.

Most of us listened to and then quickly forgot about the city's illustrious heritage. There was a reason. None of it was present in our own lives. Lowell and Lawrence were soiled, depleted cities that were barely hanging on. Our own neighborhoods might be clean and cheerful, but the city itself was our ailing uncle, too old to reinvent himself, hobbled and outmoded, unsophisticated, barely cultured—but against all odds, endearing and even lovable. Thirty miles south of us, Boston glistened with the sparkle and sass of a large metropolis. As young teenagers we occasionally ventured forth, stared up at tall buildings, and felt like bumpkins.

That was Lowell, an industrial community with no thriving industry, a farming community with no farms, sitting beside a river that was so polluted it smelled like rancid milk and glinted with a greenish, mocha, blue, or red sheen, depending on which toxic dyes had most recently been dumped into it by the vatload. Swollen dead fish floated by belly-up under its bridges. Its numerous barrooms sported large jars of pickled pigs' knuckles and served corned beef and cabbage to the declining population of mill workers, many of whom wandered in at 6 A.M. after their night shift. There was little hope for employment: the textile mills had long ago relocated to North and South Carolina and other states below the Mason-Dixon line.

The Boston Red Sox were more than our baseball team; they were our collective attitude toward the future: it stank. Like Lowell, the BoSox once had it all and lost it so long ago that almost nobody alive could remember exactly when that was. The past was a painful reminder of glory long gone, while the present was a struggle not to be pummeled by stronger, more agile competitors—like the loathed New York Yankees—and the days and years to come, well, don't hold your breath for the Sox to win the World Series any time soon. Lowell was the town that prosperity forgot.

Jack Kerouac grew up in Lowell and drank himself to death at an early age. There may not be an exact cause-and-effect relationship between those two events, but consider this passage from his novel, *Dr. Sax*, which he set in Lowell:

> There was something wet and gloomy . . . something hopeless, gray, dreary, nineteen-thirty-ish, lostish, broken not in the wind a cry but a big dull blurt hanging dumbly in a gray brown mass of semi-late-afternoon cloudy darkness and pebble grit . . . something that can't possibly come back again in America and history, the gloom of the unaccomplished mud-heap civilization.

Kerouac ran away early—to New York and Columbia University. Few in Lowell paid him much attention for years until finally he got too famous to ignore. In the mid-1980s, a park and commemorative sculptures were built to honor Jack. By then he had passed into legend; mythology and merchandising had transformed this bitterly unhappy native son into a symbol of spontaneous exuberance, and no doubt Kerouac himself would have been deeply amused to learn that Lowell was honoring his memory, since he savaged the city at every turn.

Coming along a generation after Kerouac in the '50s, I didn't share his antipathy, but I, too, picked up on the pebble grit and mud-heap gloom. One summer, at sixteen, I got a job for $1.10 an hour in a converted textile factory located where the Merrimack Canal emptied its fetid contents into the river. In this monolithic brick-faced building, which a century earlier had housed whirring looms and spindles, shoe counters were now being manufactured. A shoe counter is the stiff, rounded insert that stabilizes the heel of the shoe. Made from cellulose material, the counters were cut and molded into horseshoe wedge shapes on loudly clanking machines, dipped in a hot wax solution, then brought to us in the packing department, where they were dumped in large wire-mesh bins. Our job was to scoop up counters, insert one inside another back-to-front until half a dozen were stacked in this fashion, and then pack them in boxes. The hot wax coating stung at first; within a week I'd begun to build up calluses.

While the working conditions were less onerous than those encountered by the mill women of Lowell a century earlier, they were not as different as might be expected. Naked bulbs barely illuminated the vast interior, casting harsh shadows. We stood in the same spot all day—on wood floors embedded with generations of grime, beside metal posts that seemed to sweat grease, under small windows that opened only a few inches. In fierce summer heat with little ventilation, we were surrounded by the scent of the smoking oil used to lubricate the loud cut-and-dye machines that stamped out the counters. There was a pervasive atmosphere of listless squalor that sapped our energy and dulled our senses as soon as we entered.

I was the lucky one in our crew of teenaged packers. I was quitting in six weeks to go back to school. My pallid coworkers, also about sixteen, expected to go nowhere. They were shackled to manual labor possibly for life with no incentive to accomplish more even if they had the ambition. They viewed me, temporarily employed through a family connection to the factory owner, as a hostile alien. I had a future, an opportunity for a college education, and parents who were willing to pay for it. They had Elvis, chopped and channeled deuce coupes with spinner knobs on the steering wheel, *Confidential* magazine, pink-and-black pegged pants and duck's ass haircuts (long side strands joined to form a vertical tailfin behind the head and held in place by cement-hard Stickum hair pomade). One of my coworkers tattooed the name of his girlfriend on his inner arm with the point of a ballpoint pen. Her name was Mickie, but he misspelled it as Mickey, and when he tried to correct his mistake he jabbed the pen over the last two letters during a lunch break and punctured a vein.

That was Lowell, sinking into oblivion, circa 1957, barely more than a century after its founding—and at that time a sad testament to the power of cotton to create wealth and happiness at one moment and despair the next. As a teenager growing up with the hulks of those defunct mills staring blankly into the wind like Easter Island totems, I knew next to nothing of Lowell's illustrious past. Only years later did that persistent memory of those monolithic relics, so vivid from childhood, encourage me to investigate their origins.

I discovered that my hometown came into existence because of an act of enlightened theft. Francis Cabot Lowell, for whom the city was named after his death, stole all the mechanical designs for the complex textile machinery from intractable English mill barons. When the owners would not let Lowell leave

Manchester, England, with a license to re-create their mills in America, or allow him even to sketch a single drawing, he did the best next thing: Lowell committed the schematics of those intricate spinning machines and power looms to his photographic memory and, with the help of an industrial engineer, built them from scratch back in Massachusetts.

If there is anything I admire, it is outrageous, scandalous behavior of that sort. The story of a city that owed its life to larceny quickly became even more interesting when I looked into the details of Lowell's place in the tensions that led to the Civil War and of the tight bond between New England's mill magnates—the Lords of the Loom—and their Southern counterparts, the Lords of the Lash, who were slave-owning wealthy cotton plantation owners.

Cotton, I quickly discovered, is simply too valuable a crop to inspire only noble behavior, and too easily grown to invite self-restraint. It lends itself to greed, opportunism, hypocrisy, irrational passion, murderous rage, and episodes of brilliant creativity—all the elements I look for in a gripping tale. This one began in my own backyard and quickly grew into a chronicle of the past and present life of American cotton—a plant that revolutionized the way we live today. Someone had to write it; there were simply too many good stories to tell.

Contents

COTTON

Introduction

For a scrawny, gangling plant that produces hairs about as insubstantial as milkweed, cotton has exerted a mighty hold over human events since it was first domesticated about 5,500 years ago in Asia, Africa, and South America. Cotton rode on the back of Alexander the Great all the way from India to Europe, robed ancient Egyptian priests, generated the conflicts that led to the American Civil War, inspired Marx's and Engels' *Communist Manifesto,* fooled Columbus into thinking he had reached Asia, and made at least one bug, the boll weevil, world famous. It also created the Industrial Revolution in England and in the United States, motivated single American women to leave home for the first time in history, and played a pivotal role in Mahatma Gandhi's fight for India's independence from British colonial rule. In these pages I trace the empires cotton built and destroyed, the fortunes it created, and the revolutions it stirred up along the way as it journeyed west from India to continental Europe, then to Great Britain, and from there to the United States.

While I focus on cotton in America, it truly belongs to the world. Forty billion pounds a year grow on about seventy-seven million acres in eighty countries according to the International Cotton Advisory Committee. In Ghana, on the West African coast, mourners wrap themselves in vibrant red *kobene* cotton cloth to express their close bonds to the deceased. In Ahmadabad, India, where Gandhi held his first fast in 1918 in support of textile workers, exquisitely subtle silk-screened cotton saris hang to dry high aboveground from hundreds of bamboo racks arranged like scaffolding; in Guatemala, women gather each morning and socialize in village circles as they weave and embroider magnificently ornate blouses called *huipiles* using *cuyuscate,* naturally colored cotton that grows in soft greens, browns, yellows, and chalk grays. An entire industry in Peru is devoted to the organic cultivation of coffee-latte-hued cotton.

Just about everyone on the planet wears at least one article of clothing made from cotton at some point during the day; inevitably, by-products of the plant show up as well in something that person is doing, whether eating ice cream, changing diapers, filtering coffee, chewing gum, handling paper money, polishing fingernails, or reading a book. The source of cotton's power is its nearly terrifying versatility and the durable creature comforts it provides.

Cotton is family. We sweat in cotton. It breathes with us. We wrap our newborns in it. In fact, we pay cotton the highest compliment of all: we don't go out of our way to be nice to it. Look in your closet. The rumpled things on the floor are most probably cotton—soiled shirts and khakis, dirty housework clothes and muddied socks that rise up in dank mounds ready to be baptized with detergent and reborn in the washer, fresh and clean as new snow. Linen, silk, wool—uptown fabrics to be sure, on display in the magnificent Bayeux Tapestry woven shortly after the Norman Conquest or in priceless Aubusson rugs, but not happy to be scrubbed with sudsy hot water and churned like butter in a dryer. Those fibrous divas demand attentive coddling while cotton, the sword carrier, needs only three squares a day and a pair of shoulders to drape itself over. Cotton is the fabric wool would be if it were light enough for summer and didn't shrink to toddler-size in the dryer; it's what silk would be if it gracefully absorbed sweat, and what linen might aspire to if it didn't wrinkle on sight.

Contemporary man-made technical polyesters, all created

from petroleum by-products, have come along in their second or third generation to trump cotton as the preferred fabric for strenuous outdoor activity and gym wear. Not a problem. Cotton manufacturers responded by wedding one of the world's oldest fabrics to futuristic nanotechnology. It is there now to protect you and me from the red wine spilled on our stain-resistant nano-cotton Dockers, and as the new century unfolds, the merger of textile and technology will soon be producing bulletproof clothing as light and airy as a Hawaiian shirt. Cotton leads the way.

The common thread in all this (wayward puns, by the way, are inevitable) is cotton's extraordinary range of practical use. The 500,000 fibers in every cotton boll, or pod, that are spun and woven or knit into fabric for home furnishings, linens, industrial coverings, and apparel account for less than half of the plant's output. The seeds inside each boll, which is about the size of a walnut, make up 65 percent of the yield from a harvest.

We consume pressed cottonseed oil directly in hundreds of supermarket products—Campbell's soups, Pepperidge Farm cookies, potato chips, crackers, marinades, snacks, and salad dressing, to name but a few. Procter & Gamble created Ivory Soap from cottonseed in the nineteenth century after a man named William Fee invented a way to knock the kernels free from their hard hulls. That discovery led as well to hydrogenated shortening, or crystallized cottonseed oil better known as Crisco. In its normal state, cottonseed is poisonous to humans and all other nonruminant animals; it contains a toxic pigment, gossypol, that helps protect the plant against insects. However, chemical processing produces a protein-rich flour that is sufficiently low in free gossypol to render it suitable for human consumption. In Central America and West Africa it becomes a beverage fed to children to prevent malnutrition. Low doses of gossypol have also been used for centuries in China as a male contraceptive: it destroys the lining of tubules in the testicles where sperm are produced.

Charles Darwin could have learned everything he needed to

know about evolutionary adaptation from a cotton plant. Like Proteus, the Greek god who changed his identity at will, this swamp-loving mallow, a shrub called *gossypium,* can be as tough as braided anchor rope at one moment or as fine as the fabled sheer muslin of ancient Bengal at another. When it is not helping to clothe us, it is likely to be slipping unnoticed into the things that we use to blow each other apart. Short fibers on the cotton seed, called linters, or, less formally, "fabulous fuzz," supply the cellulose used in dynamite and other explosives, rocket propellants, shoes, handbags and luggage, book bindings, industrial abrasives, and also in plastics and fingernail polish; chemically treated and ground into pulp, linters show up in food casings for bologna, sausages, and hot dogs; they thicken ice cream and smooth makeup and find their way into lacquers, paint, and automotive parts. They are also processed into materials used in photographic and X-ray film, envelope windows, and recording and transparent tapes. The plant's discarded leaves, fibers, and stalks, which cotton growers call trash, get cleaned and become mattress stuffing for human use and barn bedding for dairy cows, while cottonseed meal feeds livestock and dairy cattle. Recycled remnants from blue-jean factories make up 75 percent of the content of United States paper currency. There are three-fourths of a pound of cotton in each pound of dollar bills. Nothing from stem to stamen goes unused in a cotton plant.

For cotton, that range of accomplishment also extends beyond the pragmatic into realms of human activity where most other plants never get past the gate: music, literature, art, pop culture, and romance. *Gone With the Wind* is as much homage to the antebellum culture of cotton as to the glory that once was Atlanta. Elsewhere in the South, at another place and time, cotton's culture also became a major contributor to the blues when slave field-hollers melded with church music and took a secular turn toward human heartbreak. Blacks fleeing the impoverished cotton fields and oppressive racism of the Mississippi Delta added a raw authentic voice to popular culture, bending their guitar strings to cry out their despair. In a lighter moment, cotton's indomitable insect foe, the boll weevil, gave rise to a host of tunes that turned tragic devastation into street entertainment. "The Boll Weevil say to the Farmer, 'You can ride in that Ford machine,' " Leadbelly sang, " 'but when I get through with your cotton, you can't buy no gasoline. You won't have no home, won't have no home.' "

By the early years of the twentieth century, cotton had insinuated itself into regional common language as well. Born in the South, if you heard anyone say she was "fair to middlin'," you knew things for her were better than just okay. While strict middling has long been the whitest, cleanest cotton that commands the best market price, colloquially "fair to middlin' " middle-grade cotton came into common use as an understated way to tell friends and neighbors the speaker was feeling on top of the world. The fiber easily embraced love-talk too. If you "cotton to" a fella, you're stuck on him. You're stuck because cotton seeds are sticky, which is why you say you cotton to him, because if you weren't stuck on him . . . well, you get the idea. And when you're in high cotton, hon, you've got the world on a string!

There are unconscious references, too, that enrich our daily conversations, whether we are following the thread of an idea, weaving a plot, spinning outrageous yarns, or even knitting our brow over an unraveling relationship. It may be that we habitually borrow from the textile crafts that transform fluffy cotton hairs into fabric simply because they mimic the way our minds work—one thought linking to another to form a complete image or idea, much the way individual warp and woof yarns intertwine to create whole cloth. One way or another, it sometimes seems, cotton is always with us.

Like T.S. Eliot's J. Alfred Prufrock, "deferential, glad to be of use / Politic, cautious, and meticulous," the fiber appears to have made its way in the world by combining protean service with humility, abetted by an outsized need to please. As many have learned the hard way, that can be a lethally deceptive disguise. No legal plant on earth has killed more people by virtue of the acrimony and avarice it provoked. No other plant ignited a civil war that sent more American men to their deaths than all other wars combined. Cotton's manufacture into cloth sentenced hundreds of thousands of young orphaned English children to labor imprisonment in the squalid, filthy cotton textile factories of nineteenth-century Manchester, England. Across the sea in the fields of the American South, it enslaved generations of uprooted Africans who were robbed of their freedom and brutalized as a matter of course.

Later, in the twentieth century, cotton crops became one of the world's most persistent and heaviest users of toxic pesticides, creating lethal environmental and health hazards that continue to plague many countries. Cotton has been responsible, too, for an ecological disaster of epic proportions in Central Asia, where rivers feeding the Aral Sea, one of the world's largest inland lakes, were diverted to provide crop irrigation and in the process brought human misery and the massive destruction of flora and fauna to a vast populated area.

At the other extreme, cotton manufacture stimulated innovations and inventions that transformed creaky, rural late-eighteenth-century England into the world's greatest industrial power before leaping the Atlantic to spur a struggling new democracy into becoming an equal among giants. Until cotton changed our country's fate, we were a pesky upstart crow with grand notions and empty pockets. After Eli Whitney invented a gin to separate green cotton's sticky seeds from its valuable hairs, or fibers, oceans of fluff erupted all over the South, and suddenly our fledgling nation owned a crucial piece of the action in international trade. As the South began to supply raw material to England's booming textile mills as well as to our own, a superpower was born. Although a colonial rebellion and a constitution gave birth to the United States, cotton added the economic muscle that any country or individual needs to achieve true independence. It was a revolutionary act all its own.

The history of cotton is filled with similar tectonic disruptions of the status quo that the fabric and fiber either instigated, accelerated, or at the very least encouraged. More revolutions are imminent as we continue to plant and export cotton seeds that carry gene-altered biotechnology to India, Brazil, and numerous other countries. Cotton is leading the way for changes (and controversies) that will redefine and dominate agriculture in the coming decades.

When we export genetically modified seeds to countries where controls over its use are difficult or impossible to enforce—most of the countries on the planet, that is—we run a serious risk of inviting ecological disaster. Third World nations like Mali and its neighbors are seizing upon American cotton and the lavish subsidies our government awards to its growers as evidence that we are as selfish and callous as our enemies maintain. Rancor over the dumping of subsidized American cotton on the world market at artificially low prices brought the 2003 World Trade Organization summit in Cancun to a screeching halt. As such pressures mount, major changes may be imminent in our trade relations as well. Yet the rewards, financial and human, for many may be worth the gamble.

Cotton always seems to force the issue, whatever the issue may be. That can be bad for nations and individuals caught up in a tangle of opposing motives and goals, but to a writer like myself it's also the stuff of compelling conflict, and it accounts for much of what drew me to the subject. I was also drawn to the extraordinary feats of imagination and ingenuity required to convert a fluffy mass of nothingness into something of substance. Looking at an opened boll of flimsy lint, the last thing you can envision is the tightly woven gold-encrusted flowing robe presented to Cortez by a Yucatan chief in 1519, or the magnificent pre-Andean textiles of Peru—or for that matter even a pair of sweat socks. Stepping across a stream in Lancashire, you would hardly guess that more than two hundred years ago a man named Richard Arkwright was able to harness its hydraulic power to drive intricate mechanical spinning machines that had never before existed and, by doing so, create the Industrial Revolution.

Whether you find the story of cotton to be a tribute to man's remarkable ability to achieve or a cautionary tale, you'll likely come away as I did with a profound respect for the power of the plant. "You dare not make war upon cotton!" South Carolina senator James H. Hammond thundered on the floor of the U.S. Senate in 1858. "No power on earth dares make war upon it. Cotton is king!" I discovered the truth to be a little more insidious. Kings are mere mortals; they die and the world keeps turning. This plant, by contrast, has eternally rewarded and punished with the haughty abandon of a capricious god. It has also stirred up more mischief than any penny-ante royal no matter how venal, and yet it remains so casually seductive in its look and feel that we are willing to forgive its sins even as we continue to pay for them. Some of us have a lot to learn from cotton.

ONE

Spun in All Directions

The trees from which the Indians make their clothes have leaves like those of a black mulberry. . . . They bear no fruit, indeed, but the pod containing the wool resembles a spring apple.

—Theophrastus (372–287 B.C.)

"They . . . brought us parrots and balls of cotton and spears and many other things which they exchanged for the glass beads and hawks' bells. They willingly traded everything they owned . . . ," the sea captain wrote with wonderment in his logbook about the natives he had encountered. Soon after that, the same fiber caught his eye again: "Here we saw cotton cloth, and perceived the people more decent, the women wearing a slight covering of cotton over the nudities."

While Christopher Columbus appreciated modesty and proved to be a keen observer of local customs, his interest in the cotton apparel and yarn he discovered among the Arawak Indians he met in the Bahamas in 1492 went far beyond cultural anthropology. Although he was the son of a skilled Italian weaver, Columbus was also a man in debt. Having persuaded King Ferdinand and Queen Isabella to finance his quest for the spices and gold of the Orient, he felt pressured to return with the bounty he promised in order to justify the crown's investment. Then, too,

Columbus had a personal stake in the venture—10 percent of the profits, governorship over the newfound lands, and the adulation that would go with his new self-promotional title: Admiral of the Ocean Sea.

The Spaniards of his day knew cotton. Arab and Moorish traders had introduced the cloth to Spain from North Africa in the eighth century. Grenada, Cordova, and Seville were its celebrated weaving centers and it was now being grown in southern areas of the country. Fustian, a blend of cotton and wool or linen, had long been a popular medieval dress fabric; "fustañeros," the weavers of Spanish fustian, organized themselves into a guild in the thirteenth century. Streets where they carried on their trade still bear their names.

Those first Muslim traders, it was widely known, often caravanned their cotton all the way from India and the West Indies. Over the centuries, travelers had confirmed its cultivation and conversion to fabric in Asia and in the Middle East (known as the Levant) as well, but it was the Asian connection that Columbus seized on to bolster his claim that he had in fact reached his fabled lands of riches: gold and cotton—where there was one, there was sure to be the other. Marco Polo's journals from the thirteenth century spoke of lavish displays of the precious metal in China and to the south. Blankets and shawls and ornamental jewelry reaching Europe via the Silk Road confirmed the presence of rare, exquisite treasures, but how to obtain them? By the end of the fifteenth century, the Turks had overrun Constantinople and eliminated the possibility of safe overland passage. That motivated Spain and Portugal to attempt to establish alternate sea routes—westward in Spain's case and southward in Portugal's, around the Cape of Good Hope. Columbus set sail to reach the opposite side of a planet he estimated to be roughly 5,000 miles in circumference at its widest point.

His extensive logbooks reveal a man willing to realign his geography to dovetail with his aspirations. For example, a few of the Taino Arawaks he and his crew came across in what is now known as the Bahamas wore small golden ornamental earrings and dressed in cotton (if they wore clothing of any sort; only married women wore a cotton-and-palm-fiber apron over their genitals, its length determined by their rank). Columbus interpreted both the apparel and decorations to mean that the Bahamas lay off the coast of fabulously bountiful India. Never mind that these Indians, as Columbus incorrectly named them, often wore vivid white and black facial markings and bore no resemblance to the clear-skinned "easterners" described by Marco Polo two centuries earlier. By the time he launched his fourth and last expedition a decade later, Columbus had ransacked Hispaniola and Cuba in a fruitless search for gold, had enslaved thousands of natives to mine for it, had spread European diseases that would soon kill millions, and had come up empty.

Setting south from Cuba on his final voyage, still looking for "a strait across the mainland that would open a way to the South Sea and the Lands of Spices," Columbus discovered approximately 12,000 pounds of cotton in a single Cuban house, as well as spindles and looms. Soon after, he came across the Maya off the coast of Honduras. Dressed in dyed shirts and breechclouts, which typically displayed wide, bright aquamarine borders and patterns of olive pigment that looped over fields of russet and mocha brown, they approached his ship in a huge canoe propelled by twenty-five paddlers laden with trade goods. To Columbus, a part-time weaver himself, these garments represented "the costliest and handsomest . . . cotton mantles and sleeveless shirts embroidered and painted in different designs and colors."

There are a dozen or more references to the fiber in Columbus's journals; cotton plays a secondary but persistent role in supporting his argument that he had reached the perimeter of his chosen destination. Ironically, he might also have used cotton to prove that he'd reached the New World if he knew of its existence, for by the late fifteenth century, cotton had long become a citizen of both the Eastern and Western Hemispheres: for thousands of years it had been cultivated in South and Central

America as well as in Asia and regions of Africa as a domesticated plant whose spun fabric was much valued for its lightness and versatility.

Columbus, who looked at exquisite Mayan textiles and saw the beauty of Indian handicrafts, can be excused for losing his bearings. He might have sailed the ocean seas with total authority, but once on land cotton took him for a wicked ride.

World Traveler

Imagine a child with a pen arbitrarily scribbling lines over a rotating global map. That more or less mimics cotton's transoceanic voyages over the more than seven million years since the first wild plants sprang to life in widespread tropical and subtropical regions of our planet. If those locales sound suspiciously vague, so be it; geneticists remain uncertain about the exact origins of Old World cottons and only a little less fuzzy about the precise location of the first wild New World varieties.

From that unpromising start, things quickly come into sharper focus. Botanists have established that wild cotton seeds, protected by their tough leathery pods, or bolls, traveled the high seas for millennia and took root many thousands of miles from "home." In an informal sort of botanical exchange program subject to the whims of nature, strains that grew in the New World eventually sailed on ocean currents back to Asia, the Middle East, and Africa, while Old World cotton genes made their way to Peru and Chile; once established, they produced varieties that continued to disperse in all directions. After a few million years four distinct species emerged whose lint—the hairs attached to the seeds—could be spun into fabric. These four, two each in the Eastern and Western Hemispheres, were first domesticated in about 3500 B.C. in Africa, western South America, India, and Mexico. None of the fifty or so remaining varieties of cotton that continue to grow up to twenty feet high in the wild bear anything approaching spinnable lint.

All domesticated species of cotton belong to one family of swamp mallow, *Gossypium malavaceae*. They grew originally as perennials, but through a process of human selection they developed what botanists refer to as an "annual habit"—flowering and fruiting within a single growing season. That represented a significant evolutionary step both for *Gossypium* and even a greater advance for homo sapiens, who discovered we could manipulate plant cycles to create cultivars (cultivated varieties) that accommodated our needs. One of the two Old World cottons, *G. arboreum,* is native to the Indus Valley in present-day Pakistan. Its cultivars spread to Nubia and Nigeria in Africa. The other, *G. herbaceum,* comes from sub-Saharan Africa and was probably first domesticated in Ethiopia or southern Arabia before spreading to Persia, Turkey, Afghanistan, Spain, and as far east as China. Contrary to popular belief, Egypt was not cotton's birthplace.

In the New World, South American cotton, or *G. barbadense,* originated along the Pacific Coast in Chile and Peru. Botanists believe the fourth domesticated species, *G. hirsutum,* first grew wild in Central America and Mexico. Within each species there are numerous varieties, all with the same general characteristics.

Whatever their differences in length and appearance, these four contributed to one of mankind's signal achievements. At approximately the same time in human history, 5,500 years ago, civilizations living as many as 10,000 miles apart and totally isolated from one another domesticated the plant and converted it into fabric. Each invented nearly identical tools—combs, bows, hand spindles, and primitive looms—to clean the fiber, then spin and weave it. That remarkable parallel discovery seems to confirm that inspired insights leap past all geographical boundaries to become the common property of mankind.

Fabricating the New World

Staple length in this mallow is everything. The difference between inch-long medium-length *G. hirsutum* fibers and those of *G. barbadense,* which reach two inches, is the difference between perfectly drinkable table wine and a celestial Chateau Lafite-Rothschild. By that standard, the Old World's short-staple cottons represent jug brands—inferior by any measure. Once their bolls open, spiraling cotton fibers twist back and forth, interlink, and cling to one another, which greatly facilitates the spinning process. But short fibers—under an inch in length—easily break apart; they also hold less moisture and tend to be more brittle, producing coarser fabric.

The Mayans and the Bahamian Arawaks that Columbus encountered were growing and spinning the two New World cotton species. Unlike their weaker Old World cousins, *hirsutum* and the elegant *barbadense* are botanical warriors and belong to a special type of hybrid called a tetraploid, meaning simply that all of the genetic structures of both parents are carried within the offspring—an unusual occurrence (most often only some of each genome shows up). As a result of their industrial-strength DNA, these tetraploids prove to be especially tough, successful, vigorous plants. They produce the hundreds of thousands of airy yet sturdy tubular cellulose fibers that lend themselves to textiles strong enough to withstand the abuses of human wear and tear.

Barbadense, which made its way to the West Indies from South America, was initially cultivated in five shades of brown as well as in yellow, russet, and lilac hues. We think of cotton as white, but that was largely a commercial decision that came about after Eli Whitney's invention of the cotton gin in 1793. White cotton—actually a pale gray in the pod—could more easily be manipulated for consistency during the bleaching process. To pre-Inca Andean civilizations, natural cotton was vibrantly multicolored, as the pigmented cotton fabrics found at an excavation at Huaca Prieta on the northern Peruvian coast reveal. Chocolate-brown cotton fragments at that pre-Columbian archaeological site date back to 3100 B.C. Peruvian fishermen preferred dark colors in order to camouflage their fishing nets and make them less detectable to fish, a practice that continues to this day.

In succeeding Andean cultures textiles were the medium for artistic expression and for relaying symbolic information about social organization, religious practices, even philosophy—much as painting, sculpture, and architecture represent the highest aesthetic achievements of Europeans. Extremely difficult to master, the primitive hand-spinning and weaving processes that created fabric were almost always the exclusive domain of women in ancient cultures, since they could be performed in a domestic setting. That allowed women to breast-feed and raise children while contributing to the community. Their ornately dyed and intricately woven textiles often incorporated geometric designs, human and animal figures, inverted motifs, and mirrored images.

Barbadense also followed these ancients into the afterlife. Elaborate cotton and related textiles often more valued than gold accompanied the dead into the grave. Unlike most ancient cotton fabrics, which have disintegrated and been robbed of their color over time in the moist climates where cotton grows, many of the Andean cottons buried with mummies have survived. The graves were dug in extremely dry sandy terrain between the mountain range and the Pacific. They tell us, if we need to be reminded, that although we may have made enormous technological strides over the past few thousand years, mechanical and digital devices have yet to capture the sophisticated, nuanced artistry of these weavers. Burial sites for the wealthiest and most revered members of the community contain long-staple cotton woven into richly embroidered mantles, tunics accented with fine lace, and vibrantly patterned shirts with border bands of bright, deep, lustrous color. Trees, plants, and insects provide the source for the vivid red, yellow, and black dyes. Resist-dyeing applied color only to selected areas, much like the batik process.

Over the centuries that followed, a dozen different societies flourished in the Andes before the arrival of the Incas in 1450. Among them were the Chancay, who blended the wool from the alpaca and llama camelids with cotton to create a heavier, sturdier cloth similar to medieval fustian. Today, Peru's *barbadense,* now highly valued as long-staple pima, continues to produce supremely strong, silky apparel and linens. Introduced to the Middle East in the early nineteenth century by the enterprising founder of modern Egypt, *barbadense* also became the progenitor for a hybrid we now know as highly prized Egyptian cotton. Although the variety is commonly assumed to have originated in that country, it actually came from South America and has far more Peruvian genes in it than any other.

By the time Columbus arrived in the late fifteenth century, there was also a profusion of the second New World species, *G. hirsutum,* better known as upland cotton, which was probably first domesticated in the coastal region of the Yucatán Peninsula. Fragments of upland dating back at least 5,000 years were found in a cave in the Mexico's Tehuacan Valley. Spanish explorers in the 1500s discovered *hirsutum* being grown and manufactured into textiles by the great Maya and Aztec civilizations. Cultivated in the Mexican and Central American lowlands, it became a highly prized commodity in the form of mantas, strips of exquisitely colored and designed cloth direct from the loom. Montezuma, the last and greatest Aztec emperor, demanded cotton mantas as tribute from his subjects. Sticky green-seeded upland would become the one that blanketed the American South and precipitated the Civil War. Today it accounts for about 95 percent of all the cotton grown and used around the world. The Maya grew it along with its longer-staple kin. Then, as now, medium-staple upland was the workhorse fiber for hammocks, clothing, wound dressings—any and all uses; its plants, which could be cultivated with less effort than long-staple varieties, also produced larger yields per acre.

Spinning Yarns

No matter what the species or variety of cotton—and there are as many as forty varieties of *hirsutum*—the trick was to devise and improve the processes required to make whole cloth out of its wispy hairs. Trying to manipulate cotton fiber by hand was a little like trying to grasp air currents.

The first step involved the use of a spindle and distaff, and it was hard work, start to finish, especially when compared to spinning the much more manageable wool and the thicker, longer flax stalks that produce linen. Tangled lint, once extracted from its bolls, had to be cleaned to remove vegetative debris, then combed, then organized into cords that could be elongated and twisted into strands of fine thread. The spindle, a round stick of wood about a foot long, was tapered at one end, its upper half ringed with stone or clay or even a potato to provide weight, balance, and momentum as it rotated to collect yarn. A notch or slit at the top funneled the strands around the spindle as it turned. The distaff, a thicker stick, supported a loose ball of the rough, cleaned cotton rope to be spun down. The spinner anchored her distaff and drew out the fiber, pulling it continuously with her left hand while winding it taut by quickly rolling the retrieving spindle against her leg with her right hand. She had to precisely coordinate her complex movements to produce an even thread. The raw fiber strands continually snapped, requiring her to stop and twist them end to end. Hours turned into days. In these cultures, spinning and weaving came to embody infinite patience and meditative concentration.

The spindle has no moving parts, the hand loom only a few; yet for all their simplicity, they are both ingenious inventions—so efficient and appropriate to the task that thousands of years later their governing principles still drive the world's textile machinery.

The first loom was probably a tree limb hanging parallel to the ground, to which vertical warp thread was attached and then anchored to the ground below by a log or similar object. That allowed horizontal weft to be interlaced through the warp and pushed tight against the previous row to form rough cloth. A sturdy frame soon replaced the branch. The next improvement was a back-strap loom that looped over the torso of the weaver, allowing her to control the tension of the spun thread by attaching one end of her loom to a fixed object and pulling away from it as she interlaced horizontal fill. Over time, inventions like the spinning jenny and power loom would minimize repetitive hand labor in industrial societies. Until the processes became fully mechanized in the late eighteenth century, they put a premium on patience and forbearance.

India's Treasures

While the two species domesticated in the New World would prove superior to India's cotton for mass producing textiles, they would not cross the Atlantic to Europe for several centuries to come. The first cotton cloth to make any impact in central and northern European countries arrived from Asia—and primarily from Calicut on India's Malabar coast, more than a thousand miles from its northwest India origins.

Archaeologists, academic gumshoes by nature, like nothing better than to unearth a tiny fragment of smelly mildewed cloth that the rest of us would not touch with a pole and proceed to deduce the habits and apparel of an entire civilization from it—with any luck, religious and mating rituals included. When excavations in 1920 along the banks of the Indus River at Mohenjo Daro, in what is now Pakistan, revealed a few shreds of dyed woven cotton stuck to a silver vase, theories about the ancient, highly advanced peoples who inhabited that area suddenly expanded to include a new realm of exploration: textiles. Able to identify the fabric and its dye, madder, the archaeologists soon realized they had discovered evidence in the far northwest corner of ancient India that Old World cotton was being harvested, woven, and dyed centuries earlier than previously determined. Experts dated the Indus remnant back at least to 2300 B.C. and probably many centuries earlier, and attributed its survival to the preserving effect of the silver salts in the vase. (Cotton is nothing if not perishable over the eons in hot, humid terrains.) Gold jewelry strewn nearby suggested that the owner had wrapped those valuables in a larger cotton cloth, perhaps hiding them as invaders approached the ancient city of Harappa. The Harappans—a name given to the Indus Valley civilization that flourished for thousands of years before mysteriously disappearing—were the first Old World cultivators of cotton; they domesticated and converted it at about the same time as the Peruvians, give or take a few hundred years.

One bust found at the site depicted a dyed shawl draped around the shoulders of a king or priest. Its pattern of intersecting discs that symbolized the unity of earth, water, and the sun proved to be the same one found on the couch in King Tutankhamen's tomb thousands of miles to the west in Egypt. Although additional clues were scant, archaeologists also linked the fabric to illustrated Harappan clay seals and cloth fragments found in Mesopotamia—now Iraq. Cotton textiles, they discovered, were a major Indus Valley export, as were ivory, copper, and gold, all carefully measured and weighed using a complex binary and decimal system.

The Harappans were at that same time domesticating sheep and cattle, ingeniously irrigating fields, and building two-story houses with sophisticated indoor plumbing. Sitting on a former floodplain whose alluvial soil was ideal for growing peas, sesame, barley, and cotton, the site was also vulnerable to the raging waters of the Indus River and its tributaries. Six times over the millennia, at least, Harappa and nearby villages were flooded and rebuilt. Although the inhabitants constructed dams, in the end the destructive and powerful forces of nature prevailed. When prosperity and calamity are never more than a heartbeat away, you know you're in cotton country.

By about 1800 B.C., after three thousand years in the area, the Harappans' economy withered and their carefully engineered villages fell into disarray. By then the continuous cultivation of cotton and food crops over many centuries may have depleted their soil. Marauding Aryans, sweeping down into northwest India from Iran, burned Mohenjo Daro to the ground and apparently subjugated its citizens. Guesswork is all that remains; there are no firsthand accounts from that period.

The Journey West

By the time Alexander the Great's armies swept through Pakistan in 325 B.C., domesticated cotton—these "nuts that produce wool," in one of his soldier's words—was so well established that the fabric provided the common dhoti or loincloths for men as well as "long shirts worn down to the knee," which were, in a modified state, also worn by women. Alexander's troops padded their saddles with this "vegetable wool" (their term for it), and traversed 6,000 miles of scorching desert and frozen tundra to show off their discovery in ancient Macedonia and Greece. By then, Indian weaving guilds had formed; they levied taxes, advanced loans, and fined delinquent members.

The earliest religious reference to weaving occurs in the *Rig Veda* hymns in 1500 B.C. Textile creation embodies the Hindu belief in a well-ordered universe, imagined as one continuous fabric. Its warp and weft pattern serves as a kind of cosmic backdrop to the dreams, illusions, and hopes we sketch on it like children drawing figures in sand. The *Atharva Veda*, in one of its passages, characterizes day and night as two sisters weaving, with the warp and the horizontal weft, or fill, symbolizing darkness and light.

Spinning and weaving also represent two essential elements of Hindu meditative practice. Repetitive behavior and single-focus concentration, both required in these handcrafts, have traditionally been valued by Hindus as paths to freeing the mind; they invite an inner tranquility that contrasts with the scattered, cluttered actions of daily life. Making fabric is said to evoke the individual's spiritual nature, helping to maintain a personal daily connection with the Divine. Because Hinduism also requires regular pilgrimages in quest of spiritual perfection and encourages interaction between isolated peoples, cotton seeds almost certainly went along by design or accident on some of these treks to southern India, where it took root and flourished.

By the time of Alexander's arrival, India was trading cotton cloth for spices and other commodities in China and Indonesia. Later, cotton trade spread to Persia, Egypt, and the Levant on its lengthy journey to Europe.

Greeks and Romans considered the fabric to be a luxury import, "an exotic cloth that had been woven by Indian craftsman for more than 1000 years," said Herodotus, who paused to admire the cotton uniforms of Xerxes' soldiers in his history of the Greek War in Persia (490–480 B.C.). Writing about eighty years after that war, the historian also beguiled his readers with tales of "wool-bearing trees with fruits of downy fleece" from the East, and more poetically created an image of a cotton cloth so delicate it resembled "webs of woven wind."

He was speaking to a people well versed in Greek mythology, in the mythology of fabric, and in its relationship to the cycles of life. Clotho, the youngest of the three Fates but one of the oldest goddesses in Greek mythology, spins the thread of human existence with her distaff; its length in the form of a string determines the life span of an individual. In another myth, Arachne, a poor, simple Greek country girl with a gift for spinning and weaving, challenges the goddess Athena to a tapestry-weaving contest.

Big mistake. In this cautionary tale about hubris, her punishment for winning is to hang forever in the air, spinning her own webs—that is, to become the world's first arachnid, or spider.

Although cotton would appear to be the fabric most suited to a hot Mediterranean climate, linen from flax plants, hemp, and wool were the common raw materials used to create cloth in those cultures, probably because their fibers were much easier to work with than cotton's fine, short strands. Silk was rare and costly. The ancient Egyptians, including their priests, primarily wore linen, a symbol to them of "divine light and purity." More than 300 yards of linen cloth have been found wrapped around mummies to help preserve them. Priests wore cotton on special occasions.

As for Rome, "there was no year in which India did not drain the Roman Empire of a hundred million sesterces," historian Pliny the Elder complained. He admired those "trees that bear wool" in North Africa but wasn't at all pleased with the high cost of imported cotton. Some of that fabric went to supply uniforms for Caesar's soldiers. The Romans called this prized cloth *carbasina,* after the Sanskrit *karpasi*.

Sixth-century Buddhist murals in the ancient caves at Ajanta depict women spinning and weaving by hand, much as they still do today. When the ruling Brahmin class no longer permitted Hindus to travel "across the black waters," to safeguard against contamination by foreign religions, Arabs saw a commercial opportunity; by the seventh century their caravans controlled the Euro-Asian overland trading routes that delivered small quantities of cotton to Venice, Europe's primary center for Asian imports.

Muslim Arab traders weren't comfortable assigning a Hindu name to this cloth. Instead they borrowed a word from their own language, "qutun," or "kutun," used to describe fine textiles. Centuries passed; cotton as a cloth was all but ignored in France, Germany, England, and the rest of northern Europe. As the subject of a widely popular fable, on the other hand, it was eagerly devoured and passed down with reverence in these countries from one generation to the next.

Fleecing the Gullible

In the Middle Ages, when a trip across the moor was the equivalent of an expedition to a remote corner of the earth, myths about strange happenings in exotic locales held listeners in thrall. One of the most exciting concerned the cotton plant, so shrouded in mystery that some chroniclers of astonishing tales made their careers embellishing accounts of its curious origins. The English knight Sir John Mandeville was no exception. In 1322 he set out to roam the known world. Returning thirty-four years later, he wrote about his adventures in a famous book, *Voyage and Travels*, much more popular in its time than Marco Polo's journal. Mandeville reported on a remarkable animal-plant he had come upon called the Vegetable Lamb.

Its cloth, Mandeville maintained, was harvested from the wool of rare and exotic baby sheep in far-off Scythia, east of Istanbul. There, he revealed, lambs grew on shrubs, each one nestled in its own downy pod. Those tender fluffy lambs could bend their necks and munch the grass below them, but when all their meadow grass was consumed, those poor little lambs gently expired and fell to earth, and the lint they left behind was gathered up and spun into a light comfortable fabric. Mandeville anglicized the Arabic lamb-fluff "kutun" to "cotton." The existence of such a miraculous, living, breathing plant proved to him that "God is marveyllous in his Werkes."

In time, philosophers, poets, and explorers would all contribute their eyewitness observations of this miraculous being and expand upon Mandeville's seductive tale to provide detailed illustrations and descriptions of a plant with a palpitating heart, dewy fur, and pursed lips. Mandeville would later be roundly discredited for inventing journeys he never took to places he'd never been, passing them off as factually accurate—but no matter; he'd created an image so powerful it survived for centuries. "The Borametz of Scythia," from the Russian word for lamb, became an eagerly read chapter in Claude Duret's *Story of Admirable Plants* in 1605. "The hunters who were in search of this creature," Duret informed his rapt reader, "were unable to capture or remove it until they had succeeded in cutting the stem by well-aimed arrows or darts, when the animal immediately fell prostrate to the earth and died." By the time a Victorian biologist, Henry Lee, added to the general confusion in 1887 in *The Vegetable Lamb of Tartary*, cotton had long since made its way into the fabric and clothing shops of Europe. Yet even at that late date Lee's book rekindled interest in the fable, which he wisely set in a remote area near Samarkand, where few were likely to trek to verify his account.

As for the nonanimal plant itself, a curious thing happened about halfway chronologically between Mandeville and Lee: cotton took a star turn as a high-fashion item—Indian chintz—and became a subject of fierce desire among the haute monde of Great Britain before beguiling France (something like a character actress waking up one morning to discover she's Marilyn Monroe). The journey of cotton from utilitarian cloth to celebrity fabric in seventeenth- and eighteenth-century Europe would change the world, stitch by stitch. Imported cotton would become a major factor in Great Britain's decision to invade India and claim that country's natural resources as its own, and lead to a series of events that would forever alter the destiny of the United States.

TWO

Star Turns

In one hall you see . . . manufacturers busily employed [spinning] those fine muslins of which are made turbans, girdles with golden flowers, and drawers, worn by females, so delicately fine as to wear out in one night.

—Francois Bernier, *Travels in the Mughal Empire, 1656–1668*

Chintz. The word itself, once applied to glazed Indian printed cotton, has long since fallen from grace. If something is chintzy you probably don't want it; it's cheap and not worth its bargain price. Yet three hundred–odd years ago Europe prized that colorful cloth from India so highly that by ripple effect it transformed rural societies into urban manufacturing powers. Enterprising English inventors, capitalizing on its sudden popularity, began to build factories to produce their own floral-print fabrics; one thing led to another, and by the final decade of the eighteenth century the Industrial Revolution was well under way. From that moment on, Western nations, including the newly formed United States, elevated industry to the status of a new religion and never looked back.

Chintz owed its creative origins to long-forgotten Indian artisans who first learned to spin, weave, dye, and paint the fabric that traveled by sea and caravan to Europe and that eventually found its way into the cushioned parlors of English ladies. The word's origin tells the story: "chintz" comes from the Hindu word *chint,* meaning variegated or multicolored. Similarly, "calico" became a term used to describe colorful dyed cloth from the southwestern Indian port of Calicut. The lingering influence of exported Indian cotton on our language is as close at hand as any clothes closet. Dungaree, gingham, khaki, madras, pajamas, sash, seersucker, and shawl are all Indian in derivation.

If Columbus had landed in India as planned, he would have found at the outdoor market stalls an inexhaustible supply of those hand-painted chintzes and vividly dyed calico prints that were to astound Europe more than one hundred years later. He, of course, set off in the wrong direction four different times, but Portuguese explorer Vasco da Gama did not; he circled the Cape of Good Hope and reached southwest India in 1498. Da Gama reported that the first Asian Indians who greeted him "pierce the ears and wear much gold in them. They go naked down to the waist, covering their lower extremities with very fine cotton stuffs. But it is only the most respectable who do this, for the others manage as best they are able." Introduced to local royalty, he carefully noted the décor: "The king was in a small court, reclining upon a couch covered with a cloth of green velvet, above which was a good mattress, and upon this again a sheet of cotton stuff, very white and fine, more so than any linen."

Da Gama sailed off with a boatload of cotton cloth, both the fine muslin he came upon in the king's court and calico, muslin's loud, saucy younger sister. Calico and chintz would become interchangeably used to describe colorful Indian cottons, but initially calico referred to unglazed fabric printed with small sedate floral or geometric designs, while chintz, glazed and shinier, featured bold, exuberant printed and painted illustrations of flowering vines, birds, animals, and similar natural motifs. Although these fabrics first made their way back in small quantities, a linen-wool blend called linsey-woolsey continued to be the preferred European fabric for most apparel.

Cotton cloth might be too lightweight for three seasons of northern European weather, but it had the advantage of being easy to clean and soft against the skin, particularly as underclothing. It would find a ready audience at length, but not until it attracted everyone's attention by making a huge splash. More than a century after Da Gama opened up Asian sea trade, England and France, as if at last awakening from a long, monochromatic slumber, seemed to jump out of bed, pull up the shades, and exclaim, "Let there be light!" Chintz fabric flooded drawing rooms, bedrooms, antichambers, clothes closets. It caught on the way all fads do, by some mysterious fermentation of popular appeal, irrational passion, and competitive insanity. Better still, these brilliantly hued fabrics—the French called them *les indiennes*—maintained their shimmering intensity when laundered. That astonished Europeans. In that era, dyed garments or coverings washed in water quickly lost their color. Not these, miraculously. Their dyes held fast, and that alone accounted for much of the hysteria that surrounded chintz.

Secrets of the Masters

From about 1660 onward, when the cotton chintz craze began to sweep across central and northern Europe, textile printers felt pressured to duplicate the permanent, colorful designs of Indian imports, though they had neither the training nor the practical experience to do so. India's master dyers had known for centuries—thousands of years, more accurately—that intense vegetable dyes will not penetrate into waxy cellulose cotton; they scatter on the surface and run. Not only did the masters have to find a way to treat raw cotton cloth to make it receptive to color, they also needed to make the dyes permeate the fibers to achieve colorfastness. Through trial and error, Indian artisans developed primitive but effective methods to do both, and they were not about to share their hard-won knowledge with foreigners. This secret wisdom, assiduously guarded, was passed from one generation to the next. It owed its science to bonding through molecular chemistry, but of course the ancient Indians knew only that they could trust the urine of goats and other barnyard animals to get the job done.

To cheat nature, dyers had to create a bridge between substances that naturally repelled one another. The first step, perfected by Indians along the eastern Coromandel coast—the center of chintz dyeing—was to repeatedly soak cotton in solutions that broke down its waxy structure. Bleaching with lemon juice or sour milk helped convert the grayish raw crop to white fiber, but that was only a beginning. From Harappa onward, dyers learned to then "animalize" their cotton fiber. The weapons of choice were buffalo milk, goat urine, and camel dung, and sometimes blood and albumen. The proteins in the animal excretions did the bulk of the work in making the fiber receptive to the dyes. Next came the mordants—from the French verb "to bite." These faintly colored metallic salts obtained from natural sources, most commonly alum and iron, were painted or block-printed onto the fabric once the design was stenciled in place with powered charcoal transferred through perforations in the artist's sketch paper. The mordant reacted with the dye to form a "lake" that permeated the animalized fiber's core, so that each color was fully absorbed into the fabric in an area saturated with the particular mordant and remained permanently fixed. The lengthy and arduous procedure could take months. The only important exception was the popular blue indigo, a pigment that chemically fixes to cotton fiber around its core and needs no chemical catalyst. Areas of prepared cloth not meant to receive a particular color were coated with beeswax, in a common practice now known as resist dyeing.

If fabrics were destined to become chintz, the mordants were applied by a fine brush, frequently in ornate designs based on the sacred Hindu Tree of Life. Painted chintzes came to be known as *kalimkari* (literally, "pen-work"). The brush was a sliver of bamboo tapered to a point, with a cloth ball soaked in dye at the lower end. The ball was squeezed to refill the tip. Like muslin spinners and weavers, kalim artists were from the lowest Indian caste. Even though they were some of the most technically precise artists in history, they remain forever anonymous. The beauty of their work in surviving *palampore* (bed covers), drapery, and wall hangings belies their lowly social status. Flower bulbs, jumping fish, birds, butterflies, and sprays of blossoming buds coil about delicate meandering branches that spread out from the sinuous tree trunk. Luminescent reds, blues, yellows, and a host of other vibrant colors spring to life in designs that are so fresh and buoyant after three hundred or more years they appear to be newly created.

The dyes themselves—chiefly madder and indigo as well as henna, annatto, turmeric, jackfruit, coral jasmine, logwood, madder, quercitron, and fustic—came originally from tree bark or seed, but more commonly from a plant's roots, or in some instances, like indigo, from its leaves. Rarer dyes came from the sea and usually went into expensive garments worn by royalty. Tyrian purple, extracted from shellfish, became the exclusive color of garments prized by Greek and Roman rulers and emperors, a practice that gave us the expression "born to the purple."

The alternate method of applying dyes to treated fabric was by wood-block print, the craft eventually taken up by Europeans. Printers transferred designs onto blocks of wood and chiseled their outlines into the wood, a delicate art all its own. Applying the same mordants as the kalimkari painters, they added a gum that would further bond the dye to the cloth (unfortunately, the gum also noticeably reduced the brilliance of the color). With their blocks—up to a foot in length—they essentially stamped designs onto the prepared fabric, or ground.

As these techniques traveled from India and Indonesia to

other countries on their journey west, they found a home in Africa, where Ivory Coast village women became adept at sending personal messages through their wax-print textile designs. One wife patterned a skirt with an attractive repeating image of small birds in flight. A closer look revealed that the birds were departing from opened cages. It was a warning from the wife to her unfaithful husband that she was about to fly the coop. Many of those wax-prints also bore titles; one was, "Condolences to my husband's mistress."

As for the Indian methods of "animalizing" cotton, they remained mysterious to most European printers until much later than might be expected—for seventy years after the arrival of chintz. Ironically, it was a man of the cloth, Jesuit Father Coeurdoux, who betrayed these fiercely guarded secrets. In 1742 the French cleric took advantage of his missionary posting on the Coromandel coast to gain the trust of Indian master dyers whom he had converted to Catholicism. They confided their secret process to him with an understanding that he would never reveal it. Coeurdoux immediately gave a detailed description in a step-by-step letter published in France. In a blink, three thousand years of clandestine artisan practice became public knowledge.

What Coeurdoux could not provide was the expertise of the Indian kalimkari painters, and it was that artistry, as well as the vibrancy of their indelible colors, that electrified Europe.

My Carpet, My Gown

"The expense of reason in a waste of shame is lust in action," Shakespeare wrote in one of his sonnets. He was speaking about sex, of course, but the erupting passion in England and France for imported chintz and calico was hardly less visceral and impulsive. By 1664, over 250,000 pieces of calico and chintz were being imported by England's East India Company. English ladies ripped apart glazed drapery, rugs, and bedcovers to convert the chintz fabric into skirts, blouses dresses, and gowns. In Molière's *Le Bourgeois Gentilhomme*, written in 1670, his nouveau-riche protagonist dresses in a flamboyant chintz robe because his tailor tells him that gentlemen wear *les indiennes* in the morning. For women, chintz was now "weare of laydes of the greatest quality," in the words of one English import company.

As this new material replaced wool in England and silk in France, it threatened traditional textile industries and incited violent protest. In 1678 an English pamphlet decried the Indian upstarts: "This trade (the woollen) is very much hindered by our own people, who do wear many foreign commodities instead of our own. . . . To remedy this it would be necessary to lay a very high impost on all such commodities as these are and that no callicos or other sort of linen be suffered to be glazened." Few listened; two years later more than a million pieces were being imported into England, and into France as well.

The French government, pressured by linen and silk manufacturers, banned imported cotton of all kind in 1686; England followed in 1700. Both bans were widely ignored, particularly by defiant French women, even at the risk of severe punishment that included death. Imported fabric bans were promoted with jingoistic fervor by patriots like Daniel Defoe of *Robinson Crusoe* fame: "About half the woolen manufacture has been entirely lost, half of the people scattered and ruined, and this by the intercourse of the East India trade," he complained. (Three centuries later President George W. Bush would be saying many of the same things about China's exports.)

About a decade later, in his *Weekly Review* in 1708, Defoe looked back with scorn:

> The general fansie of the people runs upon East India goods to that degree, that the chints and printed calicoes, which before were only made use of for carpets, quilts, etc. . . . became now the dress of our ladies; and such is the power of a mode as we saw our persons of

> quality dressed in Indian carpets, which but a few years ago their chambermaids would have thought too ordinary for them: the chints was advanced from lying upon their floors to their backs . . . : and even the queen herself at this time was pleased to appear in China and Japan, I mean china silks and calico. . . . Nor was this all, but it crept into our houses, our closets, and bedchambers . . . were nothing but calicoes or Indian stuffs; . . . in short, everything that used to be made of wool or silk.

Defoe was angry for perfectly patriotic reasons. Innocuous white cotton, turned madder red and indigo blue and shocking pink, glazed to a silky shimmer and painted with bursting scenic motifs by Indian masters, aroused so much excitement in England and elsewhere that cotton had now become a competitive fabric. Would there soon be pastures filled with unemployed sheep? One pamphleteer for wool manufacturers wrote that the threat to his trade was caused by "a tawdery, pie-spotted, flabby, ragged, low priz'd Thing call'd Calico, a Foreigner by Birth, made by a parcel of Heathens and Pagans that worship the Devil."

When the bans were finally lifted fifty years or so later, heavy duties imposed on imports had taken the play away from chintz, but by then these colorful, playful cottons had spawned an industry of domestic block printers in France and England. Once the Indian dye methods became public knowledge, print factories in Europe knocked off the original designs without hesitation, forcing thousands of Indian craftsmen out of work. Small block-print studios began to spring up around London on tributaries of the Thames, prime sources for the clear running water needed to dye and wash fabric. Textile hand-painting was a skill that eluded the English, but printing was a familiar occupation. Block printers imported raw material from India and the West Indies, and at first produced dull, clumsy imitations of the Indian originals with inferior dyes that quickly faded. In time they became more skilled and knowledgeable and produced higher quality, sometimes exquisite, fabric—but its colors and designs remained derivitive.

The French, being French, brought style to the enterprise. It was estimated that twenty million francs were spent annually on colored cottons in the early eighteenth century. By 1758, Christophe-Phillip Oberkampf, the son of a German dyer, had opened a printing factory at Jouy-le-Moutier near Versailles outside Paris. An artist as well as a shrewd businessman, he made Jouy synonymous with high fashion. Mulhouse, near the Swiss border, also became famous for its textile factories. Madame de Pompadour, mistress of Louis XV, wore Oberkampf's stunning gowns. An "inspired genius," in her words, he also eagerly embraced the latest inventions, including copper-plate printing, the automated successor to hand blocking.

Oberkampf was European cotton's first celebrity designer, a favorite also of Marie Antoinette and sought after by aristocrats all across Europe for the superior artistry of his fabrics. He survived the French Revolution even if the French royals did not, and was decorated with the Cross of the Legion of Honor by Napoleon in 1809. His citation read, in part: "M. Oberkampf began his establishment fifty years ago and naturalized in France the art of painted cotton, which had been built up in Europe from very modest beginnings . . ." Napoleon embraced him as a comrade in arms: "We will make together a rude war against the English, you by your industries and I by my armies."

Imperial Impulses

While chintz was laying the groundwork for progressive change in Europe, it was also exposing India to the threat of invasion. In addition to spices and opium, both coveted by Great Britain for their commercial value, disorganized India itself now became a desirable acquisition for its immensely profitable cotton textile industry as well as for its plentiful source for raw material. Cotton needs heat, warmth, and moisture in order to prosper. In England a balmy day is an appetizer for a main course that never arrives. The British, now importing shiploads of India's finished goods, looked to that country's robust cotton fields as well with an acquisitive glint in their eye.

Why allow these dark-skinned "oily wogs," as they were commonly denigrated, to dominate the cotton trade? Why not do the proper thing and squash those infernal Asians like bugs with Her Majesty's mighty imperial armies? Teach them a thing or two about table manners while we're at it. That was more or less the tenor of parliamentary debate that led to a vote in favor of colonizing India and making it a jewel in the British imperial crown. Lofty phrases like "moral uplift" were bandied about, but in the end, commercial self-interest drove Robert Clive, under the auspices of the royally chartered East India Company, to round up support from powerful Mughal emperors and invade Bengal for England in the Battle of Plassey in 1757. From there it was a short, crisp British strut to occupying India's cotton fields and outlawing Indian manufacture of any cotton fabric from its own raw material. India's millions of citizens in the future would have to buy their cotton goods from the British textile mills to which all that cotton would now be transported. England's export of that same cotton largely subsidized its slave trade: shipped in bulk to West Africa, Indian cotton was bartered for slaves, who were then shipped by the British to American plantations in return for sugar, cash, or both.

In a few short decades India went from a fractious but self-sustaining country to a ward of the state, a beggar at the mercy of Her Imperial Majesty's mercantile whims. The English did more than strip away one of India's primary sources of income; they also robbed the country of its soul, for the emotional link between India and its cotton ran as deep as America's patriotic connection to its amber waves of grain, or possibly even deeper. The fiber was woven into that country's history. It appeared in sacred Hindu texts dating back thousands of years. It symbolized the creation of the known world in textile hangings revered by countless generations. And while Europe was muddling through its dark ages, cotton was dazzling India. Long before chintz ever caught Europe's fancy, cotton had taken another star turn, one that transformed the fiber into an enduring symbol of Indian pride.

White Gold

"The skin of the moon removed by the executioner-star would not be so sheer. One might compare it with a drop of water if that drop fell against nature, from the fount of the sun. A hundred yards can pass through the eye of a needle, so fine is its texture. . . . It is so transparent and light that it looks as if one is in no dress at all but has only smeared the body with pure water," the fourteenth-century Persian poet Amir Khusrau rhapsodized. He was speaking of royal muslin from Dacca and Deogiri, famous for the fineness of its weave.

Cotton has rarely been sexy, but royal muslin was so erotic that men fairly sputtered when they tried to write about it. The seventeenth-century traveler Jean Baptiste Tavernier described this muslin as being so sheer that at the Grand Mogul's seraglio, "the kings and queens take great pleasure to behold the sultanesses in these shifts and see them dance with nothing else upon them."

Tavernier came late to royal muslin. Since the first century it had brought international fame to Dacca, where it was made. There are marble images of Buddha from that era that depict him draped in a fabric of exquisite lightness—an artist's tribute to both the cloth and the god. Legend has it that when a well-known courtesan first met the Buddha, she wore a richly woven semitransparent muslin sari to make certain she caught his attention. It worked.

This fabled fiber, named for the city of Mosul in Iraq, where it is alleged to have originated, came from an extinct variety of the cotton plant. The conventional Old World species, *G. arboreum* and *G. herbaceum,* both produced brittle, inferior, short-staple fibers. By contrast, the muslin that became better known as white gold was created from thread so fine that one pound of it could be stretched over two hundred miles. Young Hindu women, starting at about age thirteen, were chosen as its spinners. They had the small hands and delicate touch needed to work thread so thin it was all but invisible. Popular mythology avows that many went blind after a few years. Muslin threads were sized, or coated with gel-like rice starch, to prevent them from breaking up on the loom. Diaphanous yet strong, the highest quality muslins had an 1,800 count (the number of threads per square inch of fabric); lesser varieties were made with a 1,400 count. To put that in perspective, luxury bedsheets today are apt to boast a 400 count. Our lightest cotton fabric today is four to five times heavier than royal muslin.

The noblest work of those young female spinners and weavers of Dacca was *jamdani,* or patterned muslin. After deseeding the cotton between rollers and cleaning the fibers with the vibrating strings of a hand bow, they used the jawbone of the boalee fish to comb them parallel; once straightened and cleaned, the fibers were converted into bleached and unbleached yarns to create subtly gradated fabric that appeared to contain dancing shadows when held up to the light. *Jamdani* incorporated bright white gold, tooth white, sandal white, autumn cloud white, and autumn moon white. Floral and abstract motifs were woven into the designs as well. The Indians gave these exquisite creations evocative names like "Evening Dew," "Sweet Like a Sherbet," and "Running Water."

The English, when they arrived in force at the beginning of the nineteenth century, knew all about royal muslin as well as calico and chintz. Its superior quality presented so much of a threat to England's own cotton industry that, common lore has it, British colonial governors ordered the thumbs and index fingers of India's best spinners and weavers chopped off so that they would no longer be able to twist and manipulate its delicate threads.

Inevitably, Indian cotton had the makings of a contentious political issue. By depriving India of the fruits of its own labor, England all but guaranteed that the crop would one day come to symbolize colonial subjugation and provide a rallying point against it. When that day finally arrived in the early 1900s, a frail warrior with the heart of a lion, Mahatma Gandhi, intertwined the destinies of homespun cotton and self-rule so adroitly that he made one indistinguishable from the other. Freedom became the cotton cloth you wove and wore, a tangible protest against tyranny from abroad.

Inventing a Revolution

It wasn't arrogance alone that compelled Great Britain to annex India to its holdings; necessity had much to do with it. Chintz had created a phenomenon among the upper classes; then, as now, the general populace was eager to follow, and it did so in great numbers. Calico, a cheaper and more accessible printed version of glazed fabric, became the new, affordable fabric of choice for every shopgirl in France and in the British Isles. By the later decades of the eighteenth century, hand-block printers all over England were seeking ways to speed up the manufacturing process. Hand spinners and weavers, too, could barely keep up with the flow.

Serendipity intervened. As a result of the Age of Enlightenment in France, a major shift from passive acceptance to self-reliance was taking hold among the people of Great Britain. As science and invention continued to gain the upper hand, they spawned a new breed of inventor devoted to developing the technology of the time. Cotton was among the first commodities to benefit from these experimental inventions. As these commoners with a passion for machinery discovered they could literally take their destinies into their own hands, they turned out an endless assortment of contraptions. Most of them ended up rusting out back, but all were infused with the spirit of defiant self-expression that Diderot captured when he said, "If you forbid me to speak on religion and government, I have nothing to say."

Way ahead of the curve was an obscure barber in the hinterlands of northwest England named Richard Arkwright. He was more than willing to put his faith and trust in himself, happy to speak his mind even when silence might be the wiser choice, and as eager to experiment with machinery as any man of his time, so long as the results could be counted in coin. He had the motive; all he needed was the means, and when Arkwright met cotton, an industrial revolution was born.

THREE

The Barber from Preston

It is not from the benevolence of the butcher, the brewer, or the baker that we expect our dinner, but from their regard to their own interest.

—Adam Smith

The man primarily responsible for changing the way we earn our living in the Western world trusted no one's abilities as much as his own. The youngest of thirteen children, Richard Arkwright was born in 1732 to a poor barber in the small remote country village of Preston in northwest England; he was destined to cut hair like his father, to marry, breed, and expire into anonymity. Upward mobility was not a prominent feature of eighteenth-century England. But this plain man, "almost gross, bag-cheeked, pot-bellied . . . with an air of painful reflection, yet also of copious digestion," according to fellow Englishman Thomas Carlyle, was too ferociously ambitious and clever to live out his life in measured spoonfuls. He came along precisely at the right time to make his mark, when consumer demand for cotton goods in England and Europe vastly exceeded the output of available hand-operated machinery. In the process of trying to increase production, Arkwright created the first factory system. It transformed the production of English textiles from a rural, family-based cottage enterprise into an automated, centralized one, and succeeded on such a broad scale that soon scores of other industries, near and far, followed suit.

Most of us alive today, whose forefathers tilled land, now dwell in or near cities originally created to house industrial workers, transport raw materials and finished goods, or furnish related mercantile services and outlets. These cities are by-products of Arkwright's cotton mills; they inspired families that had eked out a living close to home for centuries to pack up and resettle in new factory towns that sprang up around them. With mill work, children and often wives became important wage earners. As dependent family members gained economic independence, the male heads of household began to lose some of their absolute authority. Politicians started to recalibrate their pitches to this new, mushrooming working class. Wars between nations, previously fought over religion and territory, now also became conflicts over commodities.

Within two short decades, cotton manufacturing would provide a source of immense wealth for England and for Arkwright himself. Tracing the entrepreneur's career from its unpromising start, one begins to fully appreciate the dimensions of his achievements and the scope of his thievery. Lionized by Victorians as the patron saint of industrial progress and vilified as a scoundrel by later pundits, Arkwright was a little of both—actually, a lot of both: genius and crook, mix and match—but he was no garden-variety reprobate. He pilfered minds, never billfolds. In that rowdier time, rough play was applauded so long as you won, and he did. Shortly before his death Arkwright was awarded a knighthood.

Apprenticing himself as a young man to a local barber (not his father), he quickly decided there was more money to be made in wigs than in snipping locks. As a young widower at twenty-five with a toddler son to provide for, he began traversing the rolling hills of Lancashire in search of women's hair that he could collect and weave into longer strands, then dye and fashion into fancy hairpieces. He mixed his own secret-formula colors, sold his proprietary dyes to other wig makers, and might have made a decent living in that trade if wigs hadn't begun to go out of style. Fashion's loss would prove to be fabric's gain.

Little is known about Arkwright's early years as a businessman, except that they evidenced an instinct for sales and a high tolerance for risk. Only a few records have survived, primarily the recollections of several acquaintances: Arkwright himself left only a few papers, none personal. One man, Thomas Ridgeway, sent Arkwright's son (also named Richard) a letter in 1799, seven years after his father's death, recalling that when Arkwright was a young man he briefly owned a public house, or pub: "His customers that had employed him in his business were generally of the better sort; he might probably have done better could he have Stooped to the vulgar, but his spirit was much superior to it. And always he seemed to have something better in view. His genius for Mechanics was observed, it was perceived in his common conversations. I well remember we had great fun with a Clock he put up, which had all the appearance of being worked by the smoke of the chimney . . . he was always thought clever in his peruke [wig] making business and very capital in Bleeding and toothdrawing. . . ."

Threads of Change

In that era, sparsely populated Lancashire was still "a wyld savage contry ferre from any habitacon," as it had been described two centuries earlier in a 1552 British government report. Its boggy moors and rock-strewn green hills remained forbidding and remote. The land was inhospitable to large farms; sheep roamed fields of green divided over time by stone fences, many of which still stand. Its inhabitants made their living largely by harvesting local wool and converting it into yarn, then cloth. Women spun, men wove. For centuries, wool took precedence over cotton: it was local and close at hand. It was also much easier and faster to spin on a wheel—a major leap forward from the hand spindle—and to weave than other fibers, including flax, because of its longer, sturdier fiber construction.

England's weaver king, Edward III, crowned in 1327, had laid the foundation for Lancashire's textile trade by urging craftsmen—spinners and weavers—to immigrate from Europe's medieval textile centers. The king convinced himself that the sheep of England would help support his costly wars against France if their fluff could be converted at home to woolen fabric, much in demand across Europe. With that in mind, he sent his emissaries off to the Netherlands and Flanders to recruit skilled workers. According to seventeenth-century historian Thomas Fuller, Edward's printed royal sales pitch quickly made the rounds:

> But oh how happy should they be, if they should come over to England bringing their mystery with them, which would provide them welcome in all places. Here they should feed on fat beef and mutton till nothing but their fullness should stint their stomach; yea, they should feed on the labours of their own hands . . . their beds should be good and their bed-fellows better.

Better bed-fellows—it works every time. Immigrant spinners and weavers settled in the North Country and established trade guilds. Wool soon provided the resource for a bountiful cottage textile industry in the pasturelands and moors of Lancashire. Not much changed for several hundred years after that until the sudden invasion of cotton calico prints. As block printers began to move their operations further north to nearby Manchester in the mid-eighteenth century to avoid London's city taxes, they brought with them all of the excitement and frenetic jockeying for position that accompanies any hot fashion trend. Chintz and unglazed cotton prints were so in vogue that legions of local wool weavers found themselves forced to convert to cotton or go hungry. They marched in protest.

Their campaign soon fizzled out, but it caught the itinerant wig-seller's attention, and Richard Arkwright began making inquiries. He discovered that James Hargreaves, a local Blackburn carpenter and weaver, had invented a mechanized spinning machine named for his daughter, Jenny, in 1764 that in a single operation efficiently combined the output of eight to ten individual women using manual spinning wheels. The story goes that Hargreaves' wife had just knocked over her hand wheel one afternoon as he walked in, and as a result its spindle began rotating vertically. For three hundred years the design of these wheels, with their horizontal spindles, had remained essentially unchanged, but Hargreaves, in a moment of inspiration, perceived that spindles could also collect threads in an upright position, which meant that if he could line them up one beside the other, he might be able to feed multiple bands of cotton fiber, called roving, through a moving carriage that clasped and pulled them into thin strands of thread before twisting them onto parallel vertical spindles, all in the same unit. Experimenting with various designs, he came up with a prototype that simultaneously stretched and twisted the roving—which was about as thick as a candle wick—into thin yarn, collected simultaneously on eight rotating spindles.

By turning the handle of the "jenny," which resembled a small bed frame fitted with a sliding, tilting rack of spindles, the weaver set the apparatus into effortless motion. Its parts retreated and advanced in precisely calibrated movements to feed roving at one end and collect twisted thread at the other.

Somewhere on his journeys, Arkwright had an opportunity to study Hargreaves' invention at close range. He apparently grasped the breakthrough advances it offered as well as the crucial problem it failed to address: the jenny could turn out reams of soft weft fill, or horizontal thread, but not sufficiently strong warp. In weaving, the vertical warp threads are strung on the loom and act as sturdy fence posts that anchor and support the weaker interwoven weft. Weavers traditionally relied on linen or wool as warp, since those heavier, longer fibers provided the necessary strength that flimsy cotton lacked. This blended cloth (called fustian), while not as lightweight or as soft as pure cotton, was much more durable and useful. However, imported Indian chintz was 100 percent cotton and all the rage. While Indian weavers had solved the tensile strength issue, mostly by hand-twisting several strands together to provide reliable warp, this solution had eluded the Europeans.

The first obstacle Arkwright needed to overcome was the insubstantial nature of cotton itself. To further complicate matters, the only available raw cotton was G. *arboreum* from India; its short brittle fibers easily splintered. Since spinning cotton is somewhat akin to pulling and twisting a bar of Turkish taffy into a ridiculously thin string without breaking it, the odds of doing that with G. *arboreum* mechanically and on a large scale were slim indeed. It was all well and good for poets to compare the fabric to webs of woven wind or to gossamer plumes of smoke, but a weaver can't make shirts or blankets out of ethereal rhyming couplets. Still, if a viable business existed here, Arkwright surmised, it lay in making cotton fabric with the muscle to withstand vigorous weaving activity.

Was Arkwright up to the task? He was born to it, actually. He told anyone who would listen that, as his name suggested, his family of boat builders descended directly from Noah. (Self-confidence was not his weakness.) In short order Arkwright blatantly swiped the best features of Hargreaves' jenny with no compensation offered to its inventor, although a patent law had been on the books in England since it was first passed under King James I in 1624. He then went about the all-but-impossible task of finding a way to make English-spun cotton cloth a worthwhile commercial fabric. Arkwright had no training in the trade and barely a farthing to his name, but others did, and what better way to put his talents to good use than to exploit all the resources within reach? He set out to raise capital, attract and organize skilled personnel, and convince a skeptical nation that cotton spinning belonged in a factory rather than in a family parlor—in his very own mill, that is, in his own Lancashire fiefdom where he would reign, if there was any justice in the world, as cotton's Imperial Majesty.

Luckily for Arkwright, technological advances were now boiling the blood of Englishmen everywhere. "Knowledge is power," Francis Bacon had declared more than a century earlier. Machines, said Bacon, will come along to liberate man and help him triumph over nature. His prediction had now developed into a national crusade. "The age is running mad after innovation," said Dr. Samuel Johnson with admiration and alarm. "All the business of the world is to be done in a new way; men are to be hanged in a new way. . . ."

Cotton happened to be the ideal meeting place for an extended conversation between mechanical experimentation and dreams of wealth and progress. The challenges to its mass production inspired a frenzy of creativity as well as the promise of wealth to the inventive industrialist who prevailed. All over the English countryside men were sequestered in their workshops busily applying cranks, gears, levers, belts, and rods to processes traditionally performed by hand.

In 1765 James Watt invented the steam engine that would in time provide vital power to factories in Great Britain and America (and elsewhere). Arkwright, when he built his first mill in 1771, instead relied on the waterwheel, a safer bet at that moment. Before he embarked on that venture he first had to decide what past and present inventions, aside from the jenny, he needed to hijack and reconfigure to his own ends.

The process called "drafting" involved a complex series of steps to elongate and interlock fibers to produce yarn. Cotton would arrive on the factory loading dock in a bale ranging in weight from 450 to 500 pounds, so compressed by a huge screw-turning device that it would crush a sturdy piece of furniture if dropped from a short height. Workers then beat the baled fibers by hand to decompress them until they separated out as fluffy tangled masses of lint still embedded with trash—broken stems, leaves, and other vegetable matter from the field.

Carding, the next step, effectively cleaned the fibers. Workers, usually women, ran a stiff nine-inch card with fine wire teeth protruding from one side through the beaten cotton until the fibers were cleaned and lined up parallel to one another in a soft tubular shape called the sliver (rhymes with diver), about as thick around as a person's arm. After that, the cotton fibers were combed, further attenuated, and loosely twisted by a series of rollers and separated into multiple bands about one-quarter-inch thick, called roving. Spinning mechanisms repeated some of these processes until the joined fibers were reduced to thin thread; they were gathered up on a bobbin that sat on the spindle rotating on a shaft. Above the spindle a whirring, bell-shaped flyer caught the thread on the horizontal feet protruding from it on opposite sides and twisted the strands tightly as they wound on the bobbin.

While the same basic steps applied as well to making linen from flax and to sheep's wool, the longer length and heavier weight of their fibers facilitated preparation and conversion. Cotton spinners working at home and operating their wheels by hand processed only one pound or so of raw cotton material every two weeks.

Where to turn for help? Arkwright wisely looked back, to half-finished inventions, to failures that contained the germ of a good idea that had been poorly executed and abandoned. He was practiced at transforming processes, such as profitably recycling snippets of hair from the floor sweepings of barbershops, and he now applied his talent and outsized ambition to the dusty mechanical discards of workshops. He might have been a minor thief, all things considered, but he was a master visionary with a gift for retooling and reassembling, and he was thoroughly merciless in his quest for glory.

Back to the Future

Roughly two decades before Arkwright launched his revolutionary factory system, one of Dr. Johnson's friends, Lewis Paul, had invented mechanical methods to speed up the preparation and spinning process. Paul, a roué, wanted mostly to lighten his workload and lessen the demands of his textile employment so that he could spend more time promenading along the boulevard in his silk embroidered frock. To reduce the drudgery he invented an automated carding machine that did the work of untangling and cleaning cotton fibers to prepare them for spinning. With the help of a wheelwright named John Wyatt, Paul also created a roller-spinner to produce strong cotton threads on multiple spindles, and eventually opened a mill to manufacture yarn. His horse-powered spinning machine, while advanced, did not function reliably. In time Paul might have overcome its defects, but he soon sabotaged the operation with his extravagant personal spending and ended up being sued by everyone—Wyatt as well as his investors—and spent much of his time in debtor's prison. His machines sat idle for two generations.

In a more logical universe, inventors like Paul and Hargreaves would first have perfected the mechanical cleaning and spinning of its fiber, then put their attention to the second phase, automated weaving, since you can't weave what hasn't been spun. But the sequence played itself out in reverse order with the early invention of the flying shuttle. In 1733, an enterprising Englishman, John Kay, studied the way that weft thread was interwoven with vertical strands, and developed an efficient mechanized system for speedy weaving long before spinning left the parlor room for the factory floor. Until Kay's invention came along, two men—a catcher and thrower—on either side of a large frame repetitiously passed a yarn-holding bobbin back and forth as it was struck by small hammers to position it over and under each warp thread; that tediously created the horizontal weave of a fabric. Kay reasoned that if you could put wheels on a shuttle fastened to the bobbin and make certain that it fed thread back and forth in a straight line, you could streamline the process and eliminate the workers. Kay's flying shuttle—named for its blinding speed—substituted two wooden launch boxes at either end of the shuttle run with picker hooks that caught and threw the thread. Both contained a string that was manipulated by the worker; pulling one string, he triggered the spring-loaded picker that sent the batten and shuttle flying across the loom with its cargo of thread; then he sent the weft back with a yank of the second string. With that invention, Kay doubled the productivity of the loom and dramatically reduced its manpower requirements.

There was a high price to pay for mechanical efficiency, Kay sadly discovered. When a local entrepreneur put a handful of his experimental models into operation around Lancashire in 1753, word quickly got out that a sinister machine was threatening to rob hand weavers of their livelihood. They formed mobs, marched across Lancashire smashing flying-shuttle looms, then stormed Kay's home in Bury, destroying everything they found—and might well have killed him if he hadn't managed to flee to France, where he died a pauper in 1781. Expecting to become wealthy, Kay had taken out a patent, but when his flying shuttle was pirated throughout the textile districts of England, offenders were lightly fined; they then simply went out and continued to use their unlicensed jerry-rigged versions of his invention. If Kay expected ethics and fair play to triumph, he woefully misjudged human nature in the orgasmic throes of greed. He was the first in a long line of disillusioned cotton inventors, including Eli Whitney, who would suffer the perils of their own genius. "I have a great many more inventions," Kay wrote in anger and despair long after his flying shuttle unleashed a turbo-charged future into textile weaving, "and the reason they have not been put forward is the bad treatment I have had from woolen and cotton factories in different parts of England."

It was only a matter of time and tenacity until spinning technology advanced to meet it. Although Arkwright saw how to improve on Lewis Paul's carding machine, which was essentially a rotating drum with fine teeth that cleaned and aligned the fibers of raw cotton fed over it, the ideal mechanism for drawing out roving to thread presented more complicated problems. There is no solution that he would likely have come upon on his own. His talent lay elsewhere—in inserting himself into a stalled project like a missing verb into a sentence fragment. Beyond recognizing the potential in another person's idea, Arkwright added his inspired business acumen and mechanical aptitude.

He saw that he would have to combine the efficiency of Hargreaves' jenny with the tensile fiber strength that could be obtained by a working version of the other Paul invention, the primitive roller-spinner. That drawing apparatus elongated cotton's short fiber strands before twisting them. Weighted rollers, synchronized with the flyer, initially did the work. If they exerted too much pressure they snapped the fibers passing between them; too little, and the finished lumpy yarn snarled the loom. When he soon came across a new and untested design created by a Warrington clockmaker named John Kay (no relation to the flying-shuttle inventor), Arkwright sensed he was closing in on a way to industrialize yarn making. Kay, for his part, had developed the crude nonfunctioning model in partnership with a man named Thomas Highs, who would later hound Arkwright and Kay for compensation, without success.

By the time Arkwright set to work with Kay to create the world's first mechanical spinner in 1768, he was already thirty-six and well into middle-age by the life expectancy standards of his day. Obsessed, driven, now remarried, and near penniless, Arkwright had long since abandoned all other possible sources of income to concentrate on building his new machine. With no collateral and no previous commercial triumphs to haul out as testimony to his shrewdness, he managed to persuade a local liquor merchant, John Smalley, to put up money for a working model, and he convinced clockmaker Kay to help him produce it. He and Kay disappeared for days into a schoolhouse room behind a garden filled with large gooseberry bushes, attempting to build a larger, working version of Kay's double-roller machine. Neighboring women heard humming noises coming from the workroom long past midnight and reported a devil inside tuning his bagpipes for a witch's reel.

In the end, Arkwright and Kay managed to create a spinning frame, loosely based on Paul's earlier machine, that was able to turn out two dozen strong threads at a time, although the first working models included only four to eleven spindles. Their ingenious solution took into account the unique properties of cotton, whose fibers are hollow and look like flattened straws under a microscope. If you stretch and twist raw cotton into thread in the same motion, you lock the short fibers together and produce frail strands; if you twist first and then stretch, the weak, locked fibers break. Arkwright and Kay realized that if the fibers were first stretched and then twisted, they would lock in a strong position and create thread that could withstand tension. Their machine—soon to become known as the water frame because it was eventually powered by river rapids—passed the lightly twisted fibers between three sets of rollers, each calibrated to speed faster than the previous set until the roving emerged as yarn; that fed at last through the rotating hooklike feet of the flyer, which gave it a final tight twist before depositing it onto awaiting spindle bobbins.

For the first time in history, these two men had mechanically produced strong cotton warp. This invention quickly became the love of Arkwright's life, as his second wife, Margaret, sadly discovered. Arkwright and she soon went their separate ways because, in the words of one nineteenth-century writer, "she, convinced that he would rather starve his family by scheming when he should have been shaving, broke some of his experimental models of machinery."

Noting that Hargreaves and others had been victimized by their own inventions as resentment swelled, Arkwright decided to skip town in advance of the vengeful mobs, and set off over the hills to relocate about thirty miles away in Nottingham. There he soon formed a partnership in 1768 with two successful silk knit-hosiery manufacturers named Jebediah Strutt and Samuel Need. He acquired the capital from them to finance a full-scale test of the new invention. Powered by horses, the water frame performed as promised. He now had the backing and the machine—and, equally important, the vision to make the most of his opportunity. Arkwright looked at wispy, delicate thread and imagined it strong enough to lash together a personal empire of monolithic textile factories. He looked at the surging foam of the Derwent River in remote Cromford and saw past the rapids to a free, efficient, and unlimited energy source. That churning water could turn a wheel and power six floors of belt-driven machinery, or so he was willing to believe. Waterwheels were a familiar source for mill power by then in England, but not for mills built on the large scale Arkwright had in mind. Fearless or reckless or both, he operated with a total disregard for cautious, unadventurous, timid, or orthodox strategies.

Force of Nature

In building his industrial dynasty, Arkwright decided there could be no other pretenders to the throne. Strutt and Need were his bank: they could stay, preferably in the background. But as for everyone else whose faith, money, and sweat equity had carried him to the brink of success, they suddenly learned just how expendable they had become. Arkwright quickly reduced his co-inventor, John Kay, to the status of glorified handyman, and dumped his original investor, the loyal John Smalley, who had put him in business way back in Preston when automated spinning was merely a pipe dream. It hardly mattered, however. Arkwright wanted him out, and Strutt, ever a gentleman, tried to mediate in a letter to Smalley:

> . . . am sorry to find matters betwixt you & Mr. Arkwright are come to such extremities & wonder he should persist in giving you fresh provocations. We cannot . . . prevent him from saying Ill-natured things nor can we regulate his actions, neither do I see that it is in our power to remove him . . . you must be sensible when some sort of people set themselves to be perverse; it is very difficult to prevent them from doing so."

No one prevented Arkwright: he grudgingly bought out Smalley's shares for £3,200 and wrote him off. Strutt himself hung on until 1781, the year that the remaining partner, Need, died. By then Arkwright had embarked on a plan for vast expansion; Strutt was content instead to build one mill of his own and tend to his silk hosiery knitting enterprise. Throughout the years that followed, Arkwright would make enemies as effortlessly as he made money; many were former business partners like Smalley, whom he abused without any recorded pangs of conscience.

Strained relations for Arkwright were simply the price one paid to realize a dream, and his dreams were more than monetary; a lifelong striver, he wanted nothing less than to be admitted into English high society, which was as imperious as any that ever arched its collective nostrils at a nouveau riche like Arkwright. Upper-crust chambers were all but impenetrable to a lowly barber's son, and yet in time they would open reluctantly to him and his newly accumulated wealth. Cotton would get him past the doorman; sheer determination would do the rest. Here was a man plagued from childhood with severe asthma, always in poor health, who continued to rise each day at 5 A.M.—until well past the age of fifty and long after becoming a multimillionaire—in order to practice his writing, grammar, and speaking skills so that he might find acceptance among the class-conscious gentry.

An excerpt from a lengthy letter he sent to his primary partner, Strutt, suggests there was plenty to improve: "When I rote to you last I had not throwoly provd the spinning; several things apening I could not acount for sinse then has proved it. . . ." Strutt himself was an entirely different proposition—poised, gracious, and to the manor born. He was also shrewd; together with Arkwright he applied in 1768 for patents on their revamped mechanized carding machine and the water frame. He too recognized that the experimental spinning frame's large size and considerable weight required more horse power than horses alone could provide. He was willing to wager that rushing water might be able to furnish sufficient power to a six-story factory filled with belt-driven machinery.

It was a foolhardy undertaking, many thought, made even more reckless by the men's decision to erect the new mill over a Derbyshire stream in Cromford, linked only by fourteen miles of poor rutted road to the nearest town, Derby, and not within easy distance of any port where cotton was off-loaded. Located at the juncture of the Derwent River and one of its smaller tributaries, Cromford was nothing more than a few sheep, a pasture, and a handful of stone structures. That's how Arkwright apparently wanted it. There would be no local mob of enraged hand-laborers to storm his operation and chase him over the hill, no indignant townspeople to placate. The proposed factory was like the baseball field in the film *Field of Dreams*: If we build it, they will come. But simply showing up in Cromford wasn't enough. Skills were required, at least for the initial building and machinery construction, because all of the mill's equipment was to be forged and die cast from the bottom up right at the site. Arkwright, Strutt, and Need placed their first advertisement for workers in the *Derby Mercury* in 1771 as construction neared completion:

> WANTED immediately, two Journeymen Clock-Makers, or others that understand Tooth and Pinion well: Also a Smith that can forge and file . . . Weavers residing at the Mill, may have good Work. There is Employment at the above Place, for Women, Children &c. and good Wages. . . .

In that simple posting one could hear the rumblings of a tectonic shift that was about to shudder across England and upend its social structure. Initially, a small stable of expensive, highly specialized metalworkers and craftsmen were needed to design and manufacture the differential components that transferred water power into an energy source for the belt-driven spinning frames, and then to build the frames and related machines. Clockmakers adept at casting bronzed gears were eagerly recruited. But once the frames were up and running, no more than a few skilled employees were required to maintain and run the machinery. The water frame was in fact designed so that all the complex processes were built into the machine itself, while its daily operation was simple enough to ensure that "there is employment at the above Place, for Women [and] Children"—in other words, cheap, unskilled factory labor.

The most revolutionary aspect of the industrial explosion across England was the way that these machines almost overnight dismantled centuries of traditional rural English family life and reassembled it to create a migratory work force. If necessity is the mother of invention, financial security might be its dowager aunt. The promise of a regular income was an alien concept to cottage-industry families, many of which were only a missing shipment of raw wool or a broken spinning wheel away from going without food. If the mill was willing to hire parents and offspring alike, families could move as a unit and remain intact. Parents were delighted their young children could find a livelihood.

By 1771, when the first mill opened at Cromford, hundreds of new employees had arrived. There would be close to a thousand within a decade. The comfortable brick row housing Arkwright built for them still stands. He also built a chapel and a plush hotel. Children as young as eight years old worked from six in the morning until seven at night with half an hour off for breakfast and forty minutes for dinner. They were educated in church on Sunday, and the mills operated twenty-three hours a day. The factory itself, built over the rapids with a massive wooden waterwheel beneath it to catch, churn, and release vast quantities of water in its horizontal vanes, called buckets, quickly became a tourist attraction. People would camp out on adjacent hills to watch the glow of the machinery at night. Among the poets it inspired was Erasmus Darwin, grandfather of Charles:

> Where the Derwent guides his dusky floods
> Through vaunted mountains and a night of woods
> The nymph *Gossypia* [cotton's botanical name] treads
> the velvet sod
> And warms with rosy smiles the watr'y gods. . . .
> From leathery pods the vegetable wool [cotton];
> With wiry teeth *revolving cards* release
> The tangled knots, and smooth the ravell'd fleece:
> Next move the *iron hands* with fingers fine,
> Combs the wire card, and forms th' eternal line;
> Slow with soft lips the *whirling can* acquires
> The tender skeins, and wraps in rising spires:
> With quicken'd pace *successive rollers* move,
> And these retain, and then extends the *rove*;
> Then fly the spokes, the rapid axles glow,
> While slowly circumvolves the labouring wheel
> below.

(From "The Botanic Garden"; italics added.)

There are many women who could only hope to be immortalized so lovingly. At Cromford, still miles from nowhere but now very much on the map, all of that action was driven, in Darwin's words, by the "pond'rous oars" of the waterwheel. Most found it all awe-inspiring. But a few nobles, like Viscount Torrington, passed through and sneered: "Every rural sound is sunk in the clamours of cotton works. These vales have lost all their beauties; the rural cot has given place to the lofty red mill; . . . the stream perverted from its course by sluices, and aqueducts, will no longer ripple and cascade. . . ." The viscount was most disturbed that the "simple peasant . . . is changed into an impudent mechanic."

In that single dismissive phrase, the viscount deftly summed up the next two hundred years of Western civilization. Arkwright, himself transformed from destitute wig maker to feudal lord-in-progress, raced about the countryside in his coach and four—that era's equivalent of a Hummer—building new cotton spinning mills as far away as Scotland, forming new partnerships, and chasing down anyone who dared try and steal his engineering designs.

It might have been predictable that a man whose personal wealth was derived from swiping intellectual property would become maniacal about protecting his own. Much of Arkwright's time was taken up defending patents, which by law expired after fourteen years, against claims that they were invalid, or in chasing down unlicensed imitators. In 1781 he sued nine competitors for infringement. The principal defendant produced John Kay and Thomas Highs to combat Arkwright's assertion that he alone was entitled to the patent rights. As a result, the patents were annulled in 1785, but by then it hardly mattered. Arkwright was so far ahead of the competition in resources and in output that pirates stood little chance against him. One thing was certain: there was big money to be made in cotton warp yarn, and now that he was making it, Arkwright wanted no rivals. His factories were his bank vaults: steal from them at your own risk. Arkwright wrote Strutt about their first Cromford mill: "Desire ward to send those other Locks and also . . . some good Latches & Catches for the out doors . . . and a large Knoker or a Bell to First door. I am Determind for the feuter to let no persons in to look at the wor[k]s except spinning."

Arkwright's earliest 1769 patent would soon expire in 1783; his 1775 carding machine patent was invalidated six years later when he brought suit against nine manufacturers for infringing on it—invalidated in part because no one could make an operable machine from the accompanying plans. That, Arkwright finally admitted, was no accident. He had purposely left out crucial components to prevent piracy. When the verdict went against him, Arkwright, furious, swore that the new factory he was then building in Manchester "shall never be worked & will sooner be left for Barricks for Soliders." This was reported by a prominent manufacturer, Matthew Boulton, who added: "It is agreed by all who know him that he is a Tyrant . . . & tis thought that his disappointment will kill him. If he had been a man of sense and reason he would not have . . . lost."

By then, after only a decade, Arkwright's hold on the cotton yarn industry was so well established that he had a virtual monopoly on it. His "fame . . . resounded throughout the land; and capitalists flocked to him, to buy his patent machines, or permission to use them," his first biographer, Edward Baines, wrote.

Arkwright, for all the grandiosity of his vision, was clever enough not to spread himself too thin. He at first attempted to introduce weaving looms as well into his Cromford mill but soon gave that up. The technology was still too primitive. So was textile printing. Both would soon advance, but not on Arkwright's nickel. Importing raw cotton from India and the West Indies and converting it into yarn for apparel, netting, covering, household furnishings, or industrial use was exclusively his domain and his passion.

Some, like Baines, a Leeds newspaper editor who wrote that era's definitive history of cotton in 1835, never forgave the man for his light-fingered boosting habits: "I found myself compelled to form a lower estimate of the inventive talent of Arkwright than most previous writers. In the investigation I have prosecuted . . . it has been shewn that the splendid inventions which even to the present day have been ascribed to Arkwright . . . belong in a great part to other and less fortunate men. In appropriating these inventions as his own, and in claiming them as the fruits of his unaided genius, he acted dishonourably, and left a stain upon his character. . . ."

Two hundred fifty years later, Christopher Charlton disagrees. As the director of the Arkwright Society in Cromford, Lancashire, Charlton naturally has a vested interest in burnishing Arkwright's tattered reputation. But it wasn't until he was called upon to turn a nonworking replica of an early water frame into a functional machine for a German exhibit that he began to appreciate Arkwright's genuine mechanical engineering gifts. Charlton and a clockmaker he hired to assist him spent months—"a thousand hours"—trying to synchronize the rotating pick-up spindle shaft that gathered the twisted yarn with the speed of the triple set of gear-driven rollers, each operating at a different velocity. "And it wasn't just his infernally intricate frame that impressed us," Charlton explains. "We occupied ourselves with the same problems he ran into in all the numerous preparation phrases and tried to come up with workable solutions. We didn't. He did."

Most impressive to Charlton were the improvements Arkwright made to his mechanical carding machine. Before his improvements, sheets of raw cotton were fed onto two rotating cylinders with protruding fine teeth and emerged ready for the first spinning stage. At the end of each cycle the machine had to be stopped while a worker ripped off the carded sheet as he might a piece of wallpaper, then hand-carried it to the first spinning frame and joined it to the previous sheet. Arduous, slow work. Arkwright came up with a modulating crank—a metal bar—and comb that pushed the cotton off the cylinders as the crank lifted up to create a single roll of carded fiber that became a continuous part of the automated production line, eliminating the stoppage. Arkwright quickly applied for patents on his mechanical improvements. The game was to stay always ahead of the pack.

"I have no doubt that Arkwright was an inventive genius," says Charlton, a retired economics historian at Nottingham University. "Also, that he was not a very pleasant man. He single-handedly pursued his own fortune without much regard for family or friends."

But then, when has an amiable disposition ever been the hallmark of a captain of industry?

Spinner of Fortune

The English cotton trade was now booming. In common usage all distinctions between glazed chintz and printed calico had been long abandoned in favor of one term, calico, to describe anything cotton and colorful. Revisions in the Calico Act allowed English textile manufacturers to turn out reams of cotton fabric without stiff taxation penalties. Arkwright's warp yarn was shipped to weavers throughout England. Between 1775 and 1783 calico output increased about 600 percent, from 56,000 to 3 million yards. Raw cotton imports increased from 4.7 million pounds in 1771 to 56 million pounds in 1800; the sizeable majority of it supplied Arkwright's mills. Consumers couldn't get enough. In 1783 a London magazine reported that "every servant girl has her cotton gown and her cotton stockings whilst . . . articles of wool be mildewed in our mercers' shops." In the end, Arkwright's factories employed about 5,000 workers; he spent millions in today's currency on new construction. "It is impossible to estimate the advantage to the bulk of the people, from the wonderful cheapness of cotton goods," Baines remarked.

Within a decade Arkwright launched more than a dozen yarn

factories. There were about two hundred in operation a decade later. Two men would soon come along who shared Arkwright's curiosity not just about the way things worked, but about the way they could work better. Both contributed their practical talents to the explosive growth of textiles.

Shy and retiring Samuel Crompton, Arkwright's opposite in most ways, was a frustrated spinner who set to work to create fine, strong yarn that could compete with Indian muslin. Ever since boyhood, Crompton had been spinning at home on the jenny for his family's needs in nearby Bolton. The jenny's abrupt movements constantly snapped weak threads. Much to his regret, Crompton spent endless hours reattaching frayed pieces by hand. For five years, to 1779, Crompton labored secretly at night, he said, "to realize a more perfect principle of spinning, and though often baffled, I as often renewed the attempt, and at last succeeded . . . at the expense of every shilling I had in the world."

Finally perfecting a carriage to replace stationary spindles with moveable ones, Crompton was able to elongate the thread that Arkwright's rollers first reduced from the thicker roving. Pulled away from the rollers like a fully extended arm, the moving spindle carriage stretched and twisted the threads. Returning, it wound the attenuated threads on its spindles. The result was a much finer yarn, strong enough for warp, and far superior to any that Arkwright's water frame could produce on its own.

Crompton's machine quickly became known as the mule, because it yoked together the best features of the water frame with those of Hargreaves' jenny. He gained local fame and a measure of prosperity but decided he could not afford a patent, so he took apart his mule at night and hid it in his loft to avoid the mobs that were smashing Hargreaves' jennies. Manufacturers got wind of his miraculous mule; he was invited to dinner with potential investors in Glasgow, but, overcome by shyness, he explained, "Rather than face up, I first hid myself and bolted from the city."

Lacking Arkwright's acquisitive instincts and the ferocious

determination he needed to create his own empire, Crompton decided to allow the mule to become public property. At length, in 1812, he received a meager £5,000 grant from Parliament for his invention. By then 4,600,000 spindles were at work on mules in Manchester, using 40,000,000 pounds of cotton annually, employing 500,000 workers. His mule was now standard equipment. Crompton, also a gifted musician and brilliant inventor but not equipped to be a businessman, used that grant to help his sons open a bleaching enterprise to prepare cotton cloth for printing. Through mismanagement, they lost all of his money, and Crompton, like Kay and Hargreaves before him, died in poverty.

Looming Triumphs

The second inventor whose work profited Arkwright was the Reverend Edmund Cartwright, a minister of the Church of England. Cartwright tells his own story best: "Happening to be in Matlock in the summer of 1784, I fell in company with some gentlemen of Manchester, when the conversation turned on Arkwright's spinning machinery. One of the company observed that, as soon as Arkwright's patent expired, so many mills would be erected, and so much cotton spun, that hands could never be found to weave it. To this observation I replied that Arkwright must set his wits to work and invent a weaving mill."

All of the Manchester gentlemen immediately agreed that this was an impractical solution. The arguments they put forth, Cartwright says, "I certainly was incompetent to answer, or even comprehend, being totally ignorant of the subject, having never at any time seen a person weave."

Not a promising start, to be sure, but Cartwright had recently seen an exhibition in London, "an automation figure which played at chess." It got him to thinking, and "sometimes afterwards . . . it struck me that, as in plain weaving according to the conception I then had of the business, there could be only three movements which were to follow each other in succession; there would be little difficulty in producing and repeating them. Full of these ideas, I immediately employed a carpenter and smithy to carry them into effect. As soon as the machine was finished I got a weaver to put in the warp which was of such material as sail cloth is made of. To my great delight, a piece of cloth, such as it was, was the production."

Cartwright, ever humble, goes on to explain that he was such a novice at all this that "his machine at first clanked and shrieked. . . . [T]he springs which threw the shuttle were strong enough to have thrown a Congreve rocket." (That same Congreve provided the "rockets' red glare" that inspired Francis Scott Key to write "The Star-Spangled Banner.") Continuing to do everything backward, Cartwright first applied for a patent in 1785, then set out to see for himself how people actually wove. ". . . [A]nd you will guess my astonishment when I compared their easy mode of operation with mine. Availing myself, however, of what I then saw, I made a loom, in its general principles, nearly as they are now made; but it was not till the year 1787 that I completed my invention, when I took out a weaving patent. . . ."

Driven by Watt's steam engine, Cartwright's power loom would soon help launch Manchester's cotton manufacturing dynasty. The functional mechanized weaving apparatus made by a complete stranger to textiles seems to confirm that the most useful background knowledge at times may be none at all. If a quiet mind cureth all, as an English philosopher once remarked, an unencumbered one seems free to entertain the impossible. Cartwright took a clean shot at a complex process and intuitively stripped it down to its essentials. Having invented the automated loom, at first powered by bulls, then by steam, the reverend went about his primary occupations—God and literature. He settled in a rectory and published legendary poems. "Few persons could tell a story so well, no man make more of a trite one," a friend remembered. Manufacturers who first looked at his loom were unimpressed, but Cartwright also continued to invent. As an Oxford don governed by scruples that never fazed Arkwright, he refused to look at any inventions he sought to make improvements on, concerned that he might unconsciously borrow an idea.

There were also the outraged mobs of hand weavers to contend with; as before, they set out to destroy what they could not adapt to or master. In 1790 they burned down a factory that contained 400 steam-powered Cartwright looms. That upset the reverend so much that he abandoned a small mill of his own and turned his attention instead to agricultural implements, inventing a three-furrow plow. His power loom did not come into general use until 1801; by 1833 there were over 100,000 in operation. Cartwright never made money from his invention, but Parliament rewarded him with a £10,000 grant; that enabled him to buy his own farm, and he become perhaps the only authentic English cotton textile inventor who did not die in poverty. He lived to eighty, a "portly, dignified old gentlemen," said a friend, "grave and polite, but full of humor and spirit."

Steaming Forward

"From the year 1770 to 1778, complete change had been effected in the spinning of yarns; that of wool had disappeared altogether, and that of linen was nearly gone; cotton, cotton, cotton had become the universal material for employment; the hand-wheels were all thrown into lumber rooms . . . ," observed early-nineteenth-century English historian William Radcliffe. He also had a few choice words to say about weaving: ". . . [E]ven old barns, cart-houses, and out-buildings of any description were repaired, windows broke through the old blank walls, and all fitted up for loom-shops . . ." But that was only the first wave of growth spurred by Arkwright's new industry. Towns became cities, and factories redefined the landscape. Nature was being pushed aside, manipulated, assaulted, or trammeled. "For desolate moors and fens . . . ," commented historian Arnold Toynbee, "we now have crowded cities with their canopies of smoke."

Steam drove that. However, until James Watt's invention was successfully applied to driving textile machinery in 1790, mills needed the power of surging water. Coal-fueled steam would be a portable commodity, freeing manufacturers to build factories away from rolling rivers, but that mobile energy source was still to come when Watt, under the patronage of Adam Smith at Glasgow University, began tinkering with a crude engine used to pump water from mines. It consumed enormous, inefficient quantities of steam and coal, Watt observed, and he used his scientific background to investigate ways to improve its construction. "I had gone for a walk on a fine Sabbath afternoon . . . ," he later wrote about his epiphany in 1765. "I was thinking of the engine at the time. I had gone as far as the herd's house when the idea came into my mind that as steam was an elastic body it would rush into a vacuum and if a connection were made between the cylinder and the exhausting vessel, it might there be condensed without cooling the cylinder. . . . I had not walked further than the Golf-house when the whole thing was arranged in my mind." He introduced a layer of steam in a jacket between the engine's inner and outer cylinders, as a way to stabilize its temperature, formed a partnership with industrialist Matthew Boulton, and obtained a patent in 1769; together they built the first commercial machine in 1777. Watt spent much of his life improving the steam engine until he made it applicable to driving the kind of heavy machinery used in textile manufacturing.

A contemporary, Lord Jeffrey, offered his eloquent appraisal of Watt's achievement: "The trunk of an elephant that can pick up a pin or rend an oak, is as nothing to it. It can engrave a seal, and crush masses of obdurate metal before it; draw out, without breaking, a thread as fine as gossamer, and lift a ship of war like a bauble in the air. . . ."

Another contemporary, speaking to a group of mechanics in

Paris, explained that "This machine represents, at the present time, the power of three hundred thousand horses, or two million of men, who should work day and night without interruption, and without repose, to augment the riches of a country. . . ." The speaker, Charles Dupin, was exhorting these French mechanics to match the output of their English counterparts in a land where, he said, more than one million of the inhabitants were employed in textile operations. "The Indies, so long superior to Europe . . . are conquered in their turn . . . so great is the power of the progress of machinery."

Mills of all sorts, at the end of the eighteenth century, began to adapt the steam engine to produce sugar, flour, lumber, iron, and pottery; in every instance wood was first used to supply the steam engine's fuel. That threatened Britain's forests and might have ended its reign but for the appearance of another apparently unrelated invention, Sir Humphrey Davy's safety lamp. That enabled English coal miners to work for prolonged periods in mines where gaslight had proved to be an explosive danger and the legendary cause of lethal accidents. As a result, miners were able for the first time to supply ample amounts of coal to run the country's steam engines. English industry surged forward as skies blackened with acrid soot.

All of these remarkable inventions were governed by unwritten laws of interdependence. The power loom had to be perfected to take full advantage of the output of the water frame and jenny; the loom required the steam engine to realize its full potential; and that engine itself would not have become fully functional without an increased supply of coal. Inspired coincidence? Perhaps. Sweat and hard work? Certainly, and presiding over it all was the godfather of meshing gears and yarn skeins, Sir Richard Arkwright. His best invention was undoubtedly himself, a gritty poor boy scrubbed and buffed to pass muster as an aristocrat.

Knight Moves

About Arkwright, Edward Baines commented, "His concerns in Derbyshire, Lancaster and Scotland were so extensive and numerous as to show at once his astonishing power of transacting business and his all-grasping spirit. . . . So unbounded was his confidence in the national wealth to be produced by it, that he would say . . . *he* would pay the national debt." That sounds like Arkwright, the kind of self-made millionaire who invites you to a splendid dinner then makes you pay for each bite of your chocolate truffle dessert with his insufferable boasting. Still, he impressed even Baines: "The most marked traits in the character of Arkwright were his wonderful ardour, energy, and perseverance. . . ." Later he adds: "To Arkwright and Watt, England is far more indebted for her triumphs than to Nelson and Wellington. Without the means supplied by her flourishing manufactures and trade, the country would not have borne up under a conflict so prolonged and exhausting."

The conflict he's referring to, the Revolutionary War, was funded in large part on the English side by cotton taxation revenues. Even so, there were problems. By 1779 France had sided with America's fight for independence and was attacking British ships; Spain blockaded Gibraltar. British textile exports declined as a result, and so did textile factory work. Arkwright kept on building. Thousands of Lancashire hand workers formed an impromptu army to bring down the spinning monopoly created by his expanding empire; mobs destroyed one of Arkwright's existing factories and marched on Cromford. Arkwright was in no mood to appease. He hired an army of 1,500 troops and brought in cannons. The mob closed in, then retreated. No one was killed.

Finally, in 1786, the king knighted him Sir Richard. As befits any feudal lord, he built himself a drafty castle above Cromford and named it Willersley. In his official portrait, done late in life, Arkwright sits beside a model of his beloved water frame. Rotund, with a bulbous nose and small piggish eyes, he wears a richly textured waistcoat over a vest that barely contains his alarming girth. Its buttons may pop off at any moment, but clearly he doesn't give a damn. He can buy your approval if he needs it. By now, in 1790, his domain extends to close to two hundred factories.

Tattered Goods

On his way to the bank, Arkwright set in motion the dynamics for vast social and economic change in England and in the United States as well as in continental Europe. Although self-absorbed, Arkwright took a genuine paternalistic interest in the welfare of his workers. The factory system that followed in his wake, however, quickly gained notoriety as England's squalid national disgrace. Manchester and numerous other cities built around the cotton textile industry became filthy, dehumanizing urban blemishes. Their mills subjected young children, orphaned "parish wards" from workhouses, to lives of enslaved labor, often at the mercy of sadistic bosses. One factory boy, Robert Blincoe—born in 1792, the same year that Arkwright died—survived to tell his story in an 1832 account that created political upheaval and put a human face on the country's labor reform movement.

Blincoe, who is generally considered to be the model for Charles Dickens's Oliver Twist, was an orphan sold into forced labor at age seven and sent to a cotton mill where children of his age were forced to work for fourteen hours a day. Blincoe wrote that the smell of untreated cotton, the hot engine oil from the frames, and the overbearing heat and noise inside the mill were instantly nauseating. The work could be dangerous as well. "Some," Blincoe explained, "had the skin scraped off the knuckles, clean to the bone; others a finger crushed or a joint or two nipped off in the cogs of the spinning frame wheels." Although Blincoe himself lost half a finger, he was ordered to continue working once the "surgeon" had stanched the bleeding. The overlookers, or bosses, he described as "a set of brutal, ferocious, illiterate ruffians alike void of understanding as of humanity." If he did not work fast enough he was cursed, pulled by the hair or beaten with a stick to "stimulate" him. "We went to the mill at five without breakfast, and worked until eight or nine, when they brought us . . . water porridge with oatcake in it . . . in a tin can. This we ate as best we could, the wheel never stopping." After a lunch break, "we worked on until eight or nine at night without bite or sup." He and the other children stood all day—there were no seats in the mills—and they often slept six or more to a windowless basement room.

Blincoe often contemplated throwing himself out a window even before turning eight, but lost his nerve.

The Factory Acts, well-intentioned reform laws on the book in England restricting work hours and regulating the treatment of children, were roundly ignored. Manchester's cotton mills employed—or enslaved—close to 200,000 children under eighteen in England by 1839. Many tried to flee. Potential delinquents frequently wore chains day and night that extended from ankle irons to their hips to prevent their running away. Crippled adults came forward to tell how, as child "doffers," their bodies had been deformed by years of stooping all day to change bobbins on the frames. Within a few short decades, cotton manufacturing, a source of immense English pride that once inspired utopian poetry, had come to symbolize the most deplorable excesses of free enterprise. Few if any held Arkwright, now dead, directly responsible, but as the godfather of the Industrial Revolution, he was eventually deemed guilty by association.

More humiliating still, generations following the Victorians forgot or ignored him. Most English children today have never heard of Arkwright, long reduced to a scrunched footnote in their school texts. On the bicentenary of Arkwright's birth in 1932, the *Manchester Guardian* smugly dismissed him as a thief and a rogue, apparently forgetting that there would have been no Manchester as a center of England's textile industry without him. However, worse insults awaited Arkwright. His first mill at Cromford was being used to store discarded toxic paint cans when Charlton first made an effort in the 1970s to salvage it as a museum to honor the man's achievements.

For better or worse—in reality, both—it was Arkwright who first put the various processes of mechanical production under one roof and who trained "human beings to renounce their desultory work habits," in the quaint phrase of a Victorian admirer. Workers no longer tended to machines between other chores; machines were now the only chore: they dictated when and where you worked; they started up with the peel of the factory whistle and demanded your attention and attendance all day long, six days a week. You manipulated their wheels, shuttles, treadles, and levers. In turn, they controlled your life.

Arkwright barely lived long enough to witness these fundamental changes; there were still more soon to come. Less than a year after his death, three thousand miles away on a sleepy plantation in Georgia, an unemployed itinerant Yankee tutor named Eli Whitney tinkered with a device to separate the embedded sticky green seeds of a cotton boll from its surrounding fibers. His ingeniously simple cotton gin—short for engine—practically overnight spurred a languishing Southern agrarian economy into becoming one of the wealthiest the world has ever known. It also set the most influential economic and political powers on three continents into frenzied motion and forced Americans to choose between money and morality—with nothing less than the life or death of their country at stake.

FOUR

Revolutionary Fiber

If the quantity of wool, flax, cotton & hemp should be encreased to ten-fold its present amount (as it easily could be) I apprehend the whole might in a short time be manufactured. Especially by the introduction of machines for multiplying the effects of labour . . .

—George Washington,
proposed address to Congress, 1789

Alone in his one-horse carriage, slogging down a rutted road in 1803 to yet another courthouse in the boiling heat of a Georgia summer, Eli Whitney must have felt like a man trapped in chains taunted by the allure of a beautiful woman who is everyone's lover but his. Wherever he looked, oceans of upland cotton undulated seductively in all directions, just out of reach. The crop offered unparalleled wealth to thousands of Southern plantation owners and their families but not to Whitney—a cruel twist of fate since his mechanical genius was solely responsible for the fortunes they accumulated. "Some inventions are so invaluable," he wrote to his father back in Massachusetts, "as to be worthless to its inventor." Whitney had devoted the previous ten years to trying to enforce the patent he and his business partner had taken out on his gin. Before Whitney's invention, a slave could pull free one pound of cotton lint a day by hand. The gin enabled her to separate out fifty pounds of raw fiber in that same time frame.

Despite the patent that Whitney and his partner, Phineas Miller, had successfully applied for in 1794, and the instantaneous popularity of the gin, they had yet to make a penny. Whitney and Miller, both transplanted Yankees, discovered that to Southern cotton farmers, a legal patent was no more intimidating than flimsy fence wire—a mere nuisance that slowed your progress only for as long as it took you to snake around it.

Ten years later Miller was near death at the young age of thirty-nine and Whitney, thirty-seven, was carrying on their futile fight against patent infringement with little hope of victory and no funds. They had unsuccessfully filed suits sixty different times in Georgia alone. All the while anyone with a few tools and some basic carpentry and handyman skills could cobble together his own crude working version of Whitney's gin and never feel any obligation to pay the licensing fee the two partners demanded. State courts from North Carolina to Georgia during that decade did nothing to stop the flow of pirated gins. At best, judges leveled meager fines; even then, Whitney and Miller received no compensation. These courthouses were situated in the heart of cotton country; juries were stacked with the planters' cronies. "I have a set of the most Depraved villains to combat and I might almost as well go to *Hell* in search of *Happiness* as apply to a Georgia-Court for Justice," Whitney wrote to a friend in 1803. By then he was at wit's end, surrounded daily by endless expanses of valuable fiber that was making everyone rich but him. "It is better not to live than to live as I have for three years past," he wrote Miller. "Toil, anxiety and disappointment have broken me down. My situation makes me perfectly miserable."

Against all odds, Whitney would live long enough to triumph as a musket maker in later life and to enjoy a measure of the riches that eluded him. But in the first decade of the nineteenth century, exhausted, penniless, and hopelessly in love with his business partner's wife, he appeared to be yet another bereft inventor cast in the mold of Hargreaves, Kay, and all the others whose genius transformed cotton from a gangly shrub into a booming automated industry and who never gained anything but grief for all their efforts.

Washington Spun Here

Early on, long before Eli Whitney and before the first muskets were drawn at Concord in 1776, the eventual leaders of the new republic realized that to be self-sustaining our citizens would have to dress themselves in native cloth grown and manufactured without any reliance on foreign powers. The alternative, they decided, was a little like borrowing clothes from your parents in order to run away from home. To George Washington, Tench Coxe, Alexander Hamilton, and their colleagues, imported British dry goods were the emperor's mantle of subjugation. They reminded one and all of our continuing economic dependence. Initially flax, spun and woven into linen, was the primary native colonial resource for fabric, along with wool. Cotton, cultivated domestically only on small acreage, might have seemed like an appendage at best, but to our first leaders it also represented an investment of hope in the country's future.

George Washington knew cotton. At Mount Vernon he spun and wove it and other clothing goods as well for twenty-eight persons in addition to his wife, Martha. In his 1768 business summary, Washington noted that about three hundred of the 1,556 yards of cloth were used for "cotton-striped, cotton plain, cotton filled, cotton-birdseye . . . and Cotton-India dimity." For his own use he spun and wove linen, wool, linsey, and "forty yards of cotton," in an effort to free himself and his family from English dependence. In that, he anticipated Mahatma Gandhi more than a century later, symbolically reducing India's massive struggle for self-rule to a single strand of cotton thread.

Later, in 1789, newly elected President Washington expressed even more passion about home industry. As he wrote in a letter, "I hope it will not be a great while before it will be unfashionable for a gentleman to appear in any other dress (except homespun). Indeed, we have already been too long subject to British prejudices. I use no porter or cheese in my family, but such as is made in America." By then transplanted English calico printers had begun to set up shop in earnest. The most renowned, John Hewson in Philadelphia, received a £200 business loan from the state of Pennsylvania treasury in that year. Hewson gained Washington's attention both as an artisan and loyal citizen. He fought for the colonies in the Revolutionary War, was captured, and escaped with his life despite a reward of fifty guineas offered "for his body, dead or alive." The British did not want Hewson as a rival to its burgeoning textile trade.

Even after the Revolutionary War had been fought and won, English cotton prints continued to flood the colonial market. In 1785, more than 350,000 yards arrived on our shores. To Hewson, his former countrymen were "savage foes." He poured all of his patriotism into his work. Martha Washington wore Hewson's cotton calicos with pride in the Federal Parade in Philadelphia in 1788, and she commissioned him to print cotton handkerchiefs displaying an image of her husband on horseback in full military dress. They were a huge hit, particularly because the medium was also the message: General Washington in miniature on cotton printed in America, a celebration of his heroism and the country's homespun ingenuity.

In the absence of machinery capable of both spinning, weaving, and mass-producing finished cloth, domestic cotton would not become the nation's leading textile until New England's new mills began to manufacture it in large quantity in the early decades of the 1800s. Still, it had established its American roots from the very first settlement at Jamestown, Virginia, in 1607. It had been planted there in the first year, and by 1621 "the plentiful coming up" of seeds sown experimentally gained attention on both sides of the Atlantic. In 1619, James Rolfe brought "twenty Negars," the first slaves, to Jamestown, an eerie precedent for things to come.

Wool from imported sheep was the basic fabric spun and woven at Jamestown and Plymouth to protect against harsh winters. William Bradford, leader of the Plymouth Colony, earned his living as a weaver blending wool and cotton into fustian. Cotton showed up as an import in Boston in 1638. Two years later, the colonial legislators of Massachusetts offered a bounty for three years of three-pence on a shilling for cotton, linen, or wool if spun or woven locally, in order to encourage domestic manufacture. Two of the governor's sons settled in the West Indies, where Sea Island cotton grew profusely on slave plantations; they shipped the raw cotton back home. "We can make dimities and fustians for our own summer cloathing," said the proud author of "New England's First Fruits" in 1642. By the late seventeenth century farmers in Virginia and the Carolinas were exporting their crops to other colonies.

Things quickly became more complicated. God-fearing Puritans set aside their bibles long enough to import slaves from Africa and to barter them for cotton and rum to West Indian plantation owners. That made New England the first group of slave-trading colonies. Its citizens wove their cotton into a variety of blended fabrics. Soon spinning wheels and looms filled the homes of New England's villagers. Spinning festivals were held on the public commons. Hemp and linen prevailed; cotton was still in short supply, although that would change in the decades to come. As England enacted its Stamp Act and other repressive measures, domestic home textile manufacturing began to take on the patriotic trappings that Washington espoused. The senior class at Harvard College in 1768 dressed entirely in native fabrics to show its contempt for harsh British taxation. On a small scale, cotton was grown as far north as New Jersey for use during the Revolutionary War.

None of the recent textile innovations in England—James

Hargreaves' jenny among them—made it across the ocean; that left the colonies to rely on hand wheels and looms that produced fabric with agonizing slowness, while Richard Arkwright in England was busy transforming an artisan craft into a mechanized industry.

Tench Coxe, who also saw that the "Cotton Spinning Mill might be brought into beneficial use in the Unites States," was a founding father with a fascination for technology; he sent an English mechanic across the seas at his own expense to make brass models of Arkwright's water frame, but the designs and models were seized, and the mechanic was forced to remain in England for three years. The English, obsessively protective of their textile inventions, kept a close watch on exports, but that didn't prevent early industrial spies from trying to smuggle innovative machinery to America as cargo purposely mislabeled "cards for cattle" and "teeth for horse-rakes" (actually a shipment of spindles).

James Madison, like Coxe and Washington, looked to the future. Madison recalled telling Coxe at a convention held in 1786 to devise means to bolster the industrial output of the country that, "There was not reason to doubt that the United States would one day become a great cotton-producing country." John Adams, however, disagreed. He argued that the United States would not be able to supply herself through her own manufacture for three hundred years. Thomas Jefferson was soon writing to a friend that "the four southernmost States make a great deal of cotton. The poor are almost entirely clothed in it in both winter and summer. In winter they wear shirts of it and outer clothing of cotton and wool mixed. In summer their shirts are linen, but their outer clothing cotton. The dress of women is almost entirely of cotton manufactured by themselves, except the richer classes, and even many of these wear a good deal of homespun cotton. It is as well manufactured as the calicoes of Europe. Those four Southern states furnish a great deal of cotton to the States north of them, who cannot make it as being too cold."

Alexander Hamilton shared the dream. "Several of these Southern colonies," he wrote, "might some day clothe the whole continent." These men knew that even with the laborious processing needed for upland cotton prior to Whitney's gin, an increasing amount of it was being grown and harvested for export and home use in the South. Many British authorities, on the other hand, were still apparently clueless. In 1784 eight bales were sent to a broker in England. They were seized at the customs house on the grounds that they violated the country's Shipping Act because they could not possibly have been harvested in the United States, their alleged country of origin, and therefore were being illegally transported in an American ship. The bales sat on the Liverpool docks for months.

Shrewd businessmen as well as public servants, the architects of American democracy, understood the importance of nurturing a resource that might one day deliver jobs and profits. Cotton provided agrarian idealists like Jefferson with visions of an agricultural economy sustained by prosperous small farmers; to industrialists like Hamilton, cotton furnished the possibility of a large-scale domestic textile manufacturing supported by native crops. Short- and medium-staple upland cotton in particular he saw as being easily adapted to machines.

All of this comprised an elevated form of wishful thinking. The Federalists knew that Britain was now in the process of mastering mass production but that she lacked the ability to grow her own crops. They knew that India provided a much less convenient source than the United States for its raw material. Cotton produced in volume might not only supply domestic needs but bring in large sums as a leading export to help support the new government. When John Cabot in 1788 built the nation's first cotton factory in Massachusetts near Boston, on the Bass River at Beverly, a Virginia weaver—President George Washington—soon showed up to take a tour. "In this manufactory, they have the new invented spinning and carding machines," he wrote in his diary, "one of which spins eighty-four threads at one time by one person. . . . In short, the whole seemed perfect, and the cotton stuffs they turn out excellent of their kind. . . ."

Cabot's machines were literally driven by horsepower under the direction of a boy, Joshua Herrick. "When the horses went too fast, Mr. Somers would call out the window, 'Hold on there! Not so fast! Slower!' and Herrick would slow up, but soon he would forget and speed up again, when again Somers would cry out, 'Hold up!' and this continued most of the day," according to one first-person account.

Washington, clearly fascinated by mechanical invention, had no way of knowing that the Beverly operation would soon fail, or that only a few months after his 1789 visit, about fifty miles south in Pawtucket, Rhode Island, a former superintendent at one of Arkwright's partner's cotton mills, Samuel Slater, would arrive from England disguised as a farmer to avoid detection by the authorities. (Fearing industrial piracy, the British refused to let anyone who worked in a cotton mill leave the country.) Slater had no models or designs on him, but in his memory he carried the blueprints for recreating Arkwright's water frame, and he set out to make his fortune from them. In that same year the Constitution was ratified, and thirteen rancorous, feuding colonies somehow managed to find enough interest in common to join forces as the United States of America and elect a textile enthusiast, George Washington, as their first leader.

Spinning a Profit

Within a year Slater built his first cotton yarn factory, also the first to use water frames; his company was financed by Moses Brown, whose family money—derived in part from slave trade—funded the conversion of Rhode Island College into Brown University. Slater had learned about Brown from a ship captain and promptly wrote to him, first offering his credentials and then making an offer Brown found hard to refuse: "Under my proposals," Slater said, "if I do not make as good yarn as they do in England, I will have nothing for my services, but will throw the whole of what I have attempted over the border," sacrificing all payment. "We shall be glad to engage in thy care and so long as thee can be made profitable to both," Brown replied. He was taking a flyer on another man's daring and ambition, and mostly on his hands-on experience. Blacksmiths and mechanics were hired. The plans called for the building of a water frame with twenty-four spindles and a requisite number of rollers and drawers, all powered by a water wheel. There were inevitable setbacks, but in late 1790 the new nation's first cotton-warp mill clambored into operation. Fashioned entirely from memory, relying on intricate machines forged for the most part on site, it miraculously performed as promised. Slater, said Andrew Jackson with great admiration, was "the father of American manufactures."

New Slater mills soon opened. Using cotton imported from Surinam and Hispaniola—now home to Haiti and the Dominican Republic—they produced yarn from long-staple varieties for chambrays, sheeting, shirts, and ginghams. It was strong enough for warp and fine enough for dress apparel, all at a significant reduction in cost. Labor-intensive homespun yarn was woven into cloth that ran as high as 40¢–50¢ a yard. Cloth produced from Slater's yarn fetched 10¢, a fraction of the cost. Cotton yarn mills now sprang up elsewhere; by 1809 eighty-seven were in operation in New England and New York. Unlike linen and wool, this fabric could be inexpensively brought to market. As a result, it would soon become America's—and the world's—cheap, practical apparel.

Cotton accounted for about 4 percent of all clothing in Europe and the United States in 1793. A century later, 73 percent of all European and American clothing was being made from cotton. Slater grew wealthy but limited himself exclusively to spinning yarn when he might have also established the country's first power-weaving operation. He was given the opportunity but declined, leaving the door open to an entrepreneur named Francis Cabot Lowell.

When Slater's Rhode Island mill opened for business, Hamilton was secretary of the treasury and Coxe his assistant secretary. These leaders' dreams of industrial glory were now, finally, taking tangible form. "The manufactory at Providence has the merit of being the first in introducing into the United States the celebrated cotton mill," Hamilton intoned, "which not only furnishes materials for that manufactory but for the supply of private families, for household manufacture." While this was all quite formal and subdued, if you listen closely you can hear Hamilton click his heels in joy: this upstart crow of a nation had ripped off the Brits at their own game.

Slater, while content to spin cotton and leave the weaving to those private families, had created a benign monster: textile factories are huge maws that ingest massive quantities of raw material to feed their ravenous machines. Brown recognized the problem. "You must shut down thy gates," he wrote Slater two years into operation, "or thee will spin all my [West Indies] farms into cotton yarn." Brown had refused to use upland cotton; the dust that adhered to its fibers resulted in an inferior yarn, he believed.

That was all about to change, and with it, the course of industry, society, and politics in the United States. "What Peter the Great did to make Russia dominant, Eli Whitney's invention of the gin has more than equaled in its relation to the power and progress of the United States," commented Thomas Macaulay, one of England's finest historians and a contemporary of Whitney. Macaulay was right up to a point. The dispiriting news about power, of course, is the damage it can do when it veers off course. In the same year that cotton exerted its greatest influence on America's economy, 1860, it brought the country to the brink of disaster.

Seeds of Change

Although Eli Whitney never figured out how to capitalize on his own genius, American cotton became a money-making force of such magnitude for so many business enterprises on two continents—from farming to manufacturing to banking to shipping to warehousing—that the governments of England and the United States shaped their policies to accommodate its needs. They even reached a kind of gentleman's agreement to stop spitting artillery in each other's direction at every opportunity. As a crop and as a textile, cotton put America on the world map, and that was due entirely to a chance meeting between a destitute young tutor and the lively widow of a revered Revolutionary War hero, Nathaniel Greene.

Graduating from Yale at twenty-seven in 1793, Whitney had accepted a tutoring job in South Carolina to pay back college loans. Meeting Katherine Greene—better known as Katy—for the first time on the sea voyage from New York, he struck up a friendship with her. Whitney might be lacking money, she saw, but hardly lacking in ideas and ambition. She learned from her plantation manager, Phineas Miller—also a Yale graduate—that Whitney had earned a reputation as an especially capable mechanical engineer, even a child prodigy. As a twelve-year-old boy raised in Westboro, Massachusetts, he took apart and reassembled his father's watch. Two years later he had developed the sophisticated molds and tools required to forge nails, and at fourteen he had set up a nail factory of his own whose profits went to his tightfisted but loving Calvinist father. At eighteen Whitney became the nation's sole manufacturer of women's hatpins. His mechanical aptitude pegged him as someone adept in the "useful arts," not exactly a socially dignified calling in that era. Still, he insisted on attending Yale, which then turned out mostly clerics and lawyers. Once out of college, Whitney was strapped for funds and, like college graduates ever since, pretty much lost at sea—in his case, literally. When he met Katy on board, he was on his way to becoming a tutor—nothing he cared to do, but the only paying job he could find.

Arriving in South Carolina, Whitney learned that his promised wages were to be cut in half. He refused the position. Katy had invited him to visit her at Mulberry Grove in nearby Savannah, Georgia, the plantation she had been awarded for her late husband's patriotic heroism, and he took her up on the invitation. Miller, who was running the plantation, was also Katy's lover. She was an extraordinarily vivacious, worldly hostess whose social contacts included George Washington and just about every other famous American of the time. There were children running about everywhere; there were acres of land to explore. Life progressed with a languorous ease and a sense of adventure Whitney had never before experienced. "I find myself in a new natural world and as for the moral world I believe it does not exist so far South," he wrote to a dear friend, Josiah Stebbins, aware that the widow Greene and young Mr. Miller were living together out of wedlock. He didn't say he missed that moral world.

Shortly after his arrival at Mulberry Grove, Katy, following her intuition, asked him to sit in on a meeting of disconsolate local cotton farmers. Their common problem, he quickly discovered, was that they had no way to remove the stubborn green seeds of upland cotton from the fibers except by hand, and so they planted it only sparingly. Each five-hundred-pound bale of upland took as many as sixteen months of a single slave's dedicated work to separate. The cost of feeding and clothing the slave for that period cut deeply into profits. The farmers' other crops—rice, tobacco, corn, and indigo—were barely providing a livelihood. Tobacco exhausted the land and could not be trusted to provide steady future income. The British were expanding their indigo plantations in Bengal, weakening the market for American dyestuff. Wetland rice required special conditions for cultivation, insuring that it would never be a widespread crop. There has been an ongoing debate ever since as to whether slavery might have died out altogether in the South without the sudden increase in the demand for field hands and other workers brought about by the explosion of planted cotton. At best, slavery was on a steep decline as a labor source and was fast becoming an economic liability.

Green-seeded cotton grew like kudzu in this climate; there was a fortune waiting to be made and no way to make it. What to do? "Gentlemen, apply to my young friend, Mr. Whitney," Katy told the assembled group of frustrated farmers. "He can fix anything!"

Ten days later Whitney came back with a working model of his cotton gin—crude, but essentially governed by mechanical design principles that are still in operation today. No engineers since have reinvented a better gin; they've simply built better versions of Whitney's original. Legend has it that Whitney's inspiration came as he was roaming the plantation grounds pondering how to solve the problem and paused to watch a cat hunt down a chicken. At the last moment the chicken fled and the cat's lunging paw came away with only a few feathers. Why then try to separate the seeds of upland cotton from its fibers? Why not instead build a device to separate the fibers from the seeds? Small difference, huge implications. If Whitney's machine could allow the fibers to be pulled away while creating a barrier that held back the seeds, the claws could exert enough force to yank the fibers free.

That elegantly uncomplicated premise led Whitney to build an apparatus that duplicated the motions of the slaves who cleaned cotton manually. He fashioned a mesh sieve or hopper with narrow slits in it running lengthwise to do the work of the hand holding the seed. On the surface of a drum rotating around the hopper he duplicated fingers pulling off the lint by attaching wire claws that protruded through the slits; these hooks grabbed the lint and wrenched it away from the seeds, which were held in check by the tight mesh. A cylindrical brush swept off the freed lint. This hand-cranked contraption, the cotton gin, was, in

Whitney's words, "an absurdly simple contrivance"—and unfortunately for Whitney, he was right: it was all too easy to copy.

Demonstrating his gin to the local planters and seeing their excitement, he knew he'd come up with a winner. So did Phineas Miller. The two young men formed a business partnership—Whitney would make the gins, Miller would license them to planters in return for one-third of their harvest, agreeing also to put up the money to build the machines. They would split all profits fifty-fifty. Katy Greene was delighted. "Mr. Whitney is a very deserving young man, and to bring him into notice was my object," she told everyone in sight. She was now engaged to Miller, who had been hired to be her children's tutor before he took over management of the plantation; they would soon marry. That left Whitney emotionally stranded. From scant but compelling evidence it appears that he both deeply loved the woman and valued his business partner's friendship. From the moment his gin became a piece of valuable farm equipment, however, Whitney had little time for personal angst. "One man and a horse will do more than fifty men with the old machine. It makes the labour fifty times less, without throwing any class of people out of business. . . . Tis generally said by those who know anything about it that I shall make a Fortune. . . . ," he wrote to his father, a proud son boasting to the man he wanted most to impress.

The gin's fortunes and Whitney's went in opposite directions from the start. After sixteen months of difficult labor in the workshop he set up in New Haven, Whitney perfected a full-sized bench model of his gin and shipped it off as part of his patent application to the newly formed federal government. His petition came to the attention of Thomas Jefferson, then secretary of state. "As I [manufacture cotton] myself, and as one of our greatest embarrassments is the cleaning of the cotton of the seed, I feel a considerable interest in the success of your invention, for family use," Jefferson wrote to Whitney, asking half a dozen tough, incisive questions about the specifics of its operation and never mentioning that mass cotton production figured prominently in the scheme of the nation's leaders to establish the financial footing of the country's independence. Jefferson, ever wily, kept it personal: "Favorable answers to these questions would induce me to engage one of them to be forwarded to Richmond for me." Whitney responded in detail, explaining how the first successful small model encouraged him to build larger working versions; in time, with Katy's assistance, the two men would become friends. In March 1794 Whitney obtained a patent.

Delighted, Whitney told his father, "I had the satisfaction to hear it declared by some of the first men in America that my machine is the most perfect & most valuable invention that has ever appeared in this country. . . . And I shall probably gain some honor as well as profit."

Whitney had every reason to expect his gin to make him rich, but there was a "new invention" loophole in the patent law first passed by Congress in 1793 just big enough for a competitive gin to squeeze through. Another inventor, Hogden Holmes, had in the interim replaced Whitney's hooks with saws. While Whitney's hooks pulverized the plant's seeds, causing their oil to leak out and clog the gin's gears with damp lint, saws did a cleaner, more efficient job, and they won Holmes a patent of his own two years later. Early on, as word got out and upland cotton farms with jerry-rigged gins suddenly began to blanket the South, Whitney and Miller were laying plans to monopolize the industry by giving their gins to growers without charge in exchange for one-third of their harvested crops. That "in-kind" arrangement had been used extensively for wheat delivered to flour mills, but to Southerners those terms seemed exorbitant: grains were a faulty model. Still, Miller and Whitney refused to renegotiate their terms. That proved to be their undoing. Possibly a more flexible business proposal might have convinced some growers to honor their patent.

The crop was in the field, with all the attendant time pressures of cultivation and harvest dictated by the growing season; cotton farmers, now planting upland widely for the first time, were not inclined to sit around and negotiate while their crops needed watering and their weeds needed chopping. They knew they could get $40 an acre for upland, while their previous crops brought in $8 or less, and so they built their own crude versions of Whitney's gin before harvest. Miller and Whitney might want to control the ginning industry, but they hadn't adequately done their homework by estimating in advance what their 33 percent of the crop would translate to in real dollars. The answer was $3 million in today's currency—for one year, 1800, alone. Southern state legislators did the math and came to the defense of their native-son growers. The governor of Georgia in 1800 railed against these two transplanted Yankees "who demand, as I am informed, $200.00 for the mere liberty of using a ginning machine. . . . Monopolies are odious in all countries, but more particularly in a government like ours. . . ."

Patriotism, the last refuge of the scoundrel, proved to be the first sanctuary of the patent pirates. Whitney and Miller gave these Southern states all the excuse they needed to wrap the American flag around their resistance, and the two paid dearly for it. By 1797 Whitney had built twenty-eight gins; none was in use. Miller continued to pump his own dwindling funds into the business—twelve thousand dollars in all. By then he and Katy had been forced to auction off Mulberry Grove. As business partners, Whitney and Miller might seem to provide a casebook study on how to turn a golden goose into a sterile cuckoo. They alienated their prospective customers, refused to license their design to other manufacturers to meet an urgent demand, laid themselves bare to inevitable bootlegs, and went broke trying to repair the damage.

Still, the partners' loss was cotton's gain. All that ultimately mattered to the care and feeding of this insensate plant was that humans now had a reason to propagate it in vast quantities, which they began to do at once. Supply could finally catch up with demand. In 1793, before Whitney's gin went into operation, the South exported 974 bales, or 487,000 pounds, of raw upland cotton to England; one short year later that quantity increased threefold to 1.6 million pounds. By 1800 the South was sending 17.8 million pounds to England; by 1805, 40 million; and by 1820, close to 128 million pounds.

Big Cotton had arrived to stay. Although Manchester's mills continued to use raw fiber imported from colonial India, they much preferred the less brittle, longer-staple, sturdier upland variety now available from America. Great Britain in turn exported about twelve million pounds of cotton fabric in 1800 and half a billion pounds by 1860. In 1790 Georgia's per capita cotton production was six pounds; it was ten times that amount by 1800 and fifty times as much by 1860. There are a slew of similar figures from all cotton-growing Southern states during the first half of the nineteenth century that point in a single direction: jaw-dropping growth as the South became the sole supplier of America's cotton mills and the primary supplier of Britain's. While Whitney's hand-cranked gin initially produced fifty pounds of lint a day, with the assistance of farm animals and steam engines, planters would eventually produce up to 1,000 pounds of cleaned, ginned upland cotton daily, and it was all spoken for.

There was one other American cotton variety that had its brief moment in the sun. Nonsticky black-seeded Sea Island cotton, *G. barbadense*, first introduced to the coast of South Carolina from the Bahamas in 1786, earned a reputation as the "cashmere of cotton." Genetically it owed its existence to the species first grown in Peru and Chile that eventually migrated to the Caribbean. Sea Island, prized for its exceptional length and satin tooth, or feel, grew only in the sandy soil of a narrowly circumscribed area—primarily on Hilton Head Island and in neighboring coastal regions. As a crop it required constant attention, produced low yields, and never became a widespread source of revenue; even so, Sea Island gained fame as the fiber of choice for Queen Victoria's exquisite lace handkerchiefs, as well as for fabled shawls and luxury shirting fashioned by British tailors. One square inch has the tensile strength of 100,000 pounds. Introduced to Egypt in 1821 by the enterprising viceroy of the Ottoman Sultan Mehemet Ali, and crossed with another variety known as Jumel, Sea Island eventually evolved into the fiber known today as Egyptian cotton. In the early twentieth century, boll weevil devastation eliminated the entire American Sea Island crop.

Since the black seeds of Sea Island cotton did not stick to its fibers, a simple roller gin, an Indian *churka,* easily deseeded it; the same device was used to deseed the common inferior Indian short-staple cotton variety, too. A vibrating primitive bow, essentially the same device that had been in use for five thousand years, cleaned it. While Sea Island was an anomaly, its success in the humid Southern climate hinted at the possibilities for upland cotton once its ginning problem was resolved. Three years after the first planting, more than a million pounds of Sea Island lint were harvested, and two million within five years. By 1793 Sea Island had reached the limits of its productivity; an acre yielded only one-half as much lint as upland at twice the cost; green-seeded upland was still too labor-intensive to replace any other crop. But that of course was the same year, as one observer remarked, in which "the skeins of this plant became involved with consequences quite as important, to say the least, as those of the Industrial Revolution."

For that momentous change in the processing of upland, which would fundamentally reshape the destiny of the United States, Whitney's gin was solely responsible. Although he provided the means, it was our Founding Fathers who offered the inspiration.

Mixed Consequences

Once Whitney's gin facilitated mass production, an immensely profitable cash crop quickly attained so much persuasive control that it threw a bolt of energy into the dying institution of slavery, which in turn triggered the escalating political, moral, religious, and economic discord that led to bloodshed. There were no circuit breakers. There were no acceptable alternatives for men and women whose capacity for rational thought in the North and South was ensnared by the wispy fibers of this puffy plant, more accurately nicknamed "white gold." Carried aloft by the winds of potential fortune, cotton muddled the sober minds of bankers, builders, planters, captains of industry, factors, legislators, shippers, and just about anyone on both sides of the Atlantic whose income was linked to the plant in the field or the fabric on the shelf. There was simply too much money to be made. In the end cousins were killing one another in remote Pennsylvania farmlands and on bluffs along the Mississippi Delta. When the Civil War ended, 1,094,446 men had been wounded and 623,026 had been killed, more than in all subsequent American wars combined.

Whitney died decades before the debacle that his invention inadvertently precipitated by setting two opposing forces in this case—profit and moral conscience—against one another. By all accounts a sensitive man, he almost certainly would have been devastated. Whitney's personal tragedy in creating more sudden wealth for more people than any other man up to that time in American history and not benefiting from it himself was at least partially offset by an eventual resolution to his lawsuits against Southern states.

After more than a decade of crying in the wilderness, Whitney finally caught the sympathetic ear of one judge, William Johnson—in Georgia, ironically—who wrote an eloquent decision in his favor: "The machine of which Mr. Whitney claims the invention . . . has suddenly become an object of infinitely greater national importance" than any previous seed-separation machine, he wrote. The judge dismissed Hogden Holmes's gin as a mere adaptation, which it was. "Every characteristic of Mr. Whitney's gin," he added, "is preserved."

States at long last, after more than a decade, began to pay attention to Whitney's claims. South Carolina agreed to pay $50,000, but after a $20,000 down payment, it reneged on the balance due. Neighboring states made their own deals. Whitney collected a total of $90,000—a handsome payment, it would seem, but in those ten years he'd run up so many unpaid bills that the money all but vanished; there were creditors to pay for equipment, laborers demanding back wages, sundry bills, and a host of lawyers to settle with. In the end, Whitney pocketed at most a few thousand dollars.

The South fared better. "Individuals who were depressed with poverty and sunk in idleness, have now suddenly risen to wealth and respectability. Our debts have been paid off. Our capitals have increased, and our lands trebled themselves in value. We cannot express the weight of the obligation which the country owes to this invention. The extent of it cannot now be seen," Judge Jackson wrote sagely.

In a sense, the conflict generated by Whitney's gin was a rehearsal for the hostilities between North and South that followed. Whitney and Miller behaved like hardheaded Northern industrialists demanding unreasonable terms of sale with no understanding of the vagaries of agriculture, an industry governed by a host of unpredictable circumstances and conditions, from weather to invasive pests. In turn, the South's greedy farmers refused to honor the financial and personal debt they owed to these men. Magnified a hundred times and greatly complicated by the moral morass of slavery, that self-absorption and conflict of wills, cultures, habits, and values eventually escalated into armed conflict.

. . .

Whitney, a successful musket manufacturer at his death in Connecticut in 1825, lived just long enough to glimpse the revolutionary changes in our nation's economy that his gin set in motion. In that year America's cotton exports to Britain alone reached 171,000,000 pounds. Two years before his death, an enterprising businessman in Massachusetts had created the first American city built solely for the purpose of producing cotton textiles from the South's upland cotton. In the meantime, the slave population of the South had more than doubled since 1793. The coming conflict between an individual's right to the pursuit of happiness in our democracy, slaves included, and the country's fixation on its cotton economy seemed inevitable to anyone who paused to consider it, and many did. But no one, in the end, was able to prevent it.

FIVE

Camelot on the Merrimack

No person can be employed by the Company whose known habits are or shall be dissolute, indolent, dishonest, or intemperate, or who habitually absent themselves from public worship or violate the Sabbath. . . .

—Posted Merrimack Manufacturing Company regulation

Lucy Larcom did not need a watch to tell the time of day. From the age of eleven, she was awakened Monday through Saturday before sunrise by the loud, insistent pealing of the Lowell factory bell, giving young women and girls in the company boardinghouse just enough time to wash, dress, and walk a short distance over the canal footbridges, through a large gate, and into the mill yard. Another bell signaled the start of their workday at 5 A.M. Once inside the long, narrow five-story redbrick factories that rose up all around them, they set about their tasks.

Men began to blow forced air into the five-hundred-pound bales of raw cotton in the adjacent picking house to decompress and disentangle their fibers. In the mill's first-floor carding room workers using fine-toothed automated combs and rollers converted sheets of loose, raw cotton fiber into slivers, long fat tubes of cleaned parallel strands. Upstairs, as loud machinery cranked into operation, Lucy and her coworkers in the spinning and weaving rooms, almost all women, filled the machines with raw material, mended broken threads, and tended to the hundreds of minute toilsome details of textile production. Below, the sun rose over the Merrimack River. It hardly mattered; they would not be allowed outside long enough to enjoy the day.

From five until seven in the morning, six days a week, these mill hands worked without food. A bell clanged at seven o'clock to announce a half-hour breakfast recess, another bell ended it, two more bells signaled the beginning and end of a half-hour lunch break at noon, and at the end of a fourteen-hour day, all but eighty or so minutes of it spent at work, a bell at seven in the evening sent them back to their boardinghouses for dinner. Afterward, if not too fatigued, they might attend a meeting of their Improvement Circles to hone educational and social skills, or they might read and gossip in the front parlor. Some went off in small groups to hear visiting speakers like eminent philosopher and writer Ralph Waldo Emerson at Lowell Hall. Housemothers enforced bedtime curfews. Contact with men was carefully monitored. Church attendance on the Sunday Sabbath was mandatory.

In the opinion of some workers, like Lucy Larcom, what you made of your time at Lowell depended on what you intended to make of the rest of your life. Mill work tested patience; it demanded vigilance; it rewarded tenacity. Early-nineteenth-century Americans held firmly to a conviction that those virtues brought you closer to God. But for some young women fresh off the farms of northern New England, most between the ages of sixteen and twenty-one, nothing but the absence of bars on windows distinguished these cotton factories from prisons. "I am going home where I shall not be obliged to rise so early in the morning, nor be dragged about by the factory bell, nor confined in the noisy room from morning till night. I shall not stay here. . . . Up before day, at the clang of a bell . . . at work in obedience to that ding-dong of a bell—just as though we were so many living machines," one mill girl complained. She wasn't alone.

Three decades after Whitney's invention, industrialization arrived in New England with the force of a sudden, turbulent storm. Winds of change blew through the multiplying cotton fields of the South, swept up mountains of lint, and deposited them into the newly minted textile factories of the North. This was all to the good—a new nation united by one cash crop grown and manufactured within its borders—but how to hire, train, organize, and regulate a large-scale industrial workforce where none yet existed in rural Massachusetts? The North improvised. Factory bells replaced cow bells: they soon sectioned off and parceled out time and labor like generals regimenting troops for combat.

For the first time in the short history of America, the mills at Lowell provided unmarried women with an opportunity to earn their own income away from home. Cotton was the catalyst. At Lowell, twenty-eight miles north of Boston, cotton created two cities—one real, the other a highly idealized fantasy bolstered by glowing accounts in the newspapers and magazines of that era. The first existed within the high thick walls of the daunting factory buildings that had rimmed the Merrimack since their construction started in the early 1820s, powered by water below and by human sweat above; the other Lowell, assembled from those same structures, took on a life of its own as the pristine embodiment of American enterprise in all its utopian glory. Three United States presidents and a host of foreign dignitaries visited, toured, and came away burbling with enthusiasm for this dynamic proof that enlightened capitalism worked. They looked on from a polite distance and saw what they wanted to, a shining City of Spindles. If they had chosen to probe further they would have discovered that inside the mills young female workers fought to maintain their mental and physical health for the better part of fourteen hours a day.

Raised on farms, Lowell's mill girls were used to hard work from a young age. They fed livestock, hoed, shoveled, raked, planted seed, swept, mended clothes, and stacked firewood, going about these and other chores at their own pace, often out of doors with rest breaks as needed. None of that applied to factory labor, they sadly discovered. Within the long, narrow brick buildings, the grinding, bone-aching tedium of mechanized production controlled their actions from dawn to dusk. Their workplace was an overheated, poorly ventilated room that filled an entire floor; it held as many as 200 machines. Screeching, clanking pulleys and levers and wheels roared with a deafening clatter; vibrations from the top-floor looms shook walls, ceilings, and floors as thousands of spring-loaded wood shuttles slammed against the side frames of looms, then back across at lightning speeds; they rattled and hammered without pause. Even at a distance, several floors below in the spinning and carding rooms, the mill girls lived with the explosive repetitive slap of those shuttles and the wheeze of whirring, straining metal wheels close by.

Only the most resourceful workers found escape. "In the sweet June weather," Lucy Larcom recalled, "I would lean out the window, and try not to hear the unceasing clash of sound inside. . . . I discovered, too, that I could so accustom myself to the noise that it became like silence to me. And I defied the machinery to make me its slave. Its incessant discords could not drown out the music of my thoughts if I would let them fly high enough."

Like Lucy, other factory women surrendered their personal freedom for the lure of independent income. In the words of one mill girl:

> Despite the toil we all agree,
> Out of the mill or in,
> Dependent on others we ne'er will be
> As long as we're able to spin.

That trade-off would motivate industrial American female labor in the decades to come. Also, until Lowell's mills opened, unmarried daughters had no way to contribute any significant earnings to their families. These new mills provided both opportunities. They offered the highest wages for working women in America, between $2 and $5 a week after deducting $1.25 for room and board, and they required few skills. Larcom was the daughter of a whaler who died young and left Lucy's mother to raise her eight children with no help. Out of necessity, Larcom and her older sister went into the cotton factories at a young age—Lucy, barely schooled at eleven, was about five years younger than almost anyone else. Mill work provided her with cash to send home and the license to dream; with scrupulous saving, she and her fellow workers had the funds for a new dress and jewelry to be seen in as they strolled up and down Lowell's shopping streets and through its groomed parks on a fine spring Sunday. More important to Lucy, a nascent writer, books from the Lowell circulating library were readily available, and she could join a group of coworkers after hours to read her latest poem, story, or essay aloud to an eager audience of boardinghouse mates.

Within the mills themselves, however, time slowed to a deadening crawl; fine hairs of cotton lint circulated through the hot air and mingled with particles of machine oil that gave off an acrid, burnt scent and seemed to coat walls, posts, and beams on every floor with a thin sheen. To maintain a high level of humidity to reduce thread breakage, managers often nailed windows closed. During summer months, temperatures soared. In balance, cotton's welfare mattered more than the workers'. Airborne lint hairs, one mill girl noted, fell as thickly "as snow falls in winter." They sometimes piled up on workers' clothing and hair; inhaled for more than twelve hours a day in the absence of fresh air, the lint caused frequent lung diseases that were difficult to diagnose and impossible to treat effectively.

One textile historian, William Moran, reports that women used nonsmoking tobacco, or snuff, as a defense against the lint they inhaled when sucking threads through the narrow end-passages of loom shuttles. They called their lip motion "the kiss of death." Doctors frequently saw female workers vomiting up little balls of cotton. Later, textile workers would be among the first diagnosed with brown lung disease, or byssinosis. Moran also notes that 70 percent of the early textile workers died of respiratory illnesses, as opposed to 4 percent of farmers in Massachusetts. Sick days meant lost wages: workers received no pay if they did not show up. Since the cost of using the factory's infirmary was deducted from your salary, many employees chose instead to work when ill, infecting others.

Then there was the work itself—sufficiently tedious to numb the mind, rarely challenging enough to stimulate it. In these mills, all of the brains were built into the machines, more technically advanced than in any other industry in America. No matter how bright you were, you were paid to tend the machines like the nannies of idiot savants who lacked all personal skills. When threads broke, as they did constantly, the machine stopped itself, but it could not repair its own severed strands. You tied a quick agile knot between loose ends as if lacing its shoe before restarting the gears in motion.

Roller frames that drew out sliver to quarter-inch roving and then drew that out into multiple threads always needed their filled bobbins of yarn, sitting on stationary spindles, replaced with empty ones. Endless rows of looms on the fourth floor, as many as two hundred to one room, had been ingeniously designed with automated beams, harnesses, and heddles. One horizontal beam that looked like a rolled carpet fed thread into the loom, and at the opposite end a take-up beam rotated slowly to collect finished cloth. Between the two, metal heddles, attached above to rows of leather harnesses, lifted and fell in precise sequence to open two planes of warp thread at once. That jawlike opening, called the shed, enabled horizontal shuttles of weft to shoot through at a right angle with terrifying velocity and noise. Through an eye at the top of each of these heddles—thick vertical wire rods—a mill girl passed a single strand of thread from the feeding beam to hold it in position throughout the weaving process. Belt-driven cams attached to shafts moved the harnesses up and down.

As technology progressed, Lowell's mills automated an earlier invention by Joseph-Marie Jacquard that allowed complex patterns to be woven into fabrics. Engineers updated his punch-card system that activated specific individually programmed harnesses to create designs. There might be twenty or more harnesses on a single loom. Much like a player piano, Jacquard looms operated by pre-ordering the operation of those mechanisms. Where holes occurred in the punch cards—later adapted to revolving chains—the warp harness was lifted; where there were no holes, that harness—and the threaded heddles attached to it—remained in place.

These power looms turned out miles of fabric at speeds no human could match, but unless you laboriously threaded individual strands through the eyeholes of their heddles every hour of every day of every week, the magnificent machines could not produce one fraction of an inch of finished cloth. You might be glorified as an "operative" in the newspapers, but in your heart of hearts you knew you had been hired to nursemaid these inanimate mechanisms so they could outperform you at your best. It wasn't surprising to hear a coworker announce that in the past week she had drawn 43,000 separate pieces of thread through heddles. That was your lot in life. Eyeholes. Threads. Hosing down a pig trough took more skill. It also offered more diversion. And there were no supervisory positions for women anywhere in these mills.

For two or three years, the length of most of the mill girls' stay at Lowell, maintaining a healthy outlook became a second occupation. Friends helped. Lucy and her close companion Harriet Hanson Robinson sometimes paused as time permitted to watch the new recruits arrive from isolated rural communities. They were "dressed in such an old-fashioned style that each young girl looked as if she had borrowed her grandmother's gown," Robinson later recalled. They clutched bandboxes that held all their worldly possessions, including love letters. "Years after, this scene dwelt in my memory, and wherever anyone said anything about being homesick, there rose before me the picture of a young girl with a sorrowful face and a big tear in each eye, clambering down the steps at the rear of the great covered wagon, holding fast to a cloth-covered bandbox." Some wept openly. The sheer size of these massive factories stretching half a mile or more down river inspired an equal measure of terror and reverence at first sight, especially when the largest structure most of them had ever seen was a hay barn.

Surrounded by soaring walls of brick, they stared up with awe at the central belfry tower that rose high into the sky like a church spire. They'd soon learn, as Larcom and Robinson had, that the bell in that tower controlled the movements of their waking hours as rigidly as phases of the moon controlled the tides. Some would make the best of a stultifying work regimen by exploring small ways to satisfy personal needs. Although all books were banned in the mills, bibles included, Lucy had managed to paste recent newspaper clippings on the bright panes near her work space—her "window gems," as she called them—and she looked up from her spindles from time to time throughout the day to stay informed about the world beyond those walls. Others escaped into their imaginations, since conversations were difficult to sustain over the thunderous racket of machinery. These tactics worked intermittently at best. After two or so years in the mills, noticeable fatigue and poor health overtook many of these workers. There was a high turnover. "The daughter leaves the farm . . . a plump, rosey-cheeked, strong and laughing girl . . . ," one female reformer wrote, "but alas, how changed!" after being tucked away inside these mills three hundred days a year. "This is a dark picture, but there are even darker realities, and these in no inconsiderable numbers."

Reformers seized on the oppressive working conditions to press their case for shorter hours and related improvements. In time these grievances combined with a depressed economy to create an irreparable rift between management and the first generation of New England's rural female employees, who were, all things considered, vastly better off than their brutally treated counterparts in Manchester, England. As one of that wave, Larcom initially spoke for many: "Certainly we mill girls did not regard our lot as an easy one, but we accepted its fatigues and discomforts as unavoidable, and could forgive them in struggling forward to what was before us." That tolerance would not last much past the first two decades. When it collapsed, Lowell's workforce swiftly and permanently changed character. By then, though, the ambition, literary output, and jaunty public demeanor of these New England-born mill girls had already cast them as stars of an idyllic fairy tale set in the magical City of Spindles.

Celebrities, politicians, and foreign princes came to survey this Camelot for themselves. Citizens at all levels were emotionally invested in the dream of an American Industrial Revolution with Lowell as its spiritual and geographic center. Lowell had to be everything that Manchester, England, that septic tank of child enslavement, was not. One visitor, Captain Basil Hall, summed up the prevailing mood in his journal: ". . . [T]he village speckled over with girls, nicely dressed and glittering with bright shawls and gay bonnets, all streaming along . . . with an air of lightness, and an elasticity of step, implying an obvious desire to get to their work."

Hard evidence might dispute that point of view, but it hardly mattered: the creative capitalists who transformed these four square miles of rock-strewn farmland into an international symbol of American manufacturing power had accomplished their goal with brilliant audacity and in the process launched the nation's Industrial Revolution. They'd built the first fiber-to-fabric textile factories, an end-to-end system that converted raw cotton to finished goods. They were feeding an inexhaustible market for cheap, lightweight, colorful cloth. More impressive still, they had done so with exceptional foresight. This group of men, called the Boston Associates, knew as much about human engineering as they did about rollers and heddles. Early on, before the first brick was mortared into place along the Merrimack River in 1821, they'd figured out how to repackage the drudgery of factory labor as the chance of a lifetime for self-improvement; they knew who to pitch and how to pitch. They knew, too, that if they succeeded they would be able to put together a diligent, dedicated workforce and gain the world's admiration for their efforts.

Fabricating an Empire

Francis Cabot Lowell, the man primarily responsible for realizing that vision, was to American cotton as Henry Ford would later be to automobiles. Although he died in 1817, a few years before the founding of the city that honored his name, Lowell had already fashioned both the machinery to produce finished cloth and the strategies that attracted an eager labor force to his factories. Shrewd and persuasive, he surrounded himself with capable industrialists and savvy financiers, and drew on family and social connections to raise capital. Lowell had something else going for him as well: he was an unapologetic thief in the noble textile tradition of Samuel Slater and Richard Arkwright. Without bold acts of industrial espionage, cotton might never have progressed in the Western world beyond primitive spinning wheels and hand looms, but then again, capitalism's darlings have often been a rogue's gallery of impudent crooks.

Lowell arrived in Manchester, England, in 1810 on a prolonged tour of Europe intended to improve his poor health. Born to a prominent Boston mercantile family, he had gained respect as a successful businessman with interests in shipping, banking, foreign exchange, and real estate since graduating from Harvard the same year that Whitney invented the gin. Through family contacts he befriended English textile industrialists and gained access to their cotton mills after meeting up with a fellow Bostonian and family friend, Nathan Appleton. As a business courtesy, the British owners instructed their managers to provide details of their operations to the visiting Americans. Together, Appleton and Lowell spent days watching the elaborate spinning frames and automated looms of Lancashire convert loose shreds of fiber into substantial sheets of cloth. Driven by belts powered by steam engines that had by then replaced waterwheels, the frames and looms operated as many as twenty-three hours a day from Birmingham to Leeds. Lowell, a man with an advanced mechanical aptitude, learned everything he could about the design specifications of these miraculous machines—the power looms in particular.

Like Slater, he committed their functional operations to memory. That may not have been his first choice, but Lowell now knew he wanted to dedicate himself to automating cotton production in America. To do so he needed to license Manchester's machinery, raise capital, and build his own empire. Yet he found himself dealing with the same feisty British textile industrialists who had squelched all previous attempts to export their technology to potential competitors. Lowell's elevated social station mattered not a whit. He was still a Yankee, and America was still an impertinent renegade nation, so ill mannered it refused to bow down to Her Majesty. This made for sticky wickets all around. For Lowell, that might have ended the quest, but hasn't forbidden fruit always been the tastiest? Rather than retreat empty-handed, he decided to memorize as many of the working parts of the Manchester power loom as possible.

Enthralled by the dance of automated harnesses, sheds, reeds, heddles, and shuttles that moved simultaneously and efficiently in contrary directions, he committed himself to a massive feat of mental gymnastics. These components were choreographed through trial and error to synchronize precisely; any fluctuations or malfunctions in their split-second timing immobilized the entire machine. But a photographic memory, Lowell soon discovered, is an industrial spy's greatest ally. So, too, are secret compartments in steamer trunks where the spy might wedge a sketch or two. Although Lowell left no record of his trip to England, the consensus is that he smuggled back notes to aid his recollections; if so, they escaped the British customs officials who doggedly searched his belongings.

Equally important, during his prolonged stay in Manchester, Lowell experienced living conditions in one of the foulest cities on earth. Black day and night with overhanging clouds of choking, eye-watering smoke from coal used to fuel steam engines, Manchester epitomized the worst excesses of unregulated free enterprise. "Everywhere heaps of debris, refuse and offal; standing pools and gutters and a stench which alone would make it impossible for a human being in any degree civilized to live in such a district," wrote another visiting foreigner also destined to change the world: Friedrich Engels, son of a German cotton manufacturer who co-owned a Manchester factory. Engels's disgust with these obscene living conditions spurred him to write an exposé that created a national scandal and then to join forces with another disillusioned foreigner, Karl Marx. Together in the *Communist Manifesto*, they created their own alternative wealth distribution system. Manchester's cluttered, cramped, cold, and damp labor housing lacked sanitation or ventilation; its rivers ran brown with raw sewerage. Thousands of malnourished and filthy orphaned children, England's "parish wards," slept six or eight to a room and worked in these cotton mills for interminable hours under overseers who often beat and chained them.

A coal historian, Barbara Freese, pointed out that inventor James Watt believed "Nature can be conquered if we can but find her weak side." Manchester, she said, proved that in "looking for nature's weak side, we found our own." A brilliant observation at the very least. At one Parliamentary hearing, a deformed woman testified that she was strapped severely from the age of six. "You are considerably deformed in a consequence of this labor?" she was asked. "Yes, I am." "You were perfectly straight and healthy before you were in a mill?" "Yes, I was as straight a little girl as ever went up and down town." "Where are you now?" "In the poorhouse."

The average life expectancy among the poor in Manchester was seventeen years. More than 57 percent of newborns were dead by the age of five. Yet despite the unimaginable misery and squalor, the city itself gained stature for its manufacturing prowess.

French writer Alexis de Tocqueville best captured Manchester's paradoxical nature: "From this foul drain the greatest stream of human industry flows out to fertilize the whole world . . . here civilization works its miracles, and civilized man is turned back almost to a savage."

Francis Cabot Lowell knew he stood little chance of transplanting Manchester's mechanized industry to the United States unless he saved the broth and discarded the scum. Devout New England's farming families would never send their offspring to live and work in such a sordid environment. That meant clean living quarters for employees, excellent sanitation, no involuntary child labor, and fair wages. Sons were simply not expendable on family farms under any circumstances; Lowell's workers in the main would have to be single daughters. Coming from a similarly strict religious upbringing, he understood the concerns of those Puritan parents he would need to convince, and he shaped his pitch to address their anxieties. They would refuse to release their daughters to sample the lustful temptations of a large city without assurances that they would not stray. On the farm, strenuous physical labor was thought to exhaust a young woman's dangerously libidinous drives. In the factories, exhausting hours of day labor would have to be marketed to doubtful mothers and fathers as an equivalent preventative. Lowell would strive for an average age of twenty-one but would accept a few younger employees under special circumstances. And no one, male or female, would be beaten or otherwise brutalized in his mills. By taking these steps, Lowell introduced the corporate paternalism that would become the signature of his textile empire; it sat comfortably with his own humanistic beliefs.

As that workforce strategy gained focus, Lowell returned home to Massachusetts from England in 1812. He embarked at once on his plan to re-create a fully functional Manchester textile factory down to the carding cylinders and shuttle bobbins. Each water-frame roller had to be weighted just so to apply the exact tension needed to draw out the sliver and then the roving into thin thread without snapping it. Slater's spinning plants in Pawtucket and elsewhere provided a map of that terrain, but water-driven power looms were an unchartered wilderness. Although boldly ambitious and larcenous in heart and mind, Lowell also knew his own limitations. Grand theft alone could not realize his grand vision. To build a functional power loom he needed the assistance of a skilled mechanical engineer, and he soon found one in Paul Moody. Together they designed and built America's first power loom, a daunting project that required months of painstaking work filled with false starts and adjustments and prototypes that should have worked but didn't until more adjustments activated some parts but mysteriously deactivated others. Lowell lent his superior mathematical skills to the project and also to devising a double-speeder in the spinning room that evenly wound loose roving onto a spool. A leading mathematician at the time, Nathaniel Bowditch, reported that the calculations required for the double-speeder were so sophisticated and advanced that he was surprised that anyone in America other than himself could have arrived at them.

In 1814, after two long years, Lowell and Moody presented the first working model of their loom. Belt-driven harnesses, rising and falling in precise rotation, opened alternating vertical rows of warp thread on heddles that separated like jaws, and as they did, a horizontal row of weft filler shot across them on shuttles, pushed tight by a sliding bar against the previous row before the process was repeated. Impossibly loud, the power loom, with its thunderous shuttle, sounded as if it was designed to smash boulders into dust. Instead, it magically intertwined fragile threads of cotton into sheets of cloth with rhythmic motions as repetitive and relentless as ocean waves cresting and breaking on a beach.

Lowell invited his friend and fellow investor Nathan Appleton to see the loom in action. "I well recollect the state of admiration and satisfaction with which we sat by the hour watching the beautiful movement of this new and wonderful machine, destined, as it evidentally was, to change the character of all textile industry," Appleton wrote years later. Understandably, they were transfixed by its seductive grace and the promise of immense wealth that it offered.

Lords of the Loom

Like Appleton, Lowell had been born into Boston's "gentry elite." Their Old Money came with a set of humanistic responsibilities attached to it and a mandate to help the less fortunate after first helping yourself. The Lowells, Cabots, Higginsons, Amorys, Appletons, Lees, Lawrences, Russells, and others—dubbed Boston Brahmins for the highest Hindu caste—were raised with a sense of noblesse oblige. Their exalted status as untitled aristocracy is neatly captured in this famous rhyme:

> . . . good old Boston, the home of the bean and the
> cod
> Where the Lowells talk only to the Cabots
> And the Cabots talk only to God.

Francis Cabot Lowell was born into that world and changed it forever. As the son of a wealthy attorney for prominent New England merchant families, he had substantial personal resources, and, more significantly, access to wealth as one of the area's mercantile upper crust. He enlisted the support of Appleton as an initial financial backer, as well as of his brother-in-law, Patrick Tracy Jackson; they incorporated their new business as the Boston Manufacturing Company and raised $400,000 from Boston's ruling class. They were going to build a plant to make clothing out of a fabric that just about nobody was wearing, except those few who could afford English imports. They were doing it in a factory that had not yet been designed, powered by water they did not own, with complex machinery they would have to build from scratch, to be operated by a labor force that did not yet exist.

Lowell proceeded with single-minded determination. The inventor of a yarn winder named Shepherd learned the hard way that Lowell, like Arkwright before him, rarely let ethics stand in his way. When he and his engineer Moody discovered that Shepherd's model made some minor but substantive improvements over theirs, Lowell offered to buy the rights to Shepherd's machine at a reduced price. Shepherd turned him down. Moody announced within Shepherd's hearing that he could incorporate those improvements without illegally infringing on Shepherd's design. "You be hanged!" said Shepherd, reluctantly accepting Lowell's offer. "Too late," said Lowell, who went off to appropriate the man's work into his own machine without payment.

With the help of merchant friends, Lowell obtained a charter to build his first cotton factory at Waltham, about ten miles up the Charles River from Boston. Its working power loom enabled the mill to integrate all of the cotton manufacturing processes under one roof to turn out cheap, coarse cloth. The $400,000 he'd raised was ten times as much as the money behind a typical Rhode Island mill. Lowell built boardinghouses for the single women employees who came off the farms. To assuage their parents, he'd had them designed to closely resemble seminary dormitories; they were supervised by older housemothers who functioned as a kind of live-in morality police force. Churches went up nearby. God was everywhere, from loom to room. Knowing the power of money as well as probity to persuade, he also paid these women a higher salary than they could earn elsewhere for their fourteen-hour workdays.

Delegating the installation of the machinery to Moody, Lowell turned to his brother-in-law, Jackson, to run the daily operations. Jackson owned a large share of the business and was heavily invested in its success. Despite inevitable delays, the Waltham mill finally opened in 1814, importing raw cotton by boat from the South. In its first year of operation, Waltham sold $412 worth of cotton fabric. That may not have been surprising, since there was only one seller of domestic cloth in Boston and any fabrics that were not made in England were considered inferior. Appleton solved the problem. He had started an importing company to bring over British goods, anticipating the end of the War of 1812. He used that company, B.C. Ward, to become the new factory's agent and to auction off Waltham cotton cloth—high quality but with no pedigree. Buyers, assuming that it had been manufactured in England, bought it and praised it. When consumers learned that Lowell and Appleton's upriver mill had created it, the currency of their factory soared.

The mill soon began to make money, and by 1823 assets would grow from $39,000 to $771,000, a staggering increase. Lowell considered the initial boom in sales "too favorable to be credible." By 1815, more than 27 million pounds of raw cotton was being shipped from Southern plantations north for manufacture. But the calculating genius in all this was Lowell himself. Wrapping an American flag tightly around his business interests, he declaimed before Congress in Washington, D.C., against the hated British. Lowell petitioned Congress to institute protectionist tariffs against English imports—"articles," he said, "that are made from very inferior materials and are manufactured in a manner calculated to deceive rather than serve the consumer." In the end he won an extraordinary victory: India cotton, manufactured in Britain and Waltham's only real competition, was taxed at 83.5 percent. His patriotic fervor did not extend to Rhode Island, however. Lowell managed to slip language into the bill that excluded Slater's rival Pawtucket mill from taking advantage of his tariff exemption.

A short time later, at the moment Lowell was on the brink of expanding his mill into a textile dynasty, his fragile health again failed. When he died young in 1817, at forty-two, he had in a few short years established the textile industry in America. Long before "vertical integration" became an MBA buzzword, Lowell practiced it by controlling cotton from raw material through the sale of finished goods; Lowell also efficiently managed the whole process with sufficient capital to carry it from concept to successful execution. Remarkably, he even found a way to bring his brother-in-law into the business without capsizing it in the process. It was that man, Patrick Tracy Jackson, in fact, who not only carried on as the head of the Waltham plant but vastly increased the family fortune by moving the mill in 1821 to a more suitable Massachusetts location about thirty miles north of Boston, where the Merrimack River fell thirty-two feet in elevation within one mile before converging with the Concord.

At that site, where the Pawtucket Falls and much swifter currents provided a far superior waterpower source, Jackson established the city named for his deceased brother-in-law. He and his Boston Associates wisely purchased a majority of shares in the adjacent Pawtucket Canal that skirted the falls, and secured surrounding farmlands and waterpower rights. Like Lowell, they envisioned an empire, not a solitary mill—as many as twenty thousand inhabitants at this location in the future, one of them speculated. Since miles of existing canals in the area could be restored to distribute water to turbines and wheels as needed, Jackson also set up an entity, the Locks and Canal Corporation, to convert them to those purposes. The river Thoreau called "a silver cascade" as it flowed from the White Mountains to the Atlantic became a dammed and diverted energy source. Steam-engine power would not come to Lowell for several decades.

In order to realize their initial vision, Jackson and his main partner, Nathan Appleton, raised enough capital—$600,000, about $6.6 million in today's dollars—for their newly incorporated Merrimack Manufacturing Company to enable them to buy the land and water rights, and build an entire city from pastureland. Each share, available only to Boston's insular upper crust, was priced at $1,000 ($11,000 in 2003 dollars). They hired a skilled, haughty young English engineer, Kirk Boott, to run the new mills at Lowell as their agent—that day's version of a chief operating officer—and geared up to manufacture close to two million yards of cotton cloth annually. They had little trouble attracting investors—the Waltham mill was returning 30 percent yearly profits.

Boott, an arrogant martinet by all accounts, laid out the town to take advantage of the countryside's scenic appeal; he constructed the boardinghouses and mills, one in his own name, and designed a Georgian mansion for himself. He also imported dyers and calico printers from England, and put them in the "English Row" of buildings beyond the mills to produce miles of colorful, popular printed fabric to be marketed through a tightly controlled group of commission houses in Boston. In the previous decades textile printing had advanced from wood block to copper plates to water-powered engraved cylinders or rollers. In this mechanized process, each sheet of soft metal was covered with wax save for where its design had been etched in acid. Printers wrapped these sheets around a massive cylinder containing a single dye color and fed reams of bleached cloth between the rollers before substituting the next color and sheet and reinserting the fabric. Correct registration was critical. If the cylinders weren't perfectly aligned they produced a blurred, inferior finished product.

There was a highly specialized art and craft to that process that went far beyond tending looms, as Boott grudgingly recognized. When the first renowned printer, John Prince, arrived at Lowell from Lancashire, he learned that his annual fee, $5,000, would not be met. Prince and his troop packed up their wagon, picked up their musical instruments, and left town playing a merry tune. They had to be intercepted. "Not even the governor makes that much money," Boott protested. "Well, can the governor of Massachusetts print?" Prince replied. Boott acquiesced.

Printers aside, Boott ran his operation until his death in 1837 with tyrannical authority and no tolerance for worker dissent. By then the city had grown in size from several hundred in 1823 to more than 18,000 inhabitants. Six thousand of them worked in its mile-long stretch of mills, all owned by the Boston Associates. Lowell was a company town as much as any West Virginia mining community. Workers lived in company houses, spent their money in company stores, and prayed in company churches. For New England, Lowell represented corporate paternalism's finest hour. It had also become the nation's principal textile center and a source of vast wealth for its investors, now dubbed Lords of the Loom. The young country's insatiable demand for cotton translated to profits that averaged 24 percent annually for twenty consecutive years, until 1845. Close to 70 percent of the mill hands were single women. In the city's first two decades, its workers descended from the same Yankee blue-blood ancestors as the owners, with one significant difference: they were poor. "There was nothing peculiar about the Lowell mill girls," said Larcom, "except that they were New England girls of the older, hardier stock."

That contrast with the lowly mill girls in Europe, scorned by society as prostitutes, contributed to Lowell's reputation as a brand new chapter in the glorious chronicle of emerging America. The Lords of the Loom had produced and directed the country's first major industrial hit, a box office sensation and a patriotic triumph. That was the Lowell that visitors from President Andrew Jackson to Davy Crockett read and heard about and traveled thousands of miles to view firsthand.

Glory Days

When Michel Chevalier, a French government official, visited Lowell, Massachusetts, and the surrounding Merrimack Valley in 1834, he portrayed a fresh, gleaming, bustling utopian town where apple orchards bloomed near steam locomotives and the buzzing of honeybees found their mechanical counterpart in the hum of spindles. Lilting bells called workers into the mills; young women on the streets were well-dressed. Chevalier commented on their comeliness—and their independence. To Chevalier, the abominable filth and stench of European factory cities made these clean and sunny sidewalks seem all the more like rainbows arched across the heavenly sky of a fabulous diorama: "A pile of huge factories, each five, six, or seven stories high and capped with a little white belfry, which strongly contrasts with the red masonry of the building. . . . By the side of these larger structures rise numerous little wooden houses, painted white, with green blinds, very neat, very snug, very nicely carpeted, with a few small trees around them, or brick houses in the English style; that is to say, simple, but tasteful without and comfortable within. . . ."

Lowell, he concluded, was "neat, orderly, quiet and prudent." During the previous decade schools and churches went up. Watchmakers and booksellers and other merchants opened shops, "every one of them as fresh and new as if the bricks had been in the mold but yesterday," Captain Basil Hall reported.

"Who shall sneer at your calling? Who shall count your vocation otherwise than noble and ennobling?" asked poet John Greenleaf Whittier rhetorically and rhapsodically about the mill hands. Lowell's "acres of girlhood," he burbled, were "the flowers gathered from a thousand hillsides and green valleys of New England. . . . Nuns of Industry, Sisters of Thrift, . . . dispensing comfort and hope and happiness around many a hearthstone. . . ."

To Anthony Trollope, Lowell was a "commercial utopia," where workers were looked after "more as lads and girls of a great seminary than as hands by whose industry profit is to be made out of capital."

Reading the published accounts of Whittier, Chevalier, and others, governments in Europe looked to Lowell as proof that industrial enterprise could prosper without grim enslavement and human suffering. Worker reform movements in Europe gained momentum. This tree-lined community became their darling: charming, productive, dignified, and above all, whistling while she worked.

By the time Chevalier arrived in 1834, there were nineteen five-story mills in operation containing 84,000 spindles and 3,000 looms. Workers now turned out an astonishing quantity of cheap cotton cloth—27,000,000 yards, or 15,698 miles, of it annually. Boat travel proved impractical at that volume; Patrick Jackson was busy laying track for a steam locomotive railroad line—one of the nation's first—to transport the finished cotton cloth to Boston and to carry raw cotton upriver from its docks. Domestic markets absorbed as much calico as they could produce. In addition, these mills were turning out considerable amounts of "Negro cloth"—coarse fabric purchased by Southern planters to clothe their slaves, who knew it as Lowell cloth.

Fearing they had sacrificed too much power to Northern industrialists, Southerners sought to build regional cotton mills for their own crops close to the fields: a sensible idea whose time would come, but not for another forty years. A few small cotton factories in South Carolina and elsewhere struggled to find a labor source; whites, no matter how poor, would not work side by side with slaves, and slaves were not trained for factory work. Planters considered it an inefficient use of resources. Competitive efforts failed. As hundreds of new plants opened in the North, producing everything from pig iron to shoe soles, the North's manufacturing clout now extended to a wide variety of finished goods and materials; it was going through one of the most prodigious industrial growth spurts in history, while the South depended ever more on its cotton, tobacco, rice, and sugar crops.

Within a decade of Lowell's founding, New England's textile aristocracy had gained so much wealth and political power that they had become uncrowned royalty. Their money built schools, hospitals, museums, churches, and parks in and around Boston, a city they intended to glorify for posterity. Their success also encouraged other investment groups to open textile mills in New Hampshire, Massachusetts, and Maine.

These Lords of the Loom lived with their clans in an exclusive Boston enclave. When Appleton and Francis Cabot Lowell's son and relatives traveled to Lowell, which they frequently did during its early years, they soon departed. They were not about to make this mill community their home; while it served its purpose as their industrial showplace and cash cow, they considered it no place to socialize or to set down roots. You did not live downstairs with the help. Sensing that scorn, American journalists first used another term, "soulless corporation," to describe Nathan Appleton's remoteness from the daily lives of his textile operatives. That may have not been entirely justified. His biographer insists that good working conditions and fair treatment of labor were moral issues for him. Perhaps. But that issue he took up at a safe retreat in Backbay Boston.

The loom lords might be admired but they were not necessarily idolized by the general public. Too much wealth and power had accumulated to a chosen few. The masses identified more readily with young women who worked an average of seventy-three hours a week in these cotton factories yet eagerly absorbed culture by attending lectures and frequenting library reading rooms, still finding time to put out a magazine of their own. Their homespun magazine, *The Lowell Offering*, began life in 1840, and a year later it became a thirty-page monthly published by a young enthusiast, the Unitarian Reverend Abel C. Thomas. Readers could subscribe for six-and-a-quarter cents an issue: "A Repository of Original Articles, Written Exclusively by Females Employed in the Mills," read the banner line on every issue."

Quite quickly, *The Offering* became an international sensation, a symbol of brave new gender equality and an ode to self-improvement. All of the contents—stories, essays, verse, journal entries—were written by the mill hands and published without heavy editing. On his travels through America Charles Dickens visited Lowell and reported to his English audience, "I am now going to state three facts which will startle a large class of readers on this side of the Atlantic very much: First there is a joint-stock piano in a great many of the boardinghouses. Secondly, nearly all of these young women subscribe to circulating libraries. Thirdly, they have gotten up among themselves a periodical called *The Lowell Offering*, . . . written exclusively by females actively employed in the mills. . . ."

Dickens thought the contents of the magazine "compared advantageously with a great many English annuals." But what astonished him most was that many of its contributors were articulate women who wrote "after arduous labors of the day" about mill life. As the reform-minded author of *Oliver Twist* and essays that exposed Britain's despicable labor conditions, he was used to looking into workers' lives through a partially opened window—or grate—and supplying them with his own imagined speech, thought, and action. Not necessary here, he noted: "It is a pleasure to find that many of [*The Offering*'s] tales are of the mills, and of those who work in them." One *Offering* contributor gave Dickens and others the lowdown on her boardinghouse mates:

> One who sits on my right hand at table is in the factory because she hates her mother-in-law. . . . The next one has a wealthy father but like many of our country farmers, he is very penurious . . . the next is here because her parents are wicked infidels . . . the next is here because she has been ill-treated in so many families she has a horror of domestic service . . . the next has left a good home because her lover, who has gone on a whaling voy-

> age, wishes to be married when he returns, and she would like more money than her father will give her. . . .

Another explained her work experience: "They set me to threading shuttles, and tying weaver's knots, and such things, and now I have improved so that I can take care of one loom. I could take care of two if only I had eyes in the back of my head."

Each issue of *The Offering* combined informal, personal glimpses of factory life with literary efforts; among the most successful were the verses and essays of Lucy Larcom, who would go on to write eight books after leaving the mills and gain national prominence as an author.

A professor at the Collège de France in Paris devoted class lectures to the Lowell mill girls' magazine; a French politician brought a bound volume of *The Offering* to show the national Chamber of Deputies. Here were working women, expressing themselves with passion and clarity: there was no precedent for that, and there was much to learn in these pages about evolutionary social progress. The leaders and scholars of Europe and America looked on with an equal measure of bafflement and respect.

But if *The Lowell Offering* began as the inspirational voice of a new female labor class, it became a flash point for escalating unrest. Its well-meaning editors—two women, after the first few issues—did not foresee that within a few years worker loyalty to mill owners would be compromised by growing distrust. Workers were being pressured to accept lowered wages and ignored in their call for a shorter work day. That effort soon became known as the Ten Hour Movement. Activist mill hands wanted to use *The Offering* as their vehicle of protest. The female editors—older, more conservative—refused to criticize current conditions. "With wages, board, &c., we have nothing to do—these depend on circumstances over which we have no control," they demurred in one editorial. And in another: "They say *The Offering* does not expose all the evils . . . attendant upon a factory life . . . and they compare us to poor caged birds, singing of the flowers which surround our prison bars, apparently unconscious that those bars exist."

Two decades after its founding, Camelot on the Merrimack was crumbling at its foundations. The rare sense of genuine camaraderie between owner and worker no longer seemed like paternalism so much as exploitation. When Ralph Waldo Emerson said, "The children of New England between 1820 and 1840 were born with knives in their brains," he was talking about these female mill workers, among others. They were too sharp to be easily manipulated.

"If in our sketches there is too much light, and too little shade . . . we have not thought it necessary to state . . . that our life was a toilsome one—for we supposed that would be universally understood, after we stated how many hours a day we tended our machines," one of *The Offering* editors, Harriet Farley, wrote in self-defense. She was responding to an inflammatory attack on *The Offering*'s compromised integrity by one of its former contributors, who called her piece "J'Accuse." It didn't help matters that the magazine received substantial funding from the textile corporation. A rival worker publication, *Voice of Industry*, now spoke more authentically to Lowell's restless female factory hands. *The Offering* lost its readership and ceased publication in 1845.

By then Lowell and the nation were in major transition. Cotton did more than mirror these social and economic convulsions: it motivated them. Locally, some factory women began wondering aloud if they could in good faith continue to earn their livelihood by the product of slavery. Antiabolitionist groups had moved from the fringe of society; they now claimed converts among leading families. Rallies erupted in the streets. Within the mills, worker organizers were threatening to take their reform demands to the state legislature. That didn't sit well with the Boston Associates. Might they not be better off not to rely so heavily on these New England farm girls with their astringent opinions and dangerous ideas? The investors turned their attention to the hordes of starving European peasants arriving daily in New York. Those foreigners desperately needed work; they would not dare to strike for better pay.

By the mid-1840s the Boston Associates had learned the hard way that cotton was above all else an agricultural crop, annually subject to nature's impulsive whims. Less than a decade previously, a Southern crop failure had coincided with a sudden slowdown in demand for raw plantation product by British textile mills, leading in part to the Panic of 1837—a collapse of the nation's credit structure. Its negative economic effects lingered on. That instability only intensified the risks of relying on unpredictable yields of a raw material dependent on weather and soil conditions. Cotton prices fluctuated severely. At the other end of the cycle, belt-tightening among consumers created a soft market for the Associates' finished goods. Feeling pressure from both sides, they welcomed cheaper immigrant factory labor. Although farm girls continued to seek employment in the mills until the mid-1850s, they became a minority in the workforce and learned quickly not to assert themselves. The owners by then were unwilling to tolerate any displays of worker agitation.

These Lords of the Loom were already doing battle with another formidable enemy—their own consciences. Slavery had made them rich. Still, the bible warned that no man had the right to lay claim to another as his property. For as long as possible, New England's textile barons had walked a fine line between religious conviction and business pragmatism. Slavery in existing states was an acceptable institution, they'd decided, but its expansion into new territories they could not abide. Now, with a massive migration west, the whole ugly mess rose up before them with such ferocious intensity that it could no longer be avoided. Whether they preferred to or not, they knew they would soon be forced to choose between principle and profit in the national arena, and between profit and paternalism locally: troublesome choices at every turn.

SIX

Looming Conflicts

Without slavery there would be no cotton, without cotton there would be no modern industry. It is slavery which has given values to the colonies, it is the colonies which have created world trade. . . .

—Karl Marx,
letter to Pavel Annenkov, 1846

Antislavery Massachusetts senator Charles Sumner took to the floor of the U.S. Senate in May 1856 to deliver a now-famous speech, "The Crime Against Kansas." In his incendiary attack on an attempt by the South to rescind its agreement not to pursue slavery in the new Kansas-Nebraska territory, Sumner quoted from Cicero, Livy, Cervantes' Don Quixote, and Milton among others. For two days his stinging indictment of a new bill that proposed that the states of Kansas and Nebraska should instead enter the Union "with or without slavery" as they chose, Sumner held the packed Senate gallery in thrall. With typical oratorical flourish, he heaped special scorn on one lawmaker from South Carolina, Andrew Pickens Butler, accusing him of choosing a mistress who "though ugly to others, is lovely to him . . . I mean the harlot Slavery." This senator, Sumner continued, "touches nothing that he does not disfigure with error. . . . He cannot ope his mouth, but out there flies a blunder." If the entire state of South Carolina should disappear, he added, civilization would lose "surely less than it has already gained by the example of Kansas."

A day later Sumner lay near death on the Senate floor. Preston Brooks, a congressional representative from South Carolina and a cousin of Butler, who was not present, had snuck up on Sumner in the nearly empty chamber as he sat making notes at his desk; seeing that Sumner's long legs were pinned under his low desk, Brooks began to club him viciously from behind with the gold head of his heavy cane. Breaking loose at last and staggering up the aisle, bleeding heavily, Sumner collapsed on the floor, unconscious. Brooks continued to beat him with full force until his cane broke. He later justified the attack by claiming that he was repaying an insult to his cousin. "If I desired to kill the Senator, why did I not do it?" he asked in his own defense. "You all admit that I had him in my power." Brooks was fined $300 and allowed to maintain his congressional seat. "Every Southern man sustains me," Brooks boasted. "The fragments of the stick are as sacred relics." Constituents were bidding high prices for pieces of his cane. A group of South Carolinians sent Brooks a new cane inscribed "Hit him again!" The *Richmond Enquirer* praised Brooks as a hero in an editorial: "These vulgar abolitionists in the Senate must be lashed into submission," said the newspaper. "We consider the act good in conception, better in execution, and best of all in consequences."

Although Charles Sumner was no friend of the conservative Boston Associates, his vicious beating profoundly altered their longstanding hands-off approach to slave ownership. It also spurred the formation of their Emigrant Aid Society, which the loom lords created in an effort to keep Kansas a free state.

Elected to the Senate in 1850 on a Free-Soil ticket, Sumner had alienated Nathan Appleton and his powerful textile partners, in particular the conservative brothers Amos and Abbott Lawrence, through his vehement abolitionist stance. Sumner derided the "conspiracy," as he saw it, between the "cotton-planters and flesh mongers of Louisiana and Mississippi" and the "cotton spinners and traffickers of New England"—the Lords of the Lash, that is, and the Lords of the Loom. By then the Associates' textile empire had spread from Lowell to Maine and several other states in New England; it also interacted with powerful cotton shipping and financial interests in New York City. Banks now estimated the value of Massachusetts manufacturing to be in the vicinity of $300 million, driven largely by cotton textiles. The Associates owned one-fifth of the nation's spindles, one-third of Massachusetts railroad mileage, and two-fifths of Boston's banking capital. Their Lowell mills alone in that year produced 65,000 miles of cotton cloth—enough to circle the globe more than twice.

These men understandably believed their personal fortunes would be jeopardized if hostilities between their slave-dependent suppliers in the South and activists in the North led to secession. No appeals to conscience, it seemed, would change their minds. But Brooks's cane did what preachers' Sunday sermons had failed to: it radicalized them. The attack on Sumner transformed him into the North's fallen hero. Until that moment Amos Lawrence hardly seemed inclined to challenge the status quo. In a letter to one Southern cotton grower, he had vowed never to interfere in the question of slavery "unless requested by my brethren of the Slave-holding States." His son, Amos A. Lawrence, considered himself honor-bound as a Whig to recognize slavery as a cornerstone of the original Union contract between states. Sumner's brutal beating shook them to the core. "We went to bed one night, old-fashioned, conservative, compromise Union Whigs," said Amos A. Lawrence, "and we waked up stark mad Abolitionists." His father offered his home to Sumner for recuperation. "I assure you no one will give you a more cordial welcome," Amos promised in his invitation, although the two men had excoriated one another for decades. Sumner accepted. Their hatchet, rather than simply being buried, was hefted in unison by both men, now arm in arm, and hurled across the Mason-Dixon Line.

The Confederates might not fire on Fort Sumter for another five years, but in the last half of the 1850s it became increasingly evident that Brooks's cane had struck the first blow in the war between the states. After decades of Northern vacillation, his unprovoked attack aroused the immensely powerful Northern cotton magnates to give more than lip service to their opposition to slavery. To anyone familiar with their business-first attitude over many years, the odds seemed long that they would do what they did: put conscience before cotton.

Cotton vs. Conscience

From the moment they established their first Massachusetts mill at Waltham, these men exercised brilliant business judgment and a self-serving, suspect logic. Led by the example of their most prominent leaders, Nathan Appleton and the Lawrence brothers, Boston Associates' original members determined that if left alone, the "peculiar institution" of slavery would die out of its own accord as plantation owners over time voluntarily freed their slaves. As preposterous a notion as that appears to be in retrospect, it became a common argument among apologists with too much to lose if their source of low-cost cotton produced by free labor were to be eliminated.

They subscribed to Alexander Hamilton's belief that "ideas of a contrariety of interests between Northern and Southern regions of the Union, are in the Main as unfounded as they are mischievous. . . . Mutual wants constitute one of the strongest links of political connection. . . ." The primary mutual want in Hamilton's mind was cotton. Barely had the ink dried on his *Report on Manufacture*, written in 1791, when, two years later, Eli Whitney turned a crank on his gin that spit cotton seeds in one direction and yanked the history of the United States in another.

The explosion of raw material from the South that followed soon enriched New England's textile aristocracy whose mills were partially responsible for driving up the number of slaves fivefold between 1800 and 1860. In that year close to four million slaves accounted for nearly 40 percent of the South's population. Seeking new arable cotton acreage, Southern growers by then had relentlessly expanded westward into virgin territories that would become Texas, Louisiana, Oklahoma, and Missouri.

Witnessing a slave auction in South Carolina as a young man, Appleton wrote passionately to a friend that he abhorred "the horrid sight of the sale of human flesh." He was sickened by the taunting cruelty of the auctioneer toward the trembling, frightened slaves being bid on and humiliated as if livestock or hogs. Later, having mounted his individual fortune on slaves' scarred backs, Appleton tried to calculate his way past that repulsion. In 1847 he argued that the abolition of slavery would cost the South "a thousand millions of dollars. Was such an amount of property ever voluntarily relinquished or annihilated?" Not by him, certainly, at least not by then. "African slavery . . . is a curse . . . I consider it a tremendous evil," he said. "But we of New England ought to be able to look at it calmly and coolly. As to the existence of slavery in the slave states, I see no reason why we of the free states should make ourselves very unhappy about it. Why not leave it to the parties immediately concerned?"

Appleton, the Lawrences, and their business partners, once lauded for bringing prosperity to New England, became objects of derision in the Abolitionist press—portrayed as unfeeling men who refused to acknowledge the pain and suffering caused by slavery. Like so many movers and shakers before and after them, they continued to put personal gain before social justice. Unlike many others, in the end they could no longer abide their own rationalizations.

The South's leaders became their willing coconspirators. "I am no opponent to manufactures or manufacturers," South Carolina senator John C. Calhoun wrote to Abbott Lawrence, "but quite the reverse. Cotton threads hold the union together: unites John C. Calhoun and Abbott Lawrence. Patriotism for holidays and summer evenings, but cotton thread is the Union." This was true both literally and figuratively: during the decades that led to the Civil War millions of Southern slaves were being clothed in coarse cotton fabrics produced in the factories of New England's mill towns.

There was an overriding social connection as well between these powerful families. Members of the North and South aristocracy recognized one another as peers, ladies and gentlemen of elevated status and breeding that transcended geographical regionalism. Pillars of their respective communities who had attained enormous wealth, they felt comfortable around each other's money, with convictions and indulgences in common—fine wine and protectionist tariffs among them. Northern bankers extended credit to Southern planters, whose sons enrolled at Yale, Harvard, and Princeton: their swallowtail coats and calfskin boots were greatly admired, said Oliver Wendell Holmes. As early as 1821, cotton was already America's leading export; its transportation, manufacture, and financing required harmonious relations between these lords. Sectionalism was bad for business, and what was bad for business was bad for America.

But then, as in the 1960s, a younger generation came along to challenge and test the beliefs of their elders. The Whig Party that the older conservative businessmen had helped create in 1834 managed to put a series of pliant presidents in the White House. By the 1850s, however, its proslavery position created an internal rupture. Young Whigs revolted. An emerging, vocal antislavery wing within the party now labeled Nathan Appleton and the Lawrences "Cotton Whigs," for their conciliatory stance. The activists formed a separate wing, the "Conscience Whigs," in opposition to the status quo.

As antislavery sentiment continued to swell within their party, the Conscience Whigs became a powerful political force and managed to block Abbott Lawrence's bid for the vice presidency in 1848. That same year, pressured by Appleton and others, the conservative Whigs nominated as their presidential candidate a Louisiana slaveholder, General Zachary "Rough and Ready" Taylor, who was as clueless about the slave debate as he was about everything else. In protest, Conscience Whigs—young and committed—bolted the party forever. Aligning themselves with Free-Soilers who wanted territories like Kansas-Nebraska to enter the Union as nonslave states in 1850, they helped elect Charles Sumner to the Senate from Massachusetts. Within the Whig Party the war was on, and the party itself, which died out in 1856, never recovered. Its conservative members became Democrats, forerunners to today's Republicans, and Conscience Whigs in 1854 helped create the new more liberal Republican Party, today's Democrats.

No one, and no country doing cotton business with the United States, remained unsullied for long. Over many decades America's growers and the United Kingdom's Lancashire mills had joined forces in a campaign, it sometimes seemed, to clothe the entire world in the first inexpensive mass-market commodity. With so much at stake, governmental skirmishes stood no chance. In the Treaty of Ghent, which ended the War of 1812, England and the United States agreed to suppress the slave trade. So much for policy positions. In reality, Baltimore builders designed faster clipper ships to carry and deliver cargoes of slaves for Liverpool's thriving slave traders. Until 1834, when Britain finally abolished slavery throughout its empire, those ships plied a centuries-old triangular route: they brought raw cotton to England from the United States and West Indies, then from England to the hand spinners and weavers of West Africa, where cotton fiber was bartered for slaves. The slaves were then shipped to the American South and British West Indies colonies, where they were forced to provide the free labor to produce more raw cotton, which was shipped back to England and Africa, endlessly repeating the cycle.

Slave-produced cotton created many of England's most prestigious banks, including the giants Barclay and Lloyds. Liverpool's towering skyline of massive Victorian commercial buildings stands as a monument to cotton's supremacy. Karl Marx, living in England during that era, noted that "direct slavery is as much the pivot upon which our present day industrialization turns as are machinery and credit."

Much of Liverpool's income from America after the early nineteenth century depended on the purchase and sale of cotton, not the sale of slaves, but even so its wealthy textile brokers understood that their primary supply would most likely be endangered in the event of a civil war. These merchants kept a close watch on unfolding events. Almost certainly they, too, sensed that every effort to isolate cotton as an American product from cotton as a hotly contested political issue was ultimately doomed to fail. The rise of the Conscience Whigs demonstrated that only too vividly. "We gain nothing by allowing any portion of these people to attend our primary, or other public, political meetings," said Abbott Lawrence about the younger firebrands. Now advanced in age but not in tolerance, his ultimate insult was to label these Conscience Whigs the "Abolitionist party." Although he lived and socialized with leading antislavery activists, including Ralph Waldo Emerson and the author of *Uncle Tom's Cabin*, Harriet Beecher Stowe, Lawrence roundly denounced the views of Abolitionists. "I believe they have done mischief to the cause of freedom in several States of the Union," he said. A Boston newspaper took him to task. Lawrence, it editorialized, "had covered old Massachusetts with shame." By consensus, Abbott was even more obsessed with money than his brother Amos, who famously announced he "always put business above friends."

One brave member of the artistic Lowell clan, fledgling writer James Russell Lowell, took them on: "There is something better than Expedience and that is Wisdom, something stronger than Compromise and that is Justice," he declared.

In a desperate attempt to expand the nation without upsetting the slave- and free-states balance, Kentucky senator Henry

Clay cobbled together the Compromise of 1850, which established a fragile truce. One of its key trade-offs assured the North that Kansas and Nebraska would enter the Union as free states. After eight months of intense, often furious, debate, Congress agreed that California would enter as a free state and that slavery would be permitted in the newly acquired Texas territory. To pacify slave-state politicians, one of the compromise bills, the Fugitive Slave Law, required citizens of a free state to assist in returning escaped slaves to their owners. They could be captured and sent back to the South with no legal right to a jury trial to plead their case.

Push was quickly coming to shove for the Cotton Whigs and their most prestigious political spokesman, Massachusetts senator Daniel Webster. Over a long and distinguished career, Webster, an elderly statesman of international repute, had come to be revered as one of America's most persuasive voices of reason, an eloquent opponent of slavery, and a fierce protector of individual rights. His words carried more weight for the average citizen than those of any president in memory. When he rose to defend Clay's proposals in the interest of maintaining national unity, the country listened.

But cotton and slavery had now changed that country, Webster learned to his dismay. On behalf of the Lords of the Loom and to advance his own agenda to end factionalism, he sought the middle ground between extremists on both sides. Extremism for any cause has always been bad for business, but renouncing it doesn't necessarily burnish the careers of politicians if mixed motives are suspected. Webster put the full weight of the loom lords' muscle behind the legality of capturing and returning escaped slaves. By seconding Clay's compromises, he imperiled his own stature and infuriated New England's influential men of letters. "The word *liberty* in the mouth of Mr. Webster sounds like the word *love* in the mouth of a courtesan," said Ralph Waldo Emerson, sufficiently enraged to call Webster a whore. As for the Fugitive Slave Law endorsed by Webster: "I will not obey it, by

God!" For John Greenleaf Whittier and others, the senator had joined the disgraced Boston Associates:

> . . . men of the North, subdued and still;
> Meek pliant poltroons, only fit
> To work a master's will.

Later, Whittier sadly lamented Webster's betrayal as "the light withdrawn which once he wore!" That same sentiment applied to the young country itself. No longer did a divine radiance illuminate these United States, so bravely adventurous as to dare to seek out a new form of government and leave tyranny behind. Many of its leaders seemed instead to have traded one form of tyranny for another—monarchy for money. Cotton was the vehicle. Greed was the engine that powered it. Whatever course the future took, it carried with it a tarnished promise that democracy might someday mean true equality, or that when pressed to choose, its preeminent elected officials would put aside personal reward to ensure the liberty of one and all. Instead the controversy over cotton exposed an ugly, self-serving expediency.

For many proud Americans, the country preached progress but practiced regression. Slavery had been outlawed in Europe for nearly two decades. In 1834 Parliament emancipated all blacks throughout the British Empire. Confederates argued that of the 11,000,000 slaves brought to the New World over the previous centuries, only a tiny fraction, fewer than 5 percent, or about 500,000, came to America. By contrast, Brazil and the West Indies accounted for nine million slaves, most of them put to work on cotton and sugar crops. That argument meant little to men and women unwilling to allow any individual from this day forward to selectively determine who was or was not entitled to "the pursuit of happiness." To them there were no asterisks in the Declaration of Independence.

Daniel Webster above all others knew that, but in short order he would become the Hubert Humphrey of his day. Little more than a century later, Senator Humphrey, once a bold, courageous civil rights liberal but now President Lyndon Johnson's vice presidential puppet, shredded his own reputation as an independent thinker by stubbornly defending the unpopular war in Vietnam. Webster, now excoriated by some as a man "whose soul has been absorbed in tariffs . . . instead of devoting himself to the freedom of the future," found himself playing to an enthused but isolated audience of wealthy bankers and businessmen in the North. They feared the South was about to secede and that an economic disaster would soon follow. While conservative newspapers lauded Webster "with a calm belief that he has placed a vexed question is a position in which it can be and must be fairly settled," he was ostracized by humanists like Emerson and Whittier whose friendship deeply mattered, and who had the ear of the people. Depressed, he died less than two years later, in 1852, as did Henry Clay.

Independent Yankees were not about to be coerced into rounding up escaped blacks for a one-way ticket back to hell. It was that simple. By 1854, the year that Senator Stephen Douglas's Kansas-Nebraska bill became law and opened up the possibility of slavery in new states, the textile lords felt their power rapidly eroding. "Harmony at all costs," their unofficial slogan, was quickly becoming "calamity at every turn": they'd lost touch with the populace. Desperate to regain their former prestige and political clout, the old-line Whigs jumped at the chance to aid a fugitive slave, Anthony Burns, in Boston. He had recently been captured by U.S. marshals under the Fugitive Slave Act, which they'd supported, and he was about to be returned to his slaveowner without trial.

Amos A. Lawrence offered to put up "any amount" to provide legal assistance for Burns on behalf of the Associates, who were slip-sliding from the Cotton to the Conscience side of town in a big hurry. But Lawrence quickly discovered that he and his partners had lost control. He petitioned the mayor of Boston, announcing that he would just as soon see the courthouse burned to the ground than see Burns sent back to the plantation. It did no good. Feelings ran high. An emotional crowd threatened to raise their fists to stop the authorities from boarding ship with Burns. Lawrence looked on. Only the presence of extra security, he told his brother, "prevented the total destruction of the U.S. Marshal and his hired assistants." He and the assembled throng watched Burns march in chains to the docked ship. Suddenly, tragically, the consequences of the Fugitive Slave Act became only too clear. Before Lawrence's eyes, cotton profits materialized as flesh and bone in bonds.

"You may rely upon it," Lawrence wrote to a friend, "that the sentiment at this time among the powerful and conservative class of men is the same as it is in the country towns throughout New England."

A noble assurance, but hardly precise. The country towns of New England stood to lose little or no income as a result of their antislavery stance. Mill towns, however, did. By now they ranged from Saco, Maine, to Manchester, New Hampshire, home to the sprawling Amoskeag mill complex; from Chicopee, Massachusetts, to Taunton to Dover to Lynn to Fall River: dozens of towns, close to a hundred mills, and close to 5,000,000 spindles in operation. Though not all were owned by the Boston Associates, the mills' investors shared a common goal: to prevent secession. Cotton Whigs stood to lose millions; despite their rhetoric they acted with caution for another two years until Brooks's attack on Charles Sumner in 1856. Rudimentary math explains why.

America's cotton mills were well on their way to producing $115,000,000 worth of cotton by 1860, or three times as much as the country imported, and every ounce of it relied wholly on slave labor. A U.S. census in 1790 counted 697,124 slaves, with almost as many in New York (21,234) as in Georgia (29,264). Despite the Constitutional ban on further importation in 1808, by 1820 there were 1,533,086 slaves, almost all now in the South, and Virginia alone accounted for 425,757. Smuggling and domestic breeding increased those numbers; they continued to swell until at the brink of the Civil War nearly three out of every eight Southerners were slaves. By then the South produced an astonishing 2.275 billion pounds of raw cotton, and the crop accounted for 60 percent of the country's exports. The South now supplied over 80 percent of the cotton manufactured in Britain, two-thirds of the world's total supply, and all of the cotton used in New England's mills.

As the country now headed toward civil war in the late 1850s, many wealthy New England mill owners, spurred by the example of the Lawrences, were no longer willing to pay for their conscience with their cotton.

Against that backdrop, western expansion "woke the sleeping tiger," as one historian, Allan Nevins, later remarked. Years earlier Thomas Jefferson had pointed out that "it is cheaper for Americans to buy new lands than to manure the old," and a restless generation of settlers was proving him right yet again. New England's textile elite, which had vowed not to interfere with slavery in states where it was sanctioned by the Constitution, held true to their word—until both houses of Congress, pressured by the South, passed Illinois Senator Stephen A. Douglas's bill that made slavery an option in the Kansas-Nebraska territory.

Since Kansas represented a new, ominous turn of events, the Boston Associates organized and funded groups of Free-Soil settlers willing to set off along the Santa Fe Trail and to take up permanent residence in the developing territory. But no sooner had these emigrants arrived than armed proslavery advocates crossed over from Missouri hoping to frighten off the intruders and prevent Free-Soil transplants from deciding the slavery vote. A proslavery "posse" sacked Lawrence, Kansas. In retribution, a fanatical Abolitionist ranger named John Brown, who would soon become a national icon, formed his first raiding party and killed five innocent proslavery settlers.

Shortly after that, Brooks attacked Charles Sumner. Many of the North's business leaders as well as its politicians viewed the beating as another contemptible Southern treachery. The Union, and Sumner himself, never fully recovered. The senator experienced severe chronic headaches and nightmares and did not return to take up his full senatorial duties for more than three years. Friends said he was never again the same man. The tenuous unity of the nation had received a crushing blow that split apart the last remaining tethers of compromise. In Boston, more than 5,000 gathered at Fanueil Hall, birthplace of the American Revolution, to denounce Brooks and the "crime against the right of free speech and the dignity of a free State."

These incidents culminated on the eve of the Civil War with John Brown's raid at Harpers Ferry, Virginia, in 1859, where he and his small rag-tag band tried to liberate slaves. Disarmed and captured in a matter of hours, Brown and his men were summarily hanged. In doing so, the South executed a zealot and gave birth to a martyr. Within weeks, Northern industrial cities held memorial services honoring his memory. In Lowell, antislavery society members pushed a John Brown funeral wagon with a solemn tolling bell along streets lined with sympathizing mill hands.

Ralph Waldo Emerson spoke for thousands: "I think we must get rid of slavery, or we must get rid of freedom."

While slavery might have become the nation's most divisive and emotionally charged issue, tariffs and states' rights, both long-simmering areas of contention, again rose up to enflame passions. When a marriage goes bad the other party can do no right. Southerners looked at the way Northern industrialists stacked the deck in Congress to insure that high tariffs on foreign imports favored domestic manufacture, and they said, in essence, "Hold on! All we get out of this arrangement is to pay more dearly for things we might be buying cheaply from England, and then you Northern swindlers come along to steal our money and keep the change." The South's planters focused on states' rights, and decided there, too, the fix was in. Since the federal government and its influential textile lobby couldn't be trusted to play fair with Southern interests as the country moved west, the time had come to return decisive power back to the states to make their own laws. Only the sorriest sort of fools would take the word of a Union government whose Constitution guaranteed the right of citizens to own property, including slaves, and who then reneged on that right as if it had been written on swamp mist.

Lives in the Balance

By the time Brooks attacked Sumner in 1856, cotton's reach had expanded far beyond the domains of the planters and textile investors. It now affected the economies of New York, Savannah, New Orleans, Charleston, and Liverpool, England, among other cities, as well as the livelihood of thousands who never planted a seed or spun a strand of fiber into yarn. They included stevedores, shopkeepers, insurance agents, factors, bankers, packers, brokers, draymen, and more than 40,000 American seamen on vessels engaged in foreign cotton trade, according to an 1852 report to Congress. Though far from the fields and mills, these laborers, retailers, commission agents, factors, and money lenders felt every ripple in the shifting tectonics of the cotton trade. The Southern states' threat to secede spelled possible wide-scale financial ruin or, at the very least, severe hardship.

New York City was especially vulnerable. More than 800,000 tons of raw cotton traveled by ship from America to Europe and other countries by the 1850s, and, through cagey business arrangements, the port of New York made itself a lucrative stop for much of that cargo on route. The port, nicknamed Sandy Hook for the hazardous five-mile sandspit at the entrance to the harbor on the New Jersey side, served as one corner of the "cotton triangle." Most of the finished cotton goods that continued to be imported to the United States from England—as much as $32.5 million as late as 1860—came through New York. The holds of those ships returned to Liverpool with raw cotton bales on-loaded in New Orleans or another Southern port. Many regularly detoured to Sandy Hook for additional cargo before crossing the Atlantic.

Cotton was New York's leading export; the South depended on New York as well for European home furnishings and high quality imported fabrics including silks and linens. The irony in all this was that although a New York stopover required ships to travel 200 miles out of their direct lane between Liverpool and Charleston, Savannah, or New Orleans, there was no logistical reason for its involvement. Trade with Europe could have been carried on directly from these ports. Nevertheless, Sandy Hook collected a heavy toll on each vessel. Wharves along the East River filled with bales being unloaded and reloaded at extra cost. "The combined income from interest, commissions, freight, insurance, and other profits was so great that, when Southerners finally awoke to what was happening, they claimed that New Yorkers with a few other Northerners were getting forty cents for every dollar paid for Southern cotton," one historian reported.

While the South did not benefit from Sandy Hook's interjection, it accommodated the inconvenience and additional expense for the same reason it shipped raw cotton hundreds of miles north to be converted to cloth in New England's mills: it lacked regional commercial and industrial resources. As an agrarian society now trying to cope with one of the greatest agricultural boom crops in history, the South occupied itself with the demands of planting, cultivating, harvesting, ginning, baling, and brokering millions of tens of thousands acres of cotton. That left it dangerously dependent on the North (or France and England) to manufacture and supply every domestic item, from tea cups to tables to chairs, from the bedsheets they slept on to the paper they wrote on. Southern states that had fought to win their independence from the British crown now relinquished it economically to the North. That became only too painfully obvious to men with vision like William Gregg, born in South Carolina, a successful tradesman who decided to do something about the imbalance.

Gregg looked around at all the poor unemployed white men in the South living close to starvation—29,000 illiterates in his state alone—and who were unwilling to degrade themselves by working alongside slaves in any capacity, even for pay. He studied the example of Lowell and other industrial cities, and decided to build South Carolina's first cotton mill at Graniteville. These poor whites, living in company boardinghouses, would provide a cheap labor force. By 1846, he'd raised $300,000 to build a two-story cotton mill (and surrounding town), first powered by waterwheel, then by a more efficient turbine. Within three years Graniteville's 8,400 spindles and 300 looms were producing a steady flow of coarse cotton cloth, employing a mix of English, German, and Scotch-Irish natives who'd been disenfranchised and cast aside by the unfair competition of free slave labor.

All of that occurred as sectional tensions mounted over slavery; clearly self-sufficiency was becoming the crucial Southern issue, politically and economically. Gregg had shown the way. Still, the Southern state governments, ruled by wealthy planters, resisted change. The largest potential investors lacked the liquid capital to fund a massive investment in industry comparable to the Boston Associates' development of Lowell. Planters were land-and-crop rich and cash-poor, habitually buying on credit. While mired in nonindustrial traditions that no longer entirely served their best interests, they proved to be ill-equipped by temperament to shift gears. They glanced at the future and retreated into the past. After the outbreak of the Civil War, blockade runners smuggled in contraband calico for a hundred times its cost in the West Indies, while Gregg, offering less expensive cotton cloth from his factory, was "condemned by the public as a heartless extortioner," one observer noted.

There was no shortage of mill labor. In the end it didn't matter. There was even a greater shortage of farsighted, agile Southern business leaders like Gregg.

Cotton's Controllers

From the mid-1840s on, positions were hardening. All the while, by relinquishing control of its crop to the North, the South was sleeping with the enemy. New York did more than ship Southern cotton; it provided much of the funding for it. Hundreds of Yankee factors from New York blanketed the South every year, working with Manhattan banking houses that had the capital to make loans. Acting as independent intermediaries, the factors advanced long credit at high interest rates against the next year's crop, usually from 7 to 12 percent, and took their cut. Southern banks played a minority role.

That put the average planter at least one year in debt to the New York money lenders whose payment came from owning a certain percentage of the planter's future harvest. Debt was chronic. It resulted primarily from the growers' need to expand their acreage and buy more slaves. That in turn gave financiers from England and New York the power to monitor their operations, squeeze out higher interest rates, and of course gave these money lenders less incentive with every new loan to take a partisan stance in the North-South slavery conflict. Bales became payment; they were quickly turned into cash as New Yorkers sold that raw cotton to Liverpool to supply Lancashire's mills. Many planters instead went directly to England for loans at better interest rates, often through British factors. But there were additional risks. British mills set the terms. Cotton might leave port at one price per pound and arrive in Liverpool a few weeks later at another, buffeted by the sizeable market fluctuations that occurred during the time in transit. There would be no instant telegraphic communication between continents until after the Civil War.

These entangled cash-and-credit business relationships flourished and established a high-risk commodities market for the crop. Subject to the vagaries of nature, cotton's fortunes rose and ebbed with each annual fall harvest. Heat, rain, and related growing variables led to an overabundance or shortfall of product, and large sums gained and lost overnight on audacious business gambles. New York speculators would buy up huge amounts of cotton in lean years and sell at high prices to Liverpool. In 1825 a man named Jeremiah Thompson, then the world's foremost cotton trader, learned that the stock on hand in Liverpool was one-third less than expected. He had his agent race to New Orleans with orders to buy as much of the coming harvest as possible. A rival agent got there first and made a killing, buying from planters who did not yet know they could sell at a premium. Thompson bought whatever he could at higher prices, only to see the prices in Liverpool tumble as the Manchester mills severely cut back their output and new cotton arrived in volume from Brazil. Thompson barely survived.

The cotton fields were in the South, but the action was in the North for speculators and businessmen. One honored tradition among Southern merchants was to pack up their families and escape the heat of summer by heading up to New York—cooler, at least by degree—for an extended stay. While the family vacationed at the beaches of Long Island in July, the merchants laid in stocks of domestic and imported goods from Northern suppliers via Sandy Hook. The interaction was as social as it was commercial, and lasted until close to the outbreak of war. By then regional pride, bolstered by accumulating antagonisms, motivated Southern ports like Charleston to attempt to break off their ties to New York and pursue direct shipping with Europe rather than succumb to Northern "merchants, ship owners, capitalists, and others . . . who drink up the life-blood of her trade," in the words of a Charleston report.

But a pattern of interdependence had been well-established for three or more decades. From New England to New York to New Orleans, cotton still bound the country together as a commercial enterprise even while propelling it into a bloody war.

Seeking a bellwether, these Englishmen might have kept a sharp eye on one industrial American city's combustible mix of pro- and antislavery agitation and worker unrest. That city, Lowell, Massachusetts, birthplace of the country's cotton revolution, now reflected the divisive passions that threatened to rupture an angry citizenry into separate nations.

Tempest on the Merrimack

If you lived in Lowell during the 1850s, injustices that once occurred at a safe distance pressed close to home. A local newspaper urged "calm reflection" rather than any threat to Southerners that "a few thousand glittering bayonets are on their way." As the North's major manufacturing center for the South's slave-produced cotton, Lowell drew fire from a steady stream of reformers; there were four separate antislavery societies now active in the city out of more than 200 throughout Northern states, many organized by clergy. Lowell's mill girls went to rallies; some marched in demonstrations. Mill managers had orders to ban Abolitionist literature. It hardly mattered. These two separate movements, worker and slave reform, surged together with the percussive force of whitewater shooting through a narrow gorge.

Emancipation quickly became the single issue that united them. Factory women who saw themselves as wage slaves were inclined to identify with field slaves. In truth there were few if any similarities. Still, emotions ran high. Noted labor reformer Charles A. Dana weighed in: "A slave too goes voluntarily to his task but his will is in some manner quickened by the whip of the overseer. The whip which brings laborers to Lowell is NECESSITY. They must have money. Is this freedom?"

Some proslavery Southerners compared the North's industrial cities to the squalid slums of England. Visiting Lowell, they usually left with a different attitude. "They cum swarmin out of the factories like bees out of a hive every direction . . . thousands upon thousands of 'em . . . all looking happy and cheerful and neat and clean and butiful as if they was boarding-school misses just from their books," William T. Thompson of the *Savannah Morning News* told his readers in 1845. To Thompson, it was an interesting sight "to one who had always thought that the opparatives as they call 'em in the Northern factories was the most miserable kind of people in the world."

The North's prolabor political party, the Democrats, was also proslavery; that complicated matters further and theoretically pitted workers against Abolitionists. Perhaps for that reason Lowell's Female Labor Reform Association never formally allied itself with the Lowell Female Anti-Slavery Society. Still, they shared a common fiery contempt for blind obedience to draconian rules and institutions. They also followed in the footsteps of a former mill worker, Sarah Bagley, who showed them all how to take power into their own hands.

Bagley had all the makings of a firebrand when she organized her labor reform association in 1844; at that time she was still on the factory floor. She proposed a shorter ten-hour work day and immediately ran afoul of the mill managers. Unfazed, she pushed on, writing broadsides that lashed out at anyone who crossed her. By then *The Lowell Offering* had clearly become the house organ of management, and it refused to publish her call to arms. Bagley promptly complained to a Boston newspaper that *The Offering* "had never been an organ through which abuses of the oppressive rules or unreasonable hours might be complained of. . . ." She enlisted the help of Democrats and Whigs alike in her effort to improve conditions for the mill girls. First-generation operatives like Lucy Larcom and Harriett Robinson had initially accepted the constraints and interminable hours as a trade-off for the opportunities they saw as worthwhile. Not so Bagley. To her, fourteen-hour days were as close to dying from natural causes as was the inevitable demise of *The Offering*. She hoped to deliver a swift kick that would snuff them out for good. In the pages of an activist mill-girl publication she helped start in 1845, *The Voice of Industry*, Bagley vented her anger. *The Voice*, as different from *The Lowell Offering* as gunpowder is to lavender potpourri, quickly became a union mouthpiece for dissatisfied female factory workers even before unions formally existed.

Bagley challenged mill managers who would not permit distribution of *The Voice* or any other labor reform literature in the mills while allowing antislavery petitions to be signed. "To my mind it is slavery," she said, to be "shut up in a close room for twelve hours a day in the most monotonous and tedious of employment. . . ." One reformer found fault with sanctioning protests against the extension of slavery "while thousands of the fair daughters and noble sons of New England are daily confined from 12 to 14 hours within the prison walls of our noisy, health-destroying and humanity degrading mills."

Persistence was everything for the reformers, as their slogan, "Try Again," indicated. "In the strength of our united influence we will show these *driveling* cotton lords, this mushroom aristocracy of New England, who so arrogantly aspire to lord it over God's heritage, that our rights cannot be trampled upon with impunity," Bagley wrote. She was clearly a piece of work, someone to be reckoned with. Articulate as well as passionate, an able public speaker, and an instinctive organizer who started chapters of the reform association in Fall River, Manchester, and numerous other mill towns, she caught the attention of William Lloyd Garrison, the country's leading antislavery proponent at that time. One decade earlier in 1834 Garrison had come close to being lynched in Boston for his Abolitionist views, with the tacit agreement of the Cotton Whigs who viewed him as their menacing enemy. Garrison survived—barely—and now he and his antislavery newspaper, *The Liberator*, were gaining respect and readership throughout the North.

Emancipation was in the air everywhere cotton was being grown or manufactured or worn or used as sheeting, bedding, canvas, coverings, sails, sacks, tents, and rope—a seemingly endless array of possibilities. That popularity put even more pressure on Northern mills to produce.

The Cotton Whigs wanted no part of Bagley's agenda, but their Democratic Party counterparts listened with interest. By 1844 two ten-hour-day petitions had been sent to the state house by Lowell workers. They contained 2,600 signatures. A similar petition had collected 1,300 signatures from workers in other towns a year earlier. One petition laid out the grievances: ". . . [I]n view of our condition—the evils already come upon us, by toiling thirteen to fourteen hours per day, confined in unhealthy apartments, exposed to the poisonous contagion of air . . . debarred from proper Physical Exercise . . . thereby hastening us on through pain, disease and privation, down to a premature grave. . . ."

The Massachusetts legislature, empowered to change the law, did what politicians habitually do at first when presented with a controversial issue: not much. Tabling immediate action, it appointed one of Lowell's representatives in the legislature, newspaper publisher William Schouler, to set up an investigative committee for the next session. That would become America's first governmental inquiry into labor conditions. Schouler appeared to expect that no self-respecting woman would testify in public at such an event: he wrote to Bagley and her co-complainants, "I would inform you that as the greater part of the petitioners are females, it will be necessary for them to make the defense, or we shall be under the necessity of laying it aside." In other words, "Stop this impertinent behavior at once or I'll drag you all before us by your braids and humiliate you in public!"

But Sarah Bagley, armed with evidence, was more than willing to be seen and heard. An eight-year veteran, Bagley grew up in the mills and spoke with authority. She pointed out that inmates at the Massachusetts state prison worked four hours less a day than the operatives and that those prisoners had a splendid library. She'd paid a visit. She confirmed the testimony of other women that illness was rampant among the factory women and that pay was being lowered to an average of $2.93 for loom operators as their hours increased. In her own situation, she said, she'd been absent one-third of her work time in the past year due to illness. The legislators tore into her. If these girls weren't working, wouldn't they be idling, wasting time? Not at all, Bagley responded. Their "intellectual, moral and religious habits would . . . be benefited." One female operative after another chronicled her work-related illnesses brought on by fatigue and prolonged exposure to cotton dust. The loom lords were ready, too, parading a long line of witnesses before the committee to assert that these mill girls enjoyed excellent health, "three meals regular," and access to culture—neglecting to mention that they got docked pay for leaving work in the evening to attend lectures.

Making an expedition to Lowell to decide for themselves, the legislative committee members sounded like horticulturists on a field trip: "Grass plots have been laid out, trees have been planted, and fine varieties of flowers in their season are cultivated in their factory grounds. . . ." They concluded that "everything about the mills and boarding-houses appeared to have for its end health and comfort." In their final report, the committee decided that although "we think there are abuses; we think that many improvements may be made . . . the remedy for long hours does not lie with us. We look for it in the progressive movement of art and science, in a higher appreciation of man's destiny, in a less love for money, and a more ardent love for social happiness and intellectual superiority." Beneath the cant and rhetorical filigree is a stern sermon to these workers, male and female both, to respect their superiors and give thanks on high that they have an opportunity to earn a living wage. These legislators owed their careers to the loom lords.

Their thorough report is filled with charts and figures on the average work day (about twelve hours) and the most common kinds of diseases (consumption and dysentery top the list). But finally it relinquishes any responsibility for improving labor conditions:

"It would be impossible to restrict the hours of labor, without affecting very materially the question of wages; and that is a matter which experience has taught us can be much better regulated by parties themselves than by Legislature."

That finding did not sit well with the Lowell reform movement, which attacked the committee for its "lack of independence, honesty, and humanity." From a contemporary perspective the committee's conclusions seem in keeping with these politicians' ardent desire to maintain the status quo. The textile mills were a primary source of revenue and employment for New England, and the vaunted families that ran them pulled many of the invisible strings that kept the political machinery running smoothly.

If Lowell's women had no vote, they had lungs, and they used them to shout out their displeasure in the *Voice of Industry*. Anticipating Gloria Steinem and the Women's Liberation Movement of the 1960s and '70s, Bagley and her supporters railed against paternalism and the subjugation that it encouraged: "Bad as is the condition for so many women, it would be so much worse if they had nothing but your boasted protection to rely upon," wrote one worker to a state legislator in an open letter in 1846. Women should not look to men, their self-styled "natural protectors," for needed help, "but to the strong and resolute of their own sex." The reformers managed to help defeat Schouler in the next election, and in their "Try Again" spirit, they continued to hold conventions and sign up ten-hour petitioners—as many as 10,000 statewide workers on a 130-foot scroll. It hardly mattered: infighting between reformist splinter groups undermined their power and legislators quickly became preoccupied with other issues—the economic implications of antislavery sentiment in particular. Once again American politicians abdicated their reform leadership to the English: Parliament in 1847 voted the Ten-Hour Act into law.

To New England's thousands of factory workers, still predominantly women, the defeat of the ten-hour reform movement sent a clear message: abandon all hope. They did. In the late 1840s they also started abandoning the mills in large numbers; their antagonism toward the loom lords is deftly captured in a poem by Lucy Larcom, who also quit the mills during this period:

> If they grind and cheat as brethren should not,
>   let us go
> Back to the music of the spinning-wheel,
> And clothe ourselves at hand looms of our own
> As did our grandmothers. . . .
> If the rule of selfishness
> Must be, invariably, mill-owners' law . . .
> Nobody should toil
> Just to add wealth to men already rich. . . .

Nobody, that is, except laborers without any other source of income. "Only a drudge," says Larcom, "will toil on with no hope." That description did not comfortably fit Sarah Bagley. She was fired for her worker organizing and left Lowell to become the nation's first female telegraph operator. Curiously, the world never heard from her after that. By 1850, half of the work force had quit. Still, Bagley's abortive effort also marks the first time in American history that women banded together to organize as reform activists.

That agitation for workers' rights took place during a time of major transition in the mills that would profoundly change the social character of the mill work. Chased by starvation and destitution brought about by the potato famine and economic depression, thousands of immigrants came to America's shores from Europe during that era. The Irish—160,000 strong—settled on the eastern seaboard, particularly in New England. Along with French, Germans, Greeks, and Italians, they began to move into the factories. The Cotton Whigs, preoccupied with the internal splintering of their party, gave up entirely on the idea of their factories as a "philanthropic college" for uneducated Yankee farm girls bent on self-improvement. Many of those female mill hands had since gone out into the world as missionaries, authors (of thirty-one books), wives, teachers, suffragettes, artists, founders of schools and libraries, and, in one instance, an acting United States Treasurer. Immigrants, the investors quickly learned, worked longer hours for less pay without protest. By 1860, on the brink of the Civil War, they comprised more than 60 percent of Lowell's workforce, with the Irish alone comprising 47 percent.

The new mill hands could not have been less like those sprightly New England girls. They were illiterate, uneducated, and dirt poor; self-improvement held far less interest for them than a full belly. They took longer to train, but they were docile, even grateful at first, and they quickly assimilated. Too quickly for the owners, perhaps—the next wave learned from their elders that strikes were a legitimate form of protest.

The mills' syndicate of autocratic Boston families during all of these years had been taking substantial operating profits in the form of regular double-digit dividends. Nepotism also ran rampant among the inner circle of Brahmins. There were relatives, competent to barely functional, at all levels of the mills' operations by now (except at a spindle or loom for twelve to fourteen hours daily). Brothers, sons, sons-in-law, cousins, nephews—all were given employment at high salaries, draining profits.

The business of cotton was suddenly on tremulous ground, as those mounting antislavery demonstrations in the streets of Lowell reflected only too clearly. No matter how much cheap foreign labor you could pack into the mills, without adequate supplies of raw material, your hired hands were worthless. This was a world the founding loom lords never made but one they contributed to by their passive acceptance of slavery until too late, and they were now about to reap the bitter harvest.

"Things fall apart," William Butler Yeats would later write in "The Second Coming." "The centre cannot hold / Mere anarchy is loosed upon the world." When the Honorable James Henry

Hammond from South Carolina stood to address the U.S. Senate in 1858, two turbulent years after Sumner's beating, he sounded like a man willing to wager that one plant was so potent no amount of moral indignation over slavery could triumph over it. He was certain, too, that no political lobby could prevent its unpaid slave-labor force from spreading west into the Kansas Territory. "Would any sane nation make war on cotton?" he demanded. "Without firing a gun, without drawing a sword, should they make war on us, we could bring the whole world to our feet. The South is willing to go on one, two, or three years without planting a seed of cotton. . . . What would happen if no cotton were furnished for three years? England would topple headlong and carry the whole civilized world with her, save the South. No, you dare not make war on cotton. No power on earth dares make war on cotton. Cotton is King."

Hammond reminded his audience that the South was a honey pot, producing so much cotton revenue for the United States through its exports to England and Western Europe that its citizens enjoyed the highest per capita income—$16.66 per head—in the world. By comparison, the North's contributions hardly registered. In his view the South was funding the Union and being treated like a pariah for its troubles. As for slavery, that was simply the highest proof of "Nature's Law," that an "inferior race" of blacks by the common consent of mankind, "a class requiring but a low order of intelligence and but little skill," yet "eminently qualified in temper, in vigor, in docility" to perform the menial duties that comprised the drudgery of life, existed so that the ruling class of whites should be free to exercise their elevated skills at "progress, civilization, and refinement." That, said Hammond, "constitutes the very mud-sill of society, and you might as well attempt to build a house in the air as to build either one or the other, except on this mud-sill."

No tent-healer promising salvation ever stimulated a more ecstatic response throughout the antebellum South. "Cotton Is King" quickly became the rallying cry from Savannah to New

Orleans. That catchphrase, borrowed by Hammond from the title of a book written by a Northerner, David Christy, boosted the South's self-confidence in its ability to strike out on its own. Economic independence spelled freedom. If its raw cotton exports ($161 million) outdistanced all of the North's ($45 million) by such a wide margin, why fear? In an 1860 reissue of his book, Christy wrote in his preface that overthrowing slavery would halt the spindles and looms of Europe and "bring ruin upon millions." Not likely, he added. There has never been "any intention on the part of political agitators to wage actual war against the slave States themselves."

But in war as in love, complacency can be the most dangerous adversary. If the Southern plantation owners were convinced that powerful Northern industrialists would remain firmly behind them regardless of their personal aversion to the expansion of slavery, they miscalculated. Just about the time Christy finished writing that, as if to remind one and all that human events often rebuff human assessment, the first secessionist shots were fired. Extremists, buttressed by deadly conflict in Kansas-Nebraska, pulled conciliators on both sides from the middle to the edge. The unseen player in all of this was England. The South gambled its future on Liverpool's boundless and incessant need for its raw product and for the flow of money that would continue to pour in to finance its military ambitions. Early in 1861, Hammond's South Carolina announced that it would no longer tolerate the encroachment of the federal government on states' rights. Shortly after that Fort Sumter—a federal garrison in the Charleston harbor—was fired upon as "a foreign entity" and went up in flames. With it went the cotton economies of the South, the North, and England. By then the first presidential candidate of the new grassroots Republican Party, Abraham Lincoln, had been elected and sworn into office. Vowing to keep the Union together at all costs, Lincoln dispatched 75,000 volunteer soldiers to the South, and the Civil War was on.

SEVEN

Southern Exposure

Yankee Doodle, now, good-bye!
We spurn a thing so rotten;
Proud independence is the cry
Of Sugar, Rice, And Cotton.

—*Eastern Clarion*, Paulding, Miss., 1861

"Mr. Paxton, I want to tell you that the thing is played out," Israel Gillespie informed A. G. Paxton. The two men were standing in the cotton field of Paxton's Greenville plantation in the Mississippi Delta in 1863. Paxton, recently returned from fighting on the Confederate front, had ordered Gillespie to spread out the harvested cotton for drying. A simple order, the kind any slave could understand and execute, yet Gillespie simply stood there and did nothing. In the past, Paxton might have sent his overseer to punish him for his flagrant refusal to obey. But there was no longer an overseer, and Israel Gillespie was no longer a slave.

Instead of taking fruitless action, Paxton turned on his heel and went back to his house to think about what Israel had said. *"The thing is played out." How could that be? Cotton that still needed cleaning and ginning and baling left to rot in the sheds. No hands to do the work, and none likely to appear. Worse yet, if that was in any way possible, no place to sell that cotton legally, baled or not. A crop worth nothing, now or anytime soon.*

What shocked Paxton, what galled him above all else, was Israel's manner. You could always count on a slave's deference and respect. Now, Gillespie was talking down to him, his owner, with the patience you use to explain things to a child who can't keep up with his lessons. *The thing is played out, Mister Paxton.* No trace of anger or revenge, just a simple statement of fact as if to say, *I'm a free man now, but you, Mister Paxton, I can see in your eyes, you trapped. No slaves left to put food on the table and lace on the ladies. No one left to call you "Massa" neither, Mr. Paxton. Truth be told, I don't 'xpect to call you or any other white man that ever again.*

Paxton was hardly alone in his desolation and confusion. All around him in the Mississippi Delta—not a true river mouth, but two hundred miles by eighty miles or so east to west of the richest, sweetest soil on earth—slaves and immigrants had carved cotton plantations like his out of dense impenetrable thickets of cypress, gum trees, and tupelo, "a chaos of vine and canes and bush," in one visitor's opinion. Toss your seeds in these alluvial deposits left behind by annual Mississippi and Yazoo River floods, from Memphis as far south as Vicksburg, and you'd likely grow more cotton than you could hoe or harvest once a rudimentary system of levees was established to help control the flooding. But now in less than two years of war these same fields lay fallow, not soon to be planted again. Changes everywhere, none for the good.

Slowly but inevitably the word spread among Delta slaves after 1863 that they'd been freed by Lincoln's Emancipation Proclamation. Delta women, left to run family plantations with their men away on the battlefront, tried to protect their cattle, corn, and other property. Many were lucky to escape with their lives. One wife, Mary Delia Montgomery, enlisted the aid of Union soldiers who arrived to burn down her plantation house in retaliation for a nearby sniper attack on a Union Mississippi

River gunboat. At her request they carried out her piano, her silver, and imported silk carpets before they set fire to her home. She and her family moved for a short time to their loom shed. As recounted by James C. Cobb in his homage to the Delta, *The Most Southern Place on Earth*, Mary Montgomery's relative, Walter Sillers, watched the house go up in flames. Later he recalled that "not often did our ladies give vent to their indignation over the action of the federals, but during the burning of this home the faces of my mother and cousin showed such anger and scorn as to deeply impress me, a boy of ten years."

Montgomery discovered that her freed slaves, like Paxton's, were short on gratitude to their former masters. Many simply stood around and watched as her house burned down. Others plundered her valuables and ran off. Letti Vicki Downs, in nearby Deer Creek, watched Union troops incite a freed slave to threaten her father's life, and heard rumors of rape and pillage. She declared her contempt for "a people who instigate a race that have been raised and cared for as children to rise and slay their owners in cold blood."

Enslaved By Cotton

The South's elite and omnipotent planter class had created an airtight argument for slavery, upon which rested the smooth efficiency of their ruling system. Within it the planter assumed total paternalistic responsibility for the health and welfare of his slaves in exchange for their free labor; he provided shelter, care, and protection for a race that in his view was feckless and childlike. Slaves were punished, often severely, for making adult decisions while rewarded for docile, subservient behavior. Without absolute dominance, planters like A. G. Paxton convinced themselves that chaos beckoned. To minimize the chances of insurrection, slaves were not allowed to learn to read or write or to marry outside the plantation; they were stripped of all legal rights or appeals. Prevented from making a free choice, slaves furthered the fiction of acquiescence by their involuntary subjugation, which explains why owners generally reacted with pained bewilderment when they celebrated their newfound liberation. Southern whites genuinely believed blacks could not fend for themselves—and few seemed ready or willing to accept that they had created the conditions that made dependent behavior a mechanism for survival. When Gillespie told Paxton "the thing is played out," he was talking about slavery, yes, but more than that he was sounding a death knell for the contrivance of paternal planters and hapless blacks. Under cotton's spell, that illusion had sustained the South for better than half a century.

In their diaries and reminiscences, many written after the collapse of the antebellum South, plantation owners and their families repeatedly looked back with nostalgic fondness to a bygone era. Reading their recollections you come away with a sense that these wealthy growers, too, were enslaved by cotton. Many seemed to become victims of the myth they perpetuated until the line blurred between harsh reality and wishful thinking.

Genuine, deep friendships developed at times between plantation family members and their slaves, yet the warden-prisoner nature of these relationships rarely if ever gets acknowledged. One diarist expressed her affection for Uncle Banks, a "friend" of the master who helped build the cotton gin; Aunt Dicey who "all the little Negroes" called "Ga Muh," and who cradled the babies of field slaves in her arms and never let anyone step over a sleeping child in the grass, for fear the child wouldn't grow any bigger; and Mamma Chaney, who "had held us all from babyhood, and rocked and soothed us to sleep by her lazy and loving pat and monotonous crooning." Georgia native Edward J. Thomas airbrushed his literary canvas in soft, warm hues:

> I remember the great big cotton house, three stories high and every window glassed, where the older women

> would sit and "pick and sort" the good cotton from the bad, where the youngsters would take the newly ginned cotton to the strong men with the iron pestles, who stood in a strong bag of stout bagging—no presses those days—until the contents were hard and fast, pestling in this bag some three hundred pounds of cotton; the horse gin, where Dick and Montezuma, the two horses, took turns with Lewis and Robin, the two mules, in pulling the lever that turned the machine that ginned the cotton; the two little black nigs who rode on the lever to keep the animals at even speed, and after a few hours, when the horses got accustomed to the noise, would fall off in the nearest corner fast asleep; the pleasant rivalry between the men and women to see who would pick the most cotton, and hence get the prize—a calico dress or hat or pair of Sunday shoes—that father would offer weekly to the one picking the most cotton. The picking season then was very long, no guano those days to hasten matters, so the cotton would not open until October, and the fields would be white until after Christmas.

The media, then as now, often saw what they chose to and ignored what they didn't. An 1854 *Harper's Magazine* reporter marveled at the way slaves could leave "the incipient stalk unharmed and alone in its glory." By July, the maturing plants produced one flower for every boll. "It is . . . of a perfect cream color. It unfolds in the night, remains in its glory through the morn—at meridian it has begun to decay. The day following its birth it has changed to a deep red, and ere, as the sun goes down its petals have fallen to earth, leaving behind enclosed in the capricious calyx a scarcely perceptible germ. This germ . . . is called a 'form,' in its more perfected state, a 'boll,' " *Harper's* rhapsodized.

Or, as captured in the lyrics of a popular Southern children's song about the cotton flower:

First day white, next day red,
Third day from my birth I'm dead;
Though I am of short duration,
Yet withal I clothe the nation.

From that fleeting display of floral splendor until the mature pods burst open and the harvest began, usually in late September, workers "laid the crop by"—that is, left it unattended; summer was reserved for other chores for slaves and for vacations in the North for the families of wealthy planters. At picking time, slaves were expected to gather up to two hundred pounds a day per person, depending on gender and size. After ginning, a press comprised of a single giant screw with a block at its business end, driven by a horse, compacted the mounds of loose fluff into dense bales that were covered with burlap and secured with thick wire.

But all of that was simply process. The power of cotton resisted such easy categorization. It cast a transfixing spell on many white southern antebellum writers who produced mounds of perfumed prose that smelled faintly of manure: "Then, on the old plantation," one typically wrote about harvest time, "swarmed forth the turbaned mammies and the wenches, shining pickaninnies and black babes in arms, with bags and huge baskets and mirth, and nimble fingers as it were predestined to the cotton pod, to live in the sunshine amid the fleecy snow, and pile up white fluffy mounds at the furrow ends, chanting melodious minor chords of song as old as Africa, the women trooping home at night-fall with poised overflowing baskets on their heads, to feasts of corn-pone and cracklin' and molasses in a blaze of a light'ood fire, within sound of the thrumming of the banjo."

All that's missing is Uncle Remus chomping merrily on a watermelon while Steppin Fetchit shuffles past, broom in hand, moaning, "Feets, don't fail me now!"

An idyllic life as portrayed—but in truth one not shared by

the overwhelming majority of Southern whites, let alone blacks, before the Civil War. These haloed soft-focus images sharply contrast with the lot of most poor yeomen cotton growers. Journalist Frederick Law Olmsted, who later gained fame as the landscape designer for New York's Central Park, traveled four thousand miles across the antebellum South, mostly in the company of common people, and published a classic book in which he describes a journey through hell. Olmsted found himself immersed in squalor, rampant poverty, and abysmal living conditions: "Nine times out of ten I slept in a room with others, in a bed which stank. . . . I found no garden, no flowers, no fruit, no tea, no cream, no sugar, no bread . . . no carpets or mats. For all that, the house is swarmed with vermin."

Time after time he arrived at a shack whose interior was blackened with soot, the slapped-together home of a struggling farmer that lacked rudimentary amenities—a leaking roof, a floor, and not much more. There might be a few slaves in rags, and if so, they too suffered from malnourishment and disease.

Cotton has always been at home in tropical and semitropical conditions that spawn dysentery, typhoid, yellow fever, malaria, salmonella, and cholera. For decades there were regular epidemics of enteric disease across the cotton belt that spared neither slave, white farmer, or master; they caused ferocious, widespread illness as they decimated digestive systems and dehydrated bodies. Thousands died. There were no cures, only wagons full of worthless patent medicines. Diaries chronicled the onset of yellow fever in grisly detail; black vomit signaled its arrival. New Orleans and Memphis, among other cities, lost thousands as yellow fever swept through. Far more pervasive, and less fatal, was malaria, which brought about alternating chills and scorching fever. Travelers reported on the general listlessness of the population; it wasn't usually apathy, it was mental and physical weakness in the aftermath of a viral or bacterial infection.

Then there was cotton itself, a plant that grew just tall

enough to inflict paralyzing backaches from bending and stooping, but not so low that it could be picked clean by kneeling comfortably between rows. Each exposed pod or boll, when mature, produced tough, leathery, jagged burrs that surrounded the embedded fiber-seed ball. To pull it free, you had to do battle with these thick vegetative fingers that held the lint and seeds firmly in place, and there were dozens of bolls per plant. The hands of a newcomer would be scraped raw after a short encounter. Slaves picked up to twelve hundred pounds of lint six days a week in sweltering heat that often topped 100 degrees.

That was the punishing reality few if any Southern aristocrats dwelled on in their memoirs or that journalists reported factually in the popular press. Both groups seemed to prefer to consider master and slave as bound together productively in the service of cotton. What was missing—until a few years before the Civil War—was a radically different take: these same scenes observed from the slaves' point of view.

No Such Thing as Rest

Northerners, even those fiercely committed to abolition, had only limited access to reliable information that separated actual events from reams of antislavery propaganda intended to inflame passions. Broadsides and emotionally charged narratives like Harriet Beecher Stowe's *Uncle Tom's Cabin* placed authenticity at the service of ideology. Stowe's melodramatic novel sold 300,000 copies in its first year, and 2,000,000 copies worldwide within two years, but no fictionalized account of slavery, however compelling, delivered the visceral impact of the first-person report by Solomon Northup. A free black man living in New York until kidnapped by slave traders, Northup barely survived a dozen brutal years of bondage on a Red River plantation in Louisiana before his rescue in 1853, and soon after that he published an avidly read chronicle of his experiences. Readers recognized this father of three and devoted husband to be articulate, bright, and trustworthy—a man committed to telling the truth:

> The hands are required to be in the cotton field as soon as it is light in the morning, and, with the exception of ten or fifteen minutes, which is given them at noon to swallow their allowance of cold bacon, they are not permitted to be a moment idle until it is too dark to see, and when the moon is full, they often times labor till the middle of the night. They do not dare to stop even at dinner time, nor return to the quarters, however late it be, until the order to halt is given by the driver.
>
> The day's work over in the field, the baskets are "toted," or in other words, carried to the gin-house, where the cotton is weighed. No matter how fatigued and weary he may be—no matter how much he longs for sleep and rest—a slave never approaches the gin-house with his basket of cotton but with fear. If it falls short in weight—if he has not performed the full task appointed him, he knows that he must suffer. And if he has exceeded it by ten or twenty pounds, in all probability his master will measure the next day's task accordingly. So, whether he has two little or too much, his approach to the gin-house is always with fear and trembling. Most frequently they have too little, and therefore it is they are not anxious to leave the field. After weighing, follow the whippings; and then the baskets are carried to the cotton house . . . and their contents stored away like hay, all hands being sent in to tramp it down. . . .
>
> Finally, at a late hour, they reach the quarters, sleepy and overcome with the long day's toil. . . . An hour before daylight the horn is blown. Then the slaves arouse, prepare their breakfast, fill a gourd with water, in another deposit their dinner of cold bacon and corn cake, and hurry to the field again. It is an offense invariably

> followed by a flogging, to be found at the quarters after daybreak. Then the fears and labors of another day begin; and until its close there is no such thing as rest. . . .

A musician raising his family in Saratoga Springs, New York, Northup joined two men in 1841 who claimed to be from a circus, signing on as a fiddler to earn money. Once the troupe reached Washington, D.C., where slavery was permitted, the men drugged, chained, robbed, and sold Northup to a slave trader. Shipped to New Orleans, he was sold in a slave market. For the next twelve years, sleeping on a plank of wood twelve inches wide and ten feet long laid on top of a dirt floor, with a stick of wood for a pillow, he worked as a slave for three different masters.

Northup was as much a stranger dropped into an alien culture as any white Northerner might have been. Far from the sophistication of New Orleans or Savannah or the gentrified pretensions of Natchez, the South he described was raw, poor, and frequently violent. It was also filthy. There were no disinfectants; country people wore ragged clothing; they washed only intermittently; there was no way of preserving food, no steady supply of pure water; all the refinements of gracious living that wealthy planters enjoyed and committed to their diaries applied to a small fraction of the populace. The overwhelming majority of the South's yeoman farmers owned five or fewer slaves, barely eked out a livelihood in most instances, and often wore their random cruelty to slaves as a badge of honor. Alcoholism was rampant. In return, slaves stole whatever they could from loutish masters, ran off, and set fire to buildings.

Growers planned their lives around the needs of the cotton crop, Northup explained. Ground was broken or "disked" in late winter after old stalks were cleared; in early or mid-April, slaves hoed the plowed land's furrows, building up beds with sloped ridges; the most skilled hoe-hand created a true row—one that ran dead straight. After that a slave girl whose apron served as a seed sack walked between the beds depositing a seed every inch or so at a rate of one hundred pounds per acre. After the third leaf appeared, slaves scraped the fields, removing weeds and grass. That soon became an arduous chore as the weather heated up and abundant weeds needed to be chopped each day in fields that often extended several miles in all directions. Cotton subjected Northup to so much suffering that the very mention of the word might be expected to provoke rage. But, no: "There are few sights more pleasant to the eye," he wrote, "than a wide cotton field when it is in the bloom. It presents an appearance of purity, like an immaculate expanse of light, new-fallen snow."

For years it appeared to Northup that cotton, beautiful or not, had condemned him to die. He prayed for death time and again. Lashed repeatedly by a succession of owners and overseers, he witnessed runaway slaves trapped, mutilated, and killed by packs of hounds; his lowest moment came when he was forced by his drunken master, Epps, to whip a young female slave, Patsey, whom Epps fancied and in a jealous rage accused of betraying him. When Northup refused to continue, Epps grabbed the lash; lacerating her mercilessly, he flayed her skin. From that day forward, Northup writes, "The burden of a deep melancholy weighed heavily on her spirits. She no longer moved with that buoyant and elastic step—there was not that mirthful sparkle in her eyes that formerly distinguished her. The bounding vigor—the sprightly, laughter-loving spirit of her youth, were gone. She fell into a mournful and desponding mood, and often times would start up in her sleep, and with raised hands, plead for mercy."

Managing to escape through the intercession of a sympathetic white Louisiana neighbor, Northup carried back with him a riveting tale of prolonged degradation and inhumanity that confirmed the worst suspicions of Northerners. It provoked a furious outcry among many political powers, including New York's governor. To readers willing to consider the larger implications of his work, Northup's account also established that the North and

South had developed two hopelessly incompatible cultures with dangerously conflicting values. Cotton of course played a pivotal role in maintaining the illusion that these regions were somehow still united in their patriotic zeal for free enterprise. Since 1800 this cash crop had provided the capital to enable the South to sustain itself as a separate entity while giving the industrial North every economic incentive not to argue the point at the risk of severing vital commercial connections. Northern business leaders fully understood the importance of cotton to the South's welfare—and in many instances, to their own as well. But it was difficult if not impossible for anyone who didn't grow up among Southerners to appreciate how profoundly the cotton culture shaped personal convictions as well as just about every other aspect of daily life.

Betting the Spread

In the 1830s a prospector stopped for the night at the home of a south Louisiana planter, who showed him some metal ingots left behind years earlier by a Spanish settler. The ingots—pure silver—excited the prospector's interest. "Rumor has it," said the planter, "that this Spaniard uncovered a rich lode of that silver in the cotton country just south of Natchez along the Mississippi, but never got around to developing it."

Off went the prospector the next day in search of his fortune, and in short order, so the story goes, he came upon the vein, still untapped and deep. All he needed to do was to convince the local gentry, and he proposed to split the profits with them for the mining rights. They turned him down. "You don't understand," said the prospector, "you're sitting on a fortune." The owners simply shook their heads. Finally one spoke up: "What planter would exchange his cotton field for a silver mine?"

That pretty much summed up the attitude of all who profited from cotton in the South. Ending slavery would spell economic ruin to everyone, from factors to dock workers; it would also dismantle a carefully constructed set of values that trumpeted self-determination, honor, chivalry, bravery, and the importance of lineage. These core beliefs, supported by cotton, formed a secular religion in the South and took on a regional significance baffling to most Northerners. The terms seemed easy enough to grasp but held special meaning, as in a secret code. Self-determination, for example, meant the end of all restrictive federal regulations. Wealthy planters and Northern industrialists might befriend each other, share an appreciation for life's luxuries, and take in the need for maintaining commercial ties despite secessionist pressures. That did not mean they fundamentally understood one another.

For the Boston Associates, cotton represented financial security. For rich planters, it represented the ultimate crapshoot. The enormous wealth accumulated by many Southern aristocrats disguised the tenuous nature of their business dealings. Dependent on the performance of a future crop, much of their money was spoken for long before it was received. Without a high-stakes gambler's fondness for risk, few planters could endure for long the anxiety of living off the revenues of a contracted harvest not yet in the ground. Once in the field those cotton bolls might produce two bales an acre in a good year, or less than half that in a drought, or flood, with the planter still held liable for meeting the terms of the agreement. The world's richest alluvial cotton soil counted for nothing if bales got damaged in transport and mishandled, as often happened. A dozen or so potential mishaps stood between the seed in the soil and the safe delivery of product.

Then, too, there were the wild price fluctuations over which planters had no control. In 1838, cotton prices averaged twelve cents a pound. In 1839 they dropped to three cents. Wealthy planters with their own gins and presses obtained the best margin for their crop. But the vast majority of smaller farmers, with little cushion, often found themselves facing insolvency.

Although the boll weevil would not arrive for another fifty years, pests presented a serious, continuing problem. There were no fungicides or insecticides. In 1845 in Louisiana's fertile Red River region, a caterpillar infestation destroyed close to the entire crop. Factors, as buying-selling agents, charged a sizeable penalty in addition to their 2.5 percent commission for every bale short of the amount deliverable by contract. Cotton agriculture might produce huge profits in good years, but it was not the sort of enterprise that provided a solid financial foundation for empire-building. That stopped few wealthy plantation owners. They and their families lived in and for the moment—not like those dour penny-pinching Yankees who'd probably pour your gift of bourbon down the sink and try to sell you back the glass bottle as a vase.

Pillars of Society

Understatement did not exist in the vocabulary of rich Southern aristocrats. They might live extravagantly, sometimes at the edge of their resources, but to express worry was to betray weakness. Nowhere were the values and lifestyles of planters' families on display more flamboyantly than in Natchez, Mississippi. Centered on the high bluffs along the southern banks of the Mississippi River and extending over five counties about 170 miles north of New Orleans, the Natchez district was to cotton agriculture as Lowell, Massachusetts, was to textile manufacture—its celebrity showcase. There all comparisons end, except one: the incalculable wealth that both produced. By 1850 more millionaires per capita lived in Natchez than anywhere else on earth. The town itself, perched majestically above a bend in the river, overlooked the steady stream of steamboats and packets that docked below, under the hill, off-loading supplies and on-loading bales for transport to the bustling port of New Orleans. Better than 600 planters in the district owned fifty or more slaves, and many built lavish Greek Revival mansions in and around the town.

In total, less than 25 percent of the South's white population owned slaves. Of the thousands of cotton growers in the South who did, about 10 percent qualified as planters. That term most commonly distinguished growers who owned twenty or more slaves, the accumulation of human property rather than acreage being the determining factor. Of 384,884 slaveowning growers in 1850, only 2,500 or so had fifty or more—about 6 percent. An even smaller handful of "great planters," the power elite of the South, owned upward of a hundred slaves; these men congregated in social centers like Natchez, vacationed in Europe, summered in the North, and formed influential alliances with the Boston Associates. They too represented an oligarchy, hand picking senators and governors or representing their states themselves in Congress. For planter families, the plantation was frequently not their first choice for a home, any more than a mill town like Lowell satisfied the Associates' social and cultural needs. Planters knew that where cotton thrived, humans suffered. Healthy crops required about four inches of rain for the first three critical months, then high heat. The fiercely hot and humid bottomlands of the Yazoo-Mississippi Delta often teemed with pests, snakes, and varmints; they broadcast disease. Some fields ranged over reclaimed swamps that hatched malarial mosquitoes; and because plantations occupied hundreds or thousands of acres—the plots worked by myriad small growers, or yeomen, were simply called farms—they isolated plantation dwellers. For planters and their wives and families, fluid social interaction was pivotally important, and many masters became absentee owners.

Planters had to be accessible to factors, bankers, merchants, and other business associates. Natchez offered an ideal blend of work and leisure; it existed in a nimbus of refinement. The elite families could surround themselves with peers, cultivate a genteel and well-lettered persona, and yet live within range of their holdings. They left the daily administrative drudgery of organizing and directing slave labor to their overseer, usually a poor white. Appearances mattered. A visiting Englishman complimented the local gentry on their *"air distingué."* Others were less impressed. Frederick Law Olmsted, traveling through in 1860, commented on "the farce of the vulgar rich," and noted that while a few planters' families conducted themselves with panache, "the number of such is smaller in proportion of the immoral, vulgar, and ignorant newly-rich, than in any other part of the United States." If many of these elite were not to the manor born, or even to the shack behind it, they were eager to invent or embellish their own genealogy. Over the previous two decades a romanticized Southern identity had emerged; it promoted the belief that regional whites descended directly from aristocratic Royalist exiles in Cromwell's England. Cotton came along to underwrite that myth in grand fashion, and Natchez gave it a magnificent theatrical venue.

Plantation houses the size of small stadiums featured such practical architectural flourishes as colonnaded porches to escape the heat, buttressed by pillars and Doric columns that bordered massive carved front doors rising to infinity. Magnolia, wisteria, and dogwood trees hung with Spanish moss cast bright plumages of color and soft shadows over the expansive surrounding manicured grounds.

Stuffed to the wainscoting with heavy dark rococo furniture, gold filigree candelabras, supple fabrics, sparkling jeweled chandeliers, marble floors strewn with silk Aubussons, ceilings framed by intricate crown molding, flocked Parisian Zuber wallpaper, and rosewood paneling, their mansions went beyond ostentation. They made a statement about the power of the inhabitants and the unchallenged rights that went with it.

The end of slavery to these planters and their families in particular meant the end to everything in life that mattered. It was never more complicated than that. In their view, the North's money-grubbing industrial vulgarians could never hope to understand the elegant, languorous pleasures of their existence. Beneath their protestations of being misunderstood and maligned by Abolitionists in their diaries and journals, these Southern arriviste aristocrats uttered a cry for mercy: *Deliver us from prying eyes, that we may walk in silks through our rows of cotton trailed by faithful slaves from here to eternity, amen.*

Cotton's War

Inevitably, that prayer fell on deaf ears. The commodity that supported these fabulous lifestyles simply depended on too many uncertain business relationships to ensure that it would provide the means to fund secession.

"Why, all we have is cotton and slaves and arrogance," Rhett Butler complains in *Gone With the Wind*. The despised Yankees, he adds, have "the factories, the foundries, the shipyards, the iron and coal mines—all the things we haven't got." Butler predicts the South's army won't last a month.

In the most optimistic projections, only a steady and reliable stream of cotton revenues throughout the Civil War gave the South any chance at success; despite the prowess of its military leaders and its prevailing certainty that Northerners lacked the heart and bravery to fight, a lack of revenue from cotton exports was sure to doom its war efforts. Other factors had already weakened the Confederates' position. Out of 9 million residents, 3.6 million were slaves, leaving an army to be drawn from men of fighting age—between fifteen and fifty—among 6.2 million white citizens. By contrast, the Union had a population of 13.3 million, and more than twice the available recruits.

Northern states also had an extensive railway system, 22,000 miles of track to transport weaponry, clothes, food, and soldiers, far in excess of the South's fragmented 9,000-plus miles. Not all of that Southern track was uniform in gauge and therefore not continuous. There were no trains to bring supplies reliably to the front. The North also boasted a robust industrial capacity ten times as strong; by 1860 it was selling more than $150 million worth of manufactured goods to Southern states, which were close to being as dependent on the North as the American colonies had once been on Britain. "From the rattle . . . of the child born in the South . . . to the shroud that covers the cold form of the dead, everything comes to us from the North," a shrewd Southerner, Albert Pike, lamented in 1855.

More devastating still, the Union established an effective navy blockade along the South's seaboard shortly after the war began. In a brief conflict it might have mattered less; but both sides compounded their problems by not preparing for the long, ferocious struggle that ensued. Fighting would be over in ninety days to six months at the outside, or so they believed. In the North, Lowell's textile aristocracy sat back complacently to await the impending Confederate surrender. Underestimating the enemy was hazardous; selling off all their raw cotton inventory to other New England and British mills was moronically shortsighted. The heirs of the original Lords of the Loom now ran these mills with their forebears' money but not with their acumen; they eagerly dumped cotton reserves at a handsome profit. Shutting down eight factories in Lowell overnight, they put 10,000 workers on the street in 1861. They planned to rehire these workers in a few months, after the war, when they received new cotton shipments from the defeated South.

"This crime, this worse than crime, this *blunder*," wrote a Lowell historian in 1868, "entailed its own punishment. . . . When the companies resumed operation, their former skilled operatives were dispersed, and could no more be recalled than the Ten Lost Tribes of Israel. . . ." With more than two-thirds of Lowell's mill workforce laid off—now employed in defense plants and rival factories in Bedford and Fall River—the community plunged into years of depression. Wool might have rescued Lowell—it was used to make army uniforms—but only one local factory, the Middlesex Mill, produced wool cloth during the war. The Associates' heirs had sabotaged their own business.

Britannia Demurs

To bring in vital funding, the Confederate states expected to continue to sell their crops to the English mills that relied on them to furnish more than 80 percent of their raw material. England's support seemed assured. It wasn't. Confederate president Jefferson Davis in 1861 sent three emissaries to Great Britain who set out to persuade the British to openly challenge the Union navy that now blockaded Southern ports and its coastlines and effectively crippled trade between Europe and the South. Still, close to one-fourth of the English population depended on the textile industry; half of England's export trade was in cotton textiles. As Foreign Minister Lord John Russell confided to U.S. Ambassador Charles Francis Adams, British merchants "would, if money were to be made by it, send their ships to hell at the risk of burning their sails." Sails, yes; sales, no, as it developed. If the Confederacy believed raw cotton was a primary defensive weapon, certain to enlist Britain's aid if only to protect its own interests, Southern states quickly learned that weapons backfire.

Cotton, "the innocent cause of all the trouble," came with baggage that alienated many English citizens. There was nothing innocent about slavery, outlawed three decades earlier in all of Britain's colonies. That gave England pause, even though the country's mills had been eager partners in crime before the war. "Can any sane man believe that England and France will consent, as is now suggested, to stultify the [antislavery] policy of half a century for the sake of an extended cotton trade?" the London *Times* railed in an 1860 editorial; shortly after South Carolina seceded in January 1861, the newspaper responded with disgust: "We cannot disguise from ourselves that, apart from all political complications, there is a right and a wrong to this question, and the right belongs, with all its advantages, to the States of the North."

A surplus of cotton at that moment in England also lessened its immediate dependency on Southern crops. Three wars in England's recent past—two in Europe, one with America in 1812—made the country wary of inviting more hostilities. Then there was its own world supremacy as a naval power; it felt an obligation to honor blockades except under provocation. Lincoln's brazenly undiplomatic secretary of state, William H. Seward, announced to the world that if Britain recognized the Confederacy, "We from that hour shall cease to be friends and become once more, as we have twice before been forced to be, enemies of Great Britain."

That overt threat got Parliament's attention, while in the South hope remained eternal. South Carolinian Mary Chesnut, who kept an exhaustive Civil War diary, wrote in 1861, "How we cling to the idea of an alliance with England or France! Without France, even Washington could not have done it."

She and the South's leading lights might have saved themselves much pain and suffering by paying more attention to her good friend William Gregg, mill owner and a voice crying in the wilderness. Long before war broke out, he said that if a conflict begins, "we shall find the same causes that produced it, making enemies of the nations which are at present the best customers for our agricultural productions."

The South's conviction that Europe would remain neutral rose from a trust in shared values as well as practical economics. England's landed gentry, that elevated class for whom physical labor of any sort was repellant as well as frightfully gauche, thumbed its nose at the North's industrial prowess. Evidently in New England even the wealthiest industrialists soiled their hands in the daily exercise of getting and spending, while the

South's landed aristocracy, accustomed to the privileges of leisure and culture, remained insulated and largely aloof from such mundane matters. These dukes and earls understood and respected obedient servants in the fields and a master race in manses, serfs below and lords above. Many among Britain's powerful upper crust campaigned energetically for the Southern cause, and for a time the Brits remained perilously close to siding with the Confederates. When Lincoln announced in September 1862 that all slaves in seceding states would be freed on the first day of 1863, however, public sentiment for the North gained momentum. Parliamentarian John Bright gave a rousing speech for the Union at St. James Hall. The Civil War, he said, "has originated from the efforts of slaveholders to break up what they themselves admit to be the freest and best government that ever existed, for the sole purpose of perpetuating their institution of negro slavery." While there may be rich men who want to disregard all that, he continued, "I leave them to their consciences."

That stance infuriated the South. For decades Britain's textile industry—to an extent, its whole economy—had lived off Southern slave labor at the same time that British Abolitionists roundly denounced its people-as-property foundation. Deplorable working conditions in England's own mills amplified the hypocrisy, in the South's view. "When you look around you, how dare you talk to us before the world of slavery? For the condition of your wretched laborers, you, and every Briton who is not one of them is responsible before God and Man." That stinging indictment came down not from Southerners but as part of an 1833 Parliamentary commission report on Lancashire's factories, which referred to English factory workers as "our slaves."

Southerners applauded the British government's willingness to focus attention on its own manufacturers' execrable maltreatment of England's mill hands. Fair enough, but a basic difference went unacknowledged: the South never appeared willing to make a similar effort at soul-searching. No state issued critiques of slavery that held every Southerner "responsible before God and Man," nor did any propose social reforms, as did the commission's report on exploitive mill owners.

In the end, cotton diplomacy failed. England remained neutral; its ships never once officially violated the embargo set up by the North, although an enterprising fleet of blockade runners carried on a brisk, thriving business in smuggling along the Southern sea coast. Cotton production in the South plummeted from 4.5 million bales in 1861 to only 300,000 by 1864 as the Union blockade on Southern shipping thwarted large transports.

England in the meantime had turned to India for its raw cotton, a painful adjustment that eventually brought about its own crisis. You might expect the British above all others to have known by then that cotton defined fair-weather friendship. Its economic importance as a commodity was often proven to outstrip its predictability as an agricultural crop. That led to a series of monetary catastrophes in the first half of the nineteenth century in Great Britain as well as the United States, complex in detail but simple enough in principle. Liverpool buyers tried to anticipate yearly market fluctuations resulting in prices that often ranged dramatically between, say, nine cents and seventeen cents a pound for cotton. When they bid too high or too low in advance of the actual harvest, speculating on supply-demand ratios that failed to match their expectations, some of the participants, from planter to factor to merchant to buyer to banker, paid dearly while others benefited. Over time, no one escaped unscathed. Because the South's financial center was in New York, planters paying for supplies on credit had little local control over their own destinies. England had even more reason, if needed, to placate the North, its loyal financial partner in producing the South's cotton.

The Confederacy's threat to break away from the Union was no closely guarded secret. It shaped the politics of the United States for nearly two decades preceding the war. Still, England's textile industry did little to develop early alternative sources for raw material. A country so devoted to sea power might easily have envisioned a successful blockade, and the possible loss of $55 million in salaries for workers in England's 2,650 textile factories. By 1860 America was furnishing 84 percent of the total cotton supplied to Lancashire. Two years later that fell to 7 percent. Increased East India production lagged much too far behind to make up the immediate difference. Relatively insignificant amounts—less than 10 percent of the total—came from Brazil and Egypt; both long-staple varieties produced finer cloth at higher cost.

Taking its cues from the North and South, Britain geared up for a short American Civil War, over in a blink—not the protracted conflict that stretched out four ruinous years at a cost of 600,000 American lives. Said Secretary of State Seward in 1861 to a Lancashire public figure, "The rebellion is already arrested. . . . it owes all the success it attained to the . . . indirect favor of British statesmen" who kept England at arm's length. Despite Seward's hollow assurance, the war between the states escalated and in the next two years England's mills suffered. Half of the Lancashire spindles fell idle. Cotton prices soared. By 1862, 485,454 of the inhabitants of Lancashire were being supported by organized charity. By 1863, one-third of Great Britain's textile factories had closed; half-a-million factory hands were out of work. "The cotton famine is altogether the saddest thing that has befallen this country in many a year," the English *Saturday Review* editorialized. "No crisis in modern times has been so anxiously watched," said the London *Times*, "nor has any European war or revolution so seriously threatened the interest of England."

In 1864 the famine ended. Supplies of inferior short-staple from India rose to meet demands, and although raw cotton fiber from Surat proved to be dirty, brittle, and highly unpopular, it served the mills' immediate needs. Still, the downturn cost the British cotton trade an estimated $350 million. Many manufacturers who never recovered went bankrupt.

Black Cotton

For the same reason that the Russians burned down Moscow to cinders rather than allow the city to fall into Napoleon's hands, cotton became one of the first casualties of the Civil War. By 1862, less than a year after the war began, Confederate soldiers, under orders to torch all cotton fields to render them worthless to the invading Union troops, had set fire to thousands of acres. Federal soldiers burned crops in the fields as well to prevent the South from harvesting and selling cotton to fund its war efforts.

Anyone looking for behavior that captures the random lunacy of armed conflict might want to examine the wholesale destruction of cotton in a conflict motivated largely by both sides' vital interest in its protection and survival. Cotton became an early version of the Vietnamese village an American general burned to the ground in order to save.

Brothers were killing brothers in the border states of West Virginia, Kentucky, and Missouri, among others, all to preserve a way of life and a crop whose moral and political significance became progressively more corrupted as the war wore on. Initially cotton and its free labor force encapsulated a rigid, race-based class structure that defined Southern culture. In theory, an inferior race toiled so the superior race could lend its intelligence and elevated morality to the advancement of society; in practice, that distinction quickly eroded as hunger became the great leveler. When Union troops closed in, desperate planters or whoever was left behind to run the plantation often committed treason in the eyes of the Confederate government by selling bales to the North instead of burning their crops for the cause. One of A. G. Paxton's neighboring Delta planters, John L. Alcorn of Coahoma County, later governor of Mississippi, proudly announced before the start of the war as he voted for secession, "I follow the army that goes to Rome!" By 1863 Alcorn was knee-deep in illegal cotton sales. For him and many other wealthy Delta planters as well as poor white farmers, self-interest trumped sectional loyalty. By 1862, when federal gunboats took control of the Mississippi River, contraband cotton trade for greenbacks was well on its way to becoming a thriving underground industry. Writing to his wife that he had come upon many "agreeable acquaintances" among "the Feds," Alcorn reported he had sold eighty bales of cotton illicitly to the Union for twelve thousand dollars, and arranged to sell an additional hundred bales. If his fields survived burning by his fellow Confederates, he wrote, within five years he expected to "make a larger fortune than ever." At one point Alcorn reports that 400 bales "were openly sold and fully fifty men on the riverbank participating" in a contraband operation, appearing to justify his own operations.

Endgame

By April 9, 1865, when Lee surrendered at Appomattox, Virginia, both loyal and opportunistic Southerners had presumably learned the same painful lessons. Wars are not won by countries or coalitions of states forced to rely on outside resources; those resources can disappear as quickly as they emerge in times of prosperity, which also explains why no agrarian society has ever won a war with an industrial one. Peace at all costs is not necessarily guaranteed by elite conservative business interests on both sides with too much to lose from conflict; infighting among these groups saps their political power as it did among Whigs. The fate of cotton might not qualify as a lesson learned—perhaps more in the realm of a reminder that risking life and limb for a commodity can quickly make a muddled mess out of the loftiest objectives. By the final stages of the war between the states, President Lincoln was approving an arrangement to send food to Confederate general Robert E. Lee's troops at Petersburg in exchange for cotton to be sent to New York. Cotton swaps became so common that some troops on both sides were alternately trading and fighting with one another.

In the end, the crop that made America a self-sufficient international power had come close to destroying it from within. Pride and greed and self-delusion all played primary roles. A royal counsel in *The Persians*, the earliest surviving play in Western literature, assures the king that:

> Defeat is impossible. Defeat is unthinkable.
> We have always been the favorites of fate.
> Fortune has cupped us in her golden palms.
> It is only a matter of choosing our desire.
> Which fruit to pick from the nodding tree.

Infused with hubris, Persia had made war on Greece and lost all at the Battle of Marathon in 490 B.C., as the playwright, Aeschylus, so poignantly dramatized. The South's nodding tree was a small gangly plant laden with white puffballs.

One of cotton's victims, ironically, was the same slave who nurtured it at every stage through his back-breaking labor. Cast adrift like leaves on a swirling stream with nowhere to go and no means to support themselves, the majority of these newly freed blacks initially chose to stay in place. At the end of the war, they had an opportunity, at least in theory, to earn wages picking and chopping cotton. The work was there but not the hard currency to pay for it, or the large thriving plantations, or the certainty that somehow cotton would provide for one and all. Out of that turmoil a new cotton-farming system emerged that would add a synonym for entrenched, color-blind poverty to the American language: sharecropping.

EIGHT

Changing Fortunes

Universal Suffrage say to all, Be ye tranquil.

—Popular Reconstruction rallying cry

A merchant who supplied tenant and sharecropping Southern cotton farmers during the years that followed the Civil War was asked if the 100 percent interest rate he was collecting might not be exorbitant. Hardly, he replied. "It's a large profit, but it is profits on the books, not in the pocket."

Farmers didn't see it that way. One Delta planter complained angrily about this "extortion, genteel swindling, legitimate larceny. . . . The negro and the poor white man make nothing; the factors and the country merchant are the divinities over the agricultural interests. . . . The landlords are out in the cold ignored even by their tenants; lands are being worn out."

For a dozen or more years after 1865, cotton fields in effect became prisoner-of-postwar camps. Once-proud growers found themselves left with ravaged plantations that had lost much of their antebellum value and now often had to be parceled and subdivided into smaller units, a humbling descent from past grandeurs. Poor whites who had vowed never to pick cotton became indentured to the soil under a system of land-lease that left them permanently in debt and not remotely in control of their own destiny. The immense war losses sustained by the South reduced their options. Forty percent of the South's cotton was now being plowed and picked by whites, the remainder by blacks for next to nothing. Many yeomen farmers were operating with handicaps. Mississippi spent one-fifth of its revenues in 1866 on artificial limbs for Confederate veterans. A journalist at a town meeting in Georgia reported that 100 out of 300 men in attendance were missing an arm or leg.

For all practical purposes the hobbled South reverted to a feudal domain with one significant difference: whites and blacks alike were now both slaves to cotton. Northerners too felt the sting. They had a stake in the rapid revival of this cash crop. New England's textile factories depended on it; Northern financiers counted on cotton to enable the South to repay prewar loans. The federal government depended on raw cotton to help maintain a favorable trade balance with Europe. A Republican senator and textile manufacturer declared that without cotton, America would be "bankrupt in every particular." If no longer king, said the *New York Times*, cotton was still "a magnate of the very first rank."

Chalk that up to wishful thinking. If magnates rode in doeskin-upholstered carriages, cotton could barely afford a ride in a buggy with a busted axle. While it once raised Southern agriculture up to the level of a natural treasure, it now reduced the farmers who grew and harvested it to peons, regardless of color. Not all, certainly, but most. A variety of factors contributed to that. Less than a decade after the end of the Civil War, economic depression gripped the United States, triggered by the collapse of the nation's banking establishment. Northern railroads had overexpanded, new bonds failed, and the Panic of 1873 sent the country's economy careening off its tracks. The waves of a financial panic broadcast out to unsettle every market. Cotton prices dropped to 14 cents a pound. That trend continued until 1877 and reduced the market value of cotton by close to 50 percent. One historian discovered that by 1875 in the region surrounding Natchez, that former crown jewel atop the cotton empire, over 150 planters had forfeited their land, all or in part, unable to cover back taxes. Black farmers suffered most; after years of struggle some were finally becoming financially independent and gaining a foothold in Southern communities. Unable to sustain their freedom, they were forced into the oppressive system of sharecropping, which defined cotton agriculture and the social structure of the rural South from that point forward.

Political upheaval proved to be as influential as the shattered economy in shaping the destiny of cotton producers during these years. In an attempt to integrate former slaves into mainstream American society, the federal government instituted a series of reforms: its Reconstruction program. Cotton played multiple roles in the unfolding drama—none that would win it any honors. It grew, but it didn't prosper. It put scraps of food on the table but, at the same time, it forced poor whites and blacks to incur crushing debts. It covered the land but destroyed its fertility. Above all, cotton's erratic and unpredictable value on the postwar world market chased off those large chunks of capital that had once bankrolled the plantation South.

Cotton was no longer a sure bet. In a corrupt, carnival atmosphere populated by carpetbaggers and scalawags (locals joining the Reconstructionists) who were busy divvying up state treasuries for themselves, politicians couldn't be trusted to act with cotton's best interests at heart. For more than a decade after the Civil War, nothing in the South worked as intended, with the possible exception of the railroad and related iron industries, both finally gaining ground after years of neglect and devastation by Union troops. All of that set the stage for tenant-farming and sharecropping, systems hastily devised to continue the cotton monoculture in the absence of money with any real purchasing power. Yankee greenbacks supported by silver or gold now outvalued Mississippi script, currency backed by cotton, at a rate of about 10 to 1. At an auction in 1866, Southern dollars with a face value of $11,088 fetched $1,663 in greenbacks.

Cotton was pretty much all that farmers knew how to do and they were determined to keep right on doing it, whether or not it drove them to an early grave. By 1880 they were producing more cotton than in prewar times, but without anything close to the profits. The crop had essentially conditioned the South to serve its needs even when its needs no longer served the South. In Mississippi in 1866, "a year of wreck," one planter who expected his plantation to gross $108,000 discovered that due to a combination of labor scarcity, falling prices, and bad weather, his total proceeds amounted to $6,564. Within months eighty-one Mississippi plantations went up for sale. Many once valued at $150,000 or more were listed for under $8,000, some for $4,000. One owner, George W. Humphreys, put up his 609-acre plantation, Sligo, for lease the following year for six bales of cotton. By 1871, seven Delta counties in the country's most fertile cotton region accounted for 1.4 million acres of land forfeited for nonpayment of back taxes.

White gold had turned to dross.

The Cord That Binds

Henry Blake didn't have to be told that; he experienced it all around him every day. However you made your deal with the merchant-landlord whose cotton acreage you tilled either as a tenant leasing the land or as a sharecropper working it for an advance against a share of the crop proceeds, you got behind and stayed behind, and that was that. The merchant owned you, and you owned nothing. Born a slave in Little Rock, Arkansas, Henry looked back late in life over those Reconstruction years and recalled that his family worked shares for a while, then they rented. When they went on shares, they couldn't make a thing.

Just the cost of overalls and something to eat. Half of everything went to the white man—the merchant that handed out food and seeds and equipment—and anyone that didn't know how to count, he'd lose. A man that did know how to count—he might lose anyhow. The white folk didn't hand you itemized statements. They just told you what you owed. You could show them your account, it didn't matter, you had to go by their account. You had to take the white man's word and notes on everything. They'd keep you in debt. They were sharp. At Christmas time you could take up to twenty dollars in food and as much as you wanted in whiskey. Henry knew exactly why: anything that kept you a slave. His sweat was in that soil he plowed staring at the backside of his mule day in day out under a fierce sun, his one hope to create a life for himself and his family as a free man. Looked at dead on, that wasn't about to happen anytime soon. Not for Henry, not so long as he owed his soul to the merchant's store. And there was no telling how long that might be, all this life for sure and maybe the next.

Furnish, they called it—the supplies provided by the country merchant. That furnishing man grabbed the planter's prewar power and sat on it like an egg till it hatched. He might even be a former planter himself, but whatever his roots, the merchant ruled as a local Caesar. He furnished essentials like clothing, molasses, sowbelly, plow points, and so forth on credit, everything you needed to get through a season; the bad part was, he kept a ledger and he jacked up the price of these items from 30 to 70 percent above their fair cash value, so when it came time to settle accounts at the end of harvest you were lucky to see a nickel from all your hard work and most likely never would.

Cotton did that after the Civil War: it created a new powerful ruling class in the South that stood behind the counter in more than 8,000 dusty rural village stores from the Delta to the Black Belt, a strip of flat fertile land that stretched 300 miles across central Alabama and northeast Mississippi. From there merchants dispensed life's necessities to poor whites and blacks who were getting further in debt, bale by bale. Will Varner, the furnisher at Frenchman's Bend in William Faulkner's *The Hamlet* was also "the largest landholder and beat supervisor in one county and the Justice of the Peace in the next and election commissioner in both. . . . He was a farmer, a usurer, a veterinarian; Judge Benbow once said of him that a milder man never bled a mule or stuffed a ballot-box. He owned most of the good land in the county and held mortgages on most of the rest. He owned the store and the cotton gin and the combined grist mill and blacksmith shop in the village proper and it was considered, to put it mildly, bad luck for a man in the neighborhood to do his trading or gin his cotton or grind his meal or shoe his stock anywhere else."

Southern lore is filled with stories from that era. In one of them a merchant shouts out an order to his bookkeeper: "Charge up another barrel of flour to all the tenants! I had another bad time at poker last night, and I can't have my wife and children paying for it."

There were no longer any solvent banks in most districts. That role fell as well to the merchant. He charged from 20 to 100 percent interest on loans, and unless you happened to be a planter of means—and only a handful remained—you could expect to pay him back his money at a 40 percent rate or higher. The crossroads storekeeper also took from one-third to one-half of your crop as payment on supplies. Under a brutal crop-lien system that some called financial bailing wire for the cotton economy, your unplanted crop was pledged for a loan of an unspecified amount advanced in dry goods and sundry items instead of money. A contemporary wrote about the sharecropper that "every mouthful of food he purchases, every implement that he requires on the farm, his mule, cattle, clothing for himself and family, the fertilizer for his land, must be bought through the merchant who holds his crop lien. . . ." The farmer, he added, "has passed into a state of helpless peonage." In the end, for all his labor, a tenant farmer or cropper was lucky to come away with a "clear receipt" that discharged his debt. As for the merchant, the apparent archvillain of the piece, he too got squeezed by the exorbitant interest he had to pay in greenbacks to Northern financiers.

Louis XIV made the classic comment about all this: "Credit supports agriculture, as cord supports the hanged."

A popular song at the time caught the mood as well. One verse ran:

> Hurrah! Hurrah!
> 'Tis queer I do declare!
> We make the clothes for all the world,
> But few have we to wear.

Many ex-slaves looked at their former lives in captivity and wondered aloud if freedom was a step forward or backward. "Time was sure better long time ago than they be now, I know it," said one former slave, Sylvia Cannon. "Colored people never had no debt to pay in slavery time. Never hear tell about no colored people been put in jail before Freedom. Had more to eat and wash then, and had good clothes all the time 'cause white folk furnish everything, everything. . . . [Freedom] let the white people down. It let us down too. Left us all to about starve to death. . . . Land went down to a dollar a acre. . . ."

"Like all the fool niggers of that time, I was smartly bit by the freedom bug for a while," said another former slave, Charlie Davenport. "It sounded powerful nice to be told: 'You don't have to chop cotton no more. You can throw that hoe down and go fishin' whensoever the notion strike you. And you can roam around at night and court girls just as late as you please.' . . . I was foolish enough to believe all that stuff. But to tell the honest truth to God, most of us didn't know ourselves was no better off. Freedom meant we could leave where we'd been born and bred. But it also meant we had to scratch for ourselves."

More Is Less

By the time Reconstruction sputtered to an end in 1877, federal intervention had made such a botch out of rehabilitating the South's economy and instilling racial parity that the failed effort was used to bolster the case for states' rights for many generations to come. Cotton didn't help much. The South's fixation on its monoculture cash crop sank everyone, black and white alike, in the same economic quicksand. If you were a Southern cotton grower during these years, whether landlord, tenant, or sharecropper, you were most probably defying the unassailable laws of supply and demand and paying a high cost for your cavalier attitude. As the price that cotton fetched went down, growers planted more, not less, adding to the glut that was already depressing the market. That made no sense, but in those turbulent times, little else did as well. The systems set in place to help get the South back on its feet were bringing it to its knees. The Freedmen's Bureau, meant to transition ex-slaves into enfranchised citizens, had gradually been reduced to handing out food rations to prevent mass starvation as its social programs were jettisoned. More than 30,000 federal troops sent into the South as peacekeepers in an occupied territory after the war failed to deter violent whites from mounting campaigns of terror. During these Reconstruction years the Ku Klux Klan came into existence. It took its name from the Greek *kuklos,* meaning "circle," as in a secret society. That circle soon became the noose of a lynching rope. Klan members lynched or shot more than a hundred blacks a year on average up through 1900.

Cotton was rapidly losing its persuasive economic clout as a valuable raw material. The cost of transporting it north was becoming more of an economic burden than ever. Despite an industrial boom in the postwar years, the operating expenses of New England's textile factories continued to escalate, and their owners looked for any and all ways to maintain margins. Peering across the Mason-Dixon Line, owners saw that the South's expanding railroad system now connected rural villages with larger cities. Also, thousands of impoverished whites who couldn't make ends meet growing cotton desperately needed work. Those and a few other regional considerations set textile owners to thinking that the South might offer untapped resources for manufacturing cotton. Maybe, then, it was time to stop bringing cotton to the mills, and time to start bringing mills to the cotton.

For the South, that decision provided a transfusion of capital and a new lease on life. It also generated the largest industrial migration in American history.

Steaming South

Cotton cloth delivered the South into the twentieth century kicking and screaming like a newborn, happy to gulp fresh air but not yet equipped to survive on its own. That happened just as progress was becoming the country's most important product.

Viewed through the narrow but sharply focused lens of cotton, America ended its protracted love affair with Romance in the final decades of the nineteenth century to begin a new flirtation with Pragmatism. Behind these uptown nouns was a down-home mentality: to find God, look no further than your crankshaft, toolbox, or accounting ledger. Gone with the wind, blown to smithereens by changing fortunes, were the last vestiges of a faith in amber waves of grain or flowing puffs of cotton to divulge the larger unifying truths shaping our existence. Expositions in Atlanta, Philadelphia, and elsewhere that celebrated the advent of mechanical wonders like the Corliss Centennial Steam Engine carried an implicit message: sign up for this automated future or grab a pitchfork and head for the hills. The circular, churning Corliss behemoth stood about three stories high and occupied an entire building at the 1876 Philadelphia Exposition, where it provided power for fourteen acres of surrounding machinery. "The lines were so grand and so beautiful, and the whole machine was so harmoniously constructed, that it had the beauty, and almost the grace, of the human form," a French emissary reported. He was speaking for a generation passionately devoted to the perfection of mankind through science and technology.

To Northern textile industrialists improved steam-power technology offered liberation. Before steam, nature set the ground rules. Hydraulic power brought cotton and wool mills to the fixed energy sources of New England, its rivers and waterfalls, and then brought towns to the mills as urban areas rose up around smokestacks and bell towers to provide for the needs of a stationary workforce. No longer. The flexibility of coal-fueled steam power enabled a southern Massachusetts port, Fall River, to replace Lowell as the center of American cotton textile manufacture after the Civil War.

Still, imperceptible but ominous cracks had begun to weaken the foundation of New England's cotton empire. Hobbled by mounting operating expenses, once-profitable mills were starting to buckle under the escalating cost of transporting raw material so far from the fields, and there was no product diversity to shore them up in times of emergency. Southern planters ruined by a cotton monoculture might have reminded the New England industrialists that one of anything can be a dangerous number in the absence of a viable alternative, but for the moment they felt secure. By 1876 Fall River's forty-three factories, all on their way to being steam-driven, housed well over one million spindles and more than 30,000 looms. They spun, wove, and printed one-sixth of New England's total cotton output, ranking second in cotton production only to Manchester, England. "It is cheaper to use steam power in the midst of a dense population, than to use water power, which often makes it necessary not only to build a factory, but a town also," a weekly newspaper editorialized. That same portability would eventually motivate the mill owners to relocate south. Not so their New England mill towns: they stayed behind and died a slow, painful death.

Ironically, other technologies first developed by Northern factory machinists also hastened the southern migration and led to New England's collapse as a textile center. The first to gain widespread use, ring spinning, replaced the laborious stop-and-start operation by which yarn was spun and then wound onto bobbins one length at a time. A ring frame smoothly and rapidly stretched, twisted, and wound yarn in a single continuous motion as it circled the bobbin; ring-spinning produced coarse thread with a tighter twist, especially suited to the strong warp in denim. Weaving soon caught up.

In 1894, J. J. Northrop perfected a bobbin changer at the Draper loom works at Hopeland, Massachusetts. Resembling the revolving cartridge battery feeder on a submachine gun, it reloaded one fresh bobbin at a time onto the loom shuttle from its ten-bobbin cylinder as the empty was ejected into a hopper. Changing bobbins, once a labor-intensive function, now became automated. Both innovations had a huge impact on cotton manufacturing, but a more profound one still on the country's shifting labor force in the decades to come. These machines incorporated the intricate operations that skilled workers had previously performed. They could now be run by low-paid, inexperienced operators fresh off the farm. That severed the last ties that bound mill owners to the North—a pattern repeated in other industries as well. The toll was brutal. Southern migration put more than 100,000 of New England's textile factory employees out of work in one early decade of the twentieth century.

Three Harvard professors later labeled such advances "disruptive technology"—the new product or improvement that "sneaks into an established market because industry leaders fail to recognize the threat it poses." This theory holds that company executives underestimate its long-term effects on the industry, and pay too much attention instead to immediate customer demands.

There was no need to convince highly competitive Northern industrialists or their Southern business partners that steam and manufacturing improvements made sense. Their objectives interlocked as naturally as two strands of cotton fiber. Not only could they build mills nearer to the raw material, but other benefits awaited as well: tax savings, inexpensive real estate, lax enforcement of child-labor laws, and a dirt-cheap, hungry, native populace eager to spin and weave from dawn until late at night. That same steam power would cut down transportation costs by running railroad locomotives from Appalachian coal mines to fuel the new textile mills in the neighboring Piedmont area of central North and South Carolina. By 1880 the South was finally shedding the parochial assumptions and beliefs that had tethered it to a dead past, and none too soon. In this piston-driven, well-oiled, modern America, industrial prowess counted for more than any other factor in defining success. There was only one way to enlist, through a factory door, and the South signed up with all the enthusiasm of true believer.

Scores of impoverished mountain whites on the brink of starvation began pouring into makeshift factory villages in central North and South Carolina, Georgia, and Alabama to supply the underpaid labor force for the rapidly expanding new mills. They left the farm for what they called "public work." Whatever its name it offered food, shelter, and a measure of self-respect. "A public job is more interesting than one-horse farming," said one worker, "because you can meet your bills." While it paid little, mill work helped feed a destitute family that had been down to two meals a day if that, and it enabled children to become wage-earners. They went into the mills as doffers as young as nine years old—although twelve was the official minimum age. The pay scale was abominable: they received only a fraction of an adult salary, which itself was far below the Northern counterpart. Adult hands earned between 40 and 50 cents a day in North Carolina's mills in the 1890s; children were paid 10 to 12 cents. But there were trade-offs—much lower living costs, and the potential for all family members to contribute. At least 90 percent of Southern working children under fifteen in 1900 were employed by cotton mills.

In 1904, a reporter wrote that he'd visited Southern cotton mills where "for ten to twelve days at a time the factory hands—children and all—were called to work before sunrise, laboring from dark to dark. . . . I have seen children eight or nine years of age leaving the factory as late as 9:30 at night and finding their way . . . through the unlighted streets of the mill villages, to their squalid homes."

That immediately evokes visions of black-sooted Manchester, and casts a long dark shadow back across the utopian optimism that created Lowell. There was no pretense of social or cultural improvement here. Another mill visitor, W. J. Cash, wrote that "[b]y 1900 the cotton-mill worker was a pretty distinct physical type in the South. . . . A dead white skin, a sunken chest and stooping shoulders were the earmarks of the breed. Chinless faces, microcephalic foreheads, rabbit teeth, goggling-dead fish eyes, rickety limbs, and stunted bodies abounded. . . . The women were . . . stringy-haired, and limp of breast at twenty, and sunken hags at thirty or forty. And the incidence of tuberculosis, of insanity and epilepsy . . . were increasing."

A frightening portrait, and fodder for social reformists. While that might have been true in some mill encampments, group photographs of Southern textile workers from that era also display generally normal-looking children and adults, no rabbit teeth in evidence. Cash's account seems exaggerated for effect. If workers had a problem, and many did, it was with the respiratory ailments that caused lint build-up in their lungs, especially in the carding rooms. As for their "squalid homes," they might not compare with the trim and tidy company boardinghouses of bygone Lowell, but they formed a tight-knit community for villagers that sustained families through hardships and provided a consistent source of recreation. The Merrimack at Lowell during this period now ran foul with raw sewage. Infants accounted for close to 50 percent of all deaths. According to one essayist, rumors spread that Lowell mothers were drowning their babies in the city's mill canals to spare them the misery of a sick, dismal childhood. The North had little to feel superior about.

Motivated by equal measures of ambition and desperation, cotton textile production began as an experiment and soon became a life support system. Spindles in the four leading Southern states increased from 422,807 in 1880 to 1,195,256 in 1890, and to 3,791,654 by 1900. The labor force grew from 16,741 to 97,559, and the capital invested from roughly $17 million to $124 million during these two decades. Between 1890 and 1900, that represented a 131.4 percent rate of expansion, as contrasted to a 12 percent rate in New England. Profits between 30 and 75 percent were common. Most factory complexes were financed by Northern investors joining forces with local lawyers, doctors, and merchants in these small southeastern towns. A community would often form a corporate entity to put up minimal seed money for a new factory and hope to attract outside finance. Northern machine builders extended credit; many New England textile magnates also expanded their factory holdings to include new branches in Alabama, the Carolinas, or Georgia.

The man largely responsible for generating this turnaround died while it was still in its infancy. Henry Woodfin Grady, the son of wealthy slave-owning planters and also the editor and a co-owner of the *Atlanta Constitution*, combined personal passion with mesmerizing oratorical gifts—a dangerous combination in demagogues but valuable in a civic-minded promoter. He pounced on locally spun cotton with evangelical fervor and famously announced in 1886 at a gathering of the elite New England Society of New York attended by J. P. Morgan and other captains of industry and finance that it heralded the arrival of the "New South."

"We have fallen in love with work," Grady promised his audience, as if the Old South was in a halfway house for recovering derelicts. "The New South is . . . stirred with the breath of a new light," he avowed. He swore too that racial equality was a done deal because now "the South, with the North, protests against injustice to this simple and sincere people." Grady got through. Money materialized, and mills began to go up in earnest.

Blacks might be simple and sincere, but they weren't invited to this party, now or later. Millwork on the machines proved to be exclusively a whites-only occupation, backed by rigid segregation laws. Blacks could find work in lumber, steel, iron, and mining, but only at the lowest level of mill employment, never as operatives. They were relegated to loading and unloading, opening, and cleaning bales—dirty, grueling, back-breaking jobs. A leading textile exponent, Frances W. Dawson, calculated that in addition to exposing hardscrabble white farming families to "elevating social influences," and "improving them in every conceivable respect," millwork would provide each laborer with just over $16 per month; Dawson didn't mention that if those earnings included all dependents, they represented monthly wages of less than one dollar per person per week. The company commissary became equivalent to the merchant's store for sharecroppers. Unpaid bills required mill hands to work long hours to settle accounts. Many never saw a dime. Problem workers were blacklisted.

Still, Grady, who died at age thirty-nine in 1889, slips into history as the patron saint of the South's newly acquired identity. Somehow he managed to deliver the gift of rejuvenating industry to a downtrodden people without threatening their agrarian values. The excitement generated by local cotton manufacturing quickly erupted into a revivalist movement. According to the messages of its apostles, delivered from church pulpits, park benches, and barbershop chairs, the same flimsy fibers that sentenced white Southerners to purgatory behind plows would now spirit them up to heaven on whirring spindles. Townspeople stood up at local community meetings to swear that building a local cotton mill was the most Christian act they could perform. One commentator reported, "Everybody wanted to be a soldier for the Lord. Farmers who were too poor to buy even a single share of mill stock pooled their pennies and bought a share in common. Workers agreed to put aside 25 cents a week toward the new enterprise. Each community tried to outdo the other." The buildings they constructed might have looked like mills, but in spirit they were houses of worship. Factory chimney smoke, said one man, was daily incense to God. "The strange mixture of practicality and evangelism," another observer remarked, ". . . made industrial accomplishment seem a fevered folk movement." It sprang, he said, from the hearts of men in whom the American dream was dying.

Cotton was the first crop to democratize greed. Once the exclusive birthright of American aristocracy, a privilege that went with rank, the self-absorbed pursuit of wealth became a commoner's mission—a religious calling, even. The goal was no longer simply enough to scrape by on, but money invested as a down payment on a luxurious future. "What is the chief end of man?" Mark Twain asked in "The Revised Catechism." "To get rich. In what way?—dishonestly if we can; honestly if we must." For cotton, galloping capitalism took the play away from the fields where it was being cultivated and relocated it in these new Southern factories. Elsewhere, Andrew Carnegie and others might be constructing the nation's new skyscraper infrastructure out of steel, a much more showy display of capitalistic muscle, but even Carnegie knew where the underclass often got their first taste of American enterprise. As a boy fresh off the boat from Scotland he worked as a doffer changing bobbins in a Pittsburgh cotton mill for $1.20 a week.

The builders of these Southern mills put practicality first. Despite the steam engine mania that now enveloped New England industry, easily available waterpower would first drive their factories. Once up, steam—and soon a brand new power source, electricity—would follow. Investors initially set out to refurbish William Gregg's tottering textile factory at Graniteville and to erect another at Kalmia, both in South Carolina. Almost forty years earlier Gregg, as clairvoyant as he was innovative, said, "Let croakers against enterprise be silenced. . . ." There were few croakers left in the pond. The new owners exhumed Graniteville, fitted it with new machinery and set it into operation.

With that, a New South was born, baptized and welcomed into the American fold. Although workers experienced many of the same conditions—excessive heat, unclean air, and machinery capable of breaking a worker's arm in a millisecond—as in New England, there were purely Southern differences as well. In the Piedmont children grew up in mills, first as underage "helpers" who would bring lunch or dinner to their mother or father and stay to help spin or weave, serving a kind of informal apprenticeship; from the age of eight or nine on they learned the trade, so that at twelve they could become fully employed with little additional training. "When I was a little fellow, my daddy was a-working at Poinsett Mill," Geddes Dodson recalled. "He was a loom fixer. . . . I'd carry his lunch every day. He'd tell me to come on in the mill, and he made me fill the batteries [with bobbins] while he run the weavers' looms. . . . See, I knew a whole lot about the mill before I ever went in one." Mills were home as well as work: children gathered and played there, or close by. And the community itself bridged the gap between supervisor and worker. New England mill agents, true to their stern Yankee roots, enforced strict clauses preventing social interaction between worker and management. But these Southerners lived and worked in the same isolated mill villages, overseer and laborer alike—family in the most cherished Southern tradition. Barely making ends meet, workers pitched in to kill hogs and harvest each other's meager crops. They held quilting bees, danced, and swapped lies together.

The culture bred its own social structure. Outsiders dismissed cotton mill people as "lintheads," a close cousin to white trash. But they operated with a sense of pride within their communities. "And they don't take nothing off of people," one former worker explained. "People come and want to give them a dirty deal, they don't take it. They just fight for theirself." In disputes with upper management, supervisors would often side with workers. Mill owners like Luther Bynum in the Piedmont, on the other hand, kept a tight rein and wanted no part of community solidarity. "Old Mr. Bynum used to go over the hill at nine o'clock and see who was up. And, if you were up, he'd knock on the door and tell you to cut the lights out and go to bed," one worker, John Wesley Snipes, recalled. Loyalty mattered, said Snipes. There was a popular saying. " 'Bynum's red mud. If you stick to Bynum, it'll stick to you when it rains.' "

"They made a rule up there one time," Mildred Edmonds, who worked at Burlington in Piedmont Heights, recalled. "They said you couldn't talk to the other one on the other side of you. Well, the boss man went around and he told all of them. He got down to me and he said, 'Mil, I'll have to tell you but I know it won't do no good. You're not supposed to talk to each other.' I said, 'God gave me a tongue and I'm going to use it.' It didn't last long. He didn't want to do it. But you see, the man above him put that on him."

Overseers weren't always so obliging. There were hat-stompers too. Hat-stomping was altogether Southern. It could scare the overalls off a child worker like James Pharis. "I seen a time when I'd walk across the road to keep from meeting my supervisor. They was the hat-stomping kind. If you done anything, they'd throw their hat on the floor and stomp it and raise hell." Icy Norman, another young doffer, remembered how his supervisor would "throw it down, and spit on it . . ." In fact there was an entire vocabulary of displeasure encoded in the act, and you'd be quick to decipher its nuances if the thudding foot was set in motion by something you'd done.

For the first few decades until growing discontent with wages and working conditions intervened, Southern mill villages provided a sense of communal belonging that had been absent for most of these tenant-farm families. They'd come home to cotton.

Bridging the Pond

The transition of the cotton textile industry during this era offers a vivid reminder that miraculous machines can also be brutally efficient. They enticed investors with their promise of increased productivity and punished veteran handcraftsmen and businessmen alike—old-line cotton merchants, factors, and classers—with their brilliant performance. Mechanical ingenuity and technology suddenly trivialized their specialized skills and experience. Men who determined the precise quality and worth of cotton fiber by feel and appearance—the classers, or graders—were about to be replaced by sophisticated instruments. So too were factors and those prosperous Liverpool merchants who for nearly a century had counted on price information delays between United States growers and English mills to manipulate the cotton market.

With the completion of the Atlantic telegraph cable linking America and Europe in 1866, cotton prices jumped the Pond in minutes, not weeks. That radically changed the whole dynamic of buying and selling raw material in England, where the Manchester mills remained the world's largest converters of fiber to cloth. Unlike brokers, who simply negotiated price and quantity, Liverpool's merchants actually took possession of cotton from Southern planters or their agents. Attuned to fluctuations in supply and demand, they warehoused their cotton until prices rose, at least in theory. (When prices dropped, so did profits.) Before the telegraph they sold to English mills whose buyers had no alternative but to wait six weeks or more—the time it took news to travel via shipping vessels—to learn whether they could get a better price. In a fingersnap of time, all that changed. Deals between factory buyer and seller could be put on the table, countered, negotiated, and closed all by Morse Code within hours.

The merchant trade was not eliminated—in fact, an annual black-tie dinner for international cotton merchants continues to this day as a tradition in Liverpool—but its unchallenged power to control the market was substantially reduced. A monopoly had been broken.

By 1880, better than 80 percent of England's cotton products were being exported to underdeveloped countries, and for the vast majority of the raw material its factories once again relied on the South. More than three-quarters of the world's rapidly expanding population was now wearing or using this versatile, light, inexpensive fabric. Practically speaking—and the industrialized West was fixated on practicality—a commodity in such high demand would be among the first to grasp at any innovation that facilitated its manufacture and distribution.

Telegraph wires hummed; miles and miles of new railroad track for the first time connected small rural Southern cotton villages with each other and larger urban areas. Everything was up to date in Yazoo City. "The telegraph is used freely, and the buyer knows hour to hour what cotton can be had in each of the interior and seaboard markets," a New England correspondent reported. "He names a price to any mill man who is in need of cotton, and if he receives an order he telegraphs forthwith to his Southern correspondent to make a trade. The staple thus being secured a bill of lading is issued."

These marveling reports of cotton's coming of age were also a cautionary tale to New Englanders set in their ways. The South was oddly blessed by its lack of an existing infrastructure. It had nothing really to protect by way of historical precedent. New mills, new rail lines, and new communications bounded forward with no obstructive traditions or standing competition to overcome. By the time steam engines and telegraph lines became familiar sights in the late nineteenth century, business centers for cotton trading had also taken up permanent residence in Memphis and New Orleans, transforming these cities into international players. Exchanges in both set the rules, connected buyers, sellers, and futures traders through humming wires, and buzzed with the electric charge of men conducting important business right at the source. The South had finally begun to capitalize in every respect on its largest cash crop.

Making the Grade

Lolling around New Orleans while visiting relatives in the cotton trade in the winter of 1873, the French impressionist Edgar Degas accompanied his cousin to the city's bustling Cotton Exchange. The haughty, cultured son of a wealthy banker, Degas was more at home painting ballerinas than businessmen, but cotton seemed to be the sole subject of interest in this stifling city, which was enduring an abnormally hot and humid winter. At first Degas couldn't wait to escape. The sidewalks of Carondelet, home to the city's banks, "were perpetually crowded with cotton factors, buyers, brokers, weighers, reweighers, classers, pickers, pressers and samplers," one journalist noted.

Another visitor, Edward King, said, "In the American quarter, during certain hours of the day, cotton is the only subject spoken of; the pavements of all the principal avenues in the vicinity of the Exchange are crowded with smartly-dressed gentlemen, who eagerly discuss crops and values, and who have a perfect mania for preparing and comparing the estimates at the basis of all speculations in the favorite staple; with young Englishmen, whose mouths are filled with the slang of the Liverpool market; and with the skippers of steamers from all parts of the West and South-west, each worshiping at the shrine of the same god." Levees, lined with bales piled fifty feet high, resembled walled cities.

To relieve his boredom, Degas studied the behavior of the

exchange's cotton traders at close range, and decided they offered up an intriguing subject. Always a brilliant draftsman, he set about to capture a defining moment at the Exchange, focusing on men "occupied with a table covered with a precious material," as he wrote to a fellow painter, Tissot. That material, of course, was cotton. They were dissecting a beaten bale of loose fiber with the meticulous care Degas lavished on the minute details of a ballet class; he'd found his way into their world. If he had been able to add sound, the viewer would have heard a tape clerk shouting aloud the latest telegraphed prices above the general commotion.

Each sampled bale, like the one spread open on a large table behind the inspector, revealed its true value only when laid bare to close inspection—as if autopsied by skilled coroners. The men gathered around it in Degas' rendering have developed a sharp eye and expert hand; "leaf" impurities, staple length, texture, and discolorations all mattered. Business first, of course, but the Exchange also served as a de facto social club, as Degas shows us. Just to the rear of the seated inspector, another man casually reads the daily newspaper opened before him, paying no attention to the surrounding hubbub. By 1873 a million bales a year, about one-third of the country's total, were passing once again through New Orleans from newly planted fields in Texas, Mississippi, Louisiana, and Alabama on their way to England and continental Europe.

In Degas' finished oil on canvas, *The Cotton Exchange in New Orleans*, an elegant gentleman in top hat, bow tie, and frock coat dominates the foreground as he sits alone and inspects the hunk of the raw cotton he pulls apart in his hands. He seems to be trying to read his future in it, and perhaps he is, for cotton pricing by then had evolved into a sophisticated system of grading that determined merit and price. The man is a classer—a still-thriving breed at that time. Classing remained one crucial aspect of the cotton trade that continued to defy advances in technology until after the turn of the century. Degas' painting captured one moment in the process of determining a bale's market value, but without an accompanying cheat-sheet, no one could be expected to unravel (in every sense) this arcane art and science. Just as a professional tea taster brings a variety of talents to bear in judging the balance and mouth-feel of Ali San oolong or Ceylon Silver Tips, a cotton classer mustered a highly developed repertoire of sensory skills. Until the USDA with its instruments stepped in to take control of the inspection process for a fee, independent classers set up shop in New Orleans and the trading center of Memphis, 400 miles upriver at the head of the Mississippi Delta. Memphis had become the nation's largest inland cotton market and the world's largest cotton merchandizing city. In a single season it handled more than 400,000 bales valued at $16 million.

The cotton being wholesaled and traded came in two flavors—spot and futures. Spot cotton referred to actual bales in the warehouse or on a wharf, at some point (or spot) along the way, while futures defined cotton bought and sold for future delivery at some defined place and time. Classers determined the value of spot cotton, and nobody took their judgments lightly. For years there was a legend on Front Street's Cotton Row in Memphis that Cain killed Abel because Abel classed Cain's cotton as Low Middling, when to Cain it was Strict Low Middling.

The difference between these two classifications, while unintelligible to anyone outside the industry, represented more than small change to sellers and buyers. For white cotton, there are nine grades that range from premium Middling Fair to Good Ordinary, the jug wine of the cotton industry. Most American cotton is graded—or classed—as Good Middling, about four steps up from Strict Low Middling. Each rung on the quality ladder marks a substantial price adjustment when negotiating thousands of pounds of fluff. Bales are sampled using a one-inch staple and a midlevel grade, Middling, as median measurement standards.

Character, length, and color determine the assigned grade. Character refers to the strength, uniformity, cleanliness, smoothness, and body of the fiber. A high percentage of leaf and trash in a bale lowers its value. Fineness, or micronaire, matters too. A less mature, coarser fiber also results in a discount charged on the price paid to the grower. Before dials and wavering needles replaced human faculties, a classer with experienced, sensitive fingers would draw a handful of cotton, as in Degas' oil, and pull apart fibers to test for strength and staple length. "The classer takes a tuft of cotton, . . . grasps both ends and breaks the fiber," one observer reported, impressed that even the most skilled judge could estimate the cotton's strength. The classer would also call, or grade, for color purity. White cotton, creamy and consistent, was most prized; in descending order, he would determine if the bale was spotted, tinged, yellow-stained, or gray—that is, exposed to rain or fog for too long a time, like unintentionally botrytised grapes. As for staple length, it varied between very short—not over three-quarters of an inch—and extra-long, about one and three-eighths inches, as in Egyptian cotton. More than 83 percent of American cotton up through the mid-twentieth century was medium staple, or an inch long, with a small fraction measured above and below that in length.

Classers needed optimal lighting and humidity to do their work; in the decades before the era of the 1,000-watt incandescent bulb and air conditioner, humid weather and bright sun served as their most trusted allies. The Memphis Cotton Exchange opened in 1873 with oversized skylights inserted into the ceiling of its top floor at various angles to admit the sun's brightest rays at different times of day. Later the government recommended they be set at 23 degrees north for the truest light. No one classed cotton in overcast weather. "Four days of rain was the damnedest excuse to get drunk that a man ever had," a Front Street man with fifty-one years of experience told Memphis reporter John Branston a few years ago. Food commodities like corn or wheat might be sold by the ton, but not cotton: each bale was graded by hand and assigned a corresponding value. Each could be traced back to its grower. Buyers rejected inaccurately classed bales.

Aesthetics, of course, played a role in all of this, but anticipated factory performance counted as well. Inferior raw material pushed up mill costs. Spinning machines had to be adjusted for short-staple fiber, which constantly snapped and shut down operation; nonuniform and brittle fibers with insufficient moisture content, or body, also botched up the works. Lancashire quickly became dependent on Southern cotton again after the Civil War primarily because the supply of dirty, short-stapled East Indian cotton fouled their machines, caused shut-downs, and produced poor quality flimsy cloth.

As the 1880s progressed, the flamboyant, free-wheeling merchants who hired these cotton classers were also on their way to becoming an extinct breed; sober men representing American conglomerates like Hohenberg and Dunavant eventually replaced them—a step forward perhaps for orderliness and stability but not for the independent spirit of brash American assertiveness and daring that Europeans loved to mock yet desperately envied. The expanding railroad network that connected the rural South to the coast encroached on steamboat cargo traffic; that left the Mississippi still muddy but less mighty. Cotton exchanges in New Orleans and Memphis lost some of their former prestige as buyers and sellers bypassed them to deal directly from field to factory.

A century later the steady decline of the Memphis cotton trade has emptied Front Street's merchant warehouses; today only a few dilapidated, abandoned structures remain. Outside the Cotton Exchange Building on Front and Union a bronze marker reads, in part:

There are hearts as soft as fluff
That assume a manner gruff—
Just to hide their real fine stuff, on Front Street.

Piedmont Power

While more than 40 percent of the South's cotton was growing west of the Mississippi by 1890, in Louisiana, Texas, and the Indian territory that would become Oklahoma, its textile industry remained firmly rooted in the central Piedmont district of the Carolinas. In 1894 Columbia Mills in South Carolina became the first large factory in America whose machines were driven entirely by electricity—self-generated in this case. Soon other electric-powered factories were being built as well, in a new, more practical single-story design made possible by eliminating the main shaft of the waterwheel or steam engine. Mills could now be erected within the vicinity of highways and rail transportation, with better lighting and fewer fire hazards. Almost fifteen years into the South's resurgence, the region had become the dominant producer of coarse goods, while Fall River continued to maintain its hold on fine cotton fabric—for the moment. In that same year, 1894, the Merrimack Mills in Lowell, which had made its reputation on coarse cloth, invested $600,000 in building a Southern plant. So too did textile corporations in Chicopee Falls and Canton, Massachusetts. Southern cotton mills now exported coarse cloth to Asia, a booming market, and they soon developed a fine-fabric industry as well. Their investors instituted vertical integration. They opened bleacheries to prepare "gray goods"—straight off the loom and unfinished—for printing, and then sent it to their own local printing plants. The South was not only coming of age, it was defining the new age of American cotton textiles.

Enterprising upstarts replaced New England's cadre of cotton textile aristocrats who'd accumulated their fortunes on the backs of farm girls and indigent foreigners. Early on in New England so much capital had been required to launch a textile empire, as in Lowell, that only the established wealthy elite needed apply. Now cunning and ambition helped bankroll success, and no one had more of those qualities than two traveling salesmen, the brothers Moses and Ceasar Cone. They'd spent their early years drumming for their immigrant father's Baltimore wholesale grocery business throughout the southeast. When money was tight, mill commissary stores would barter yarn and cloth for coffee and sugar and staples. The brothers saw an opportunity and asked tough, sensible questions: What happens when the cloth leaves your factory? Who negotiates the price? Who finds the clients? How do you finance and collect payment?

It quickly became clear to Moses and Ceasar, both about thirty, that in contrast with Northern operations, this fledgling industry lacked organization, structure, and a strong sales and marketing arm. The Cones were hungry, also savvy. They knew that new Southern factories with their ring frames and Draper looms could turn out cloth at a faster rate and lower cost than New England's mills. Britain, relying on dated machinery, was falling behind. India's forty or so mills presented no real competition, because the ruling English continued to suppress that country's textile industry as a threat to Lancashire. There had also been a dramatic shift in British exports since 1820, when 60 percent of its finished cotton went to Europe and America.

As the Cones and others recognized, that greatly increased the American market for domestic manufacture—especially when more than a half million immigrants a year were now arriving at Ellis Island, and all had to be clothed and provided with bedding and linens. No need to look any further than their father, Herman Kahn, who had come across on a boat from Bavaria, Germany, in 1846 with little more going for him than bad English and great expectations. Herman, ever adaptable, quickly anglicized Kahn to Cone and opened his grocery business, H. Cone & Sons, in 1870.

The first wave of Southern mills were turning out reams of ugly, cheap, thick cotton plaids and coarse sheeting. The Cones quickly realized both were losing propositions. As selling agents they needed a higher class of product like chambrays, cheviots, and dress shirts, each carrying a quality guaranteed. By 1890 Moses and Ceasar were ready to make their play. They sold the declining grocery business and put the proceeds into the Cone Export and Commission Company, headquartered in New York, where they set up shop as agents for Southern mills. They spent $22 on tables, their first expense, and opened for business. They also had a personal stake in the mills' success, having invested $50,000 of their own money in a plaid fabric manufacturing plant in Asheville, North Carolina—the C. E. Graham Company. The few mills that accepted their advice and refined their fabric selection prospered within months. Word got around. In short order Moses and Ceasar convinced 90 percent of the Southern mills—about fifty in number—to upgrade the quality of their product and make Cone Export their selling agent on a 5 percent commission basis.

They were soon selling more product than their mills could turn out, but as Herman had taught them well, in every challenge there exists an opportunity. Way out west a former dry goods salesman, Levi Strauss, also Bavarian, was making a killing converting denim yardage into something called blue jeans. What was thick indigo-dyed twill created for if not to be ring-spun? Never were a yarn and machine more perfectly matched. Locomotives now carried freight from ocean to ocean. You could spindle in North Carolina and sell in San Francisco. Work clothes, said Ceasar, are vital in commerce, and they are so poorly made at present as to be a detriment and a failure. If the Cones applied the same quality standards to these products as they insisted on for their high-end fabrics, they could turn out dependable, durable goods. "Lasting work clothes are an economic necessity," Ceasar announced. Nobody argued.

Men of action as well as words, the brothers took the leap:

they converted some of the land they had purchased in Greensboro, North Carolina, in the heart of the Piedmont, into a denim mill. It was near the cotton fields and near gins, so they called themselves the Proximity Manufacturing Company. Denim worked out so well that three years later, in 1899, they were able to build a second mill, White Oak, and branched out to soft flannel. They called the new flannel factory Revolution Mills out of respect to its advanced machinery. A Charlotte newspaper wrote at the time that the brothers had "energy, sense, and pluck." Those three words described the ideal new twentieth-century American man, embodied by Teddy Roosevelt and the durable-denim Cones, who now became the world's largest manufacturers of indigo-dyed twill. When Moses died at fifty-one in 1908, the Cones had established an empire. Ten years later, for the first time in American history, Southern factories produced more cotton cloth than their counterparts in the North. On a graph, the Northern textile decline after that resembles the face of Mt. Everest. By 1957, 80 percent of the nation's spindles, and 82 percent of its looms had moved south.

Denim on its own did not build the South's cotton mills, but its dominance exactly fit the place and time. Cotton was about the least pretentious cloth on earth: weave it strong and tough and it asked nothing back but a bath now and again. Wherever civilization hadn't yet steamrolled over the American wilderness cotton cloth set the mood for discourse and commerce. Keep it simple, don't bend corners, lock eyes when you speak, and button up where it matters. Maybe late arrivals like the Cone brothers and Strauss saw it all a little more clearly than blue-blood Yankees ensconced within their lace doilies and stiff celluloid collars. Good table manners and worsted topcoats counted for little when chasing ponies or treeing a coon. Pants that didn't tear, though—they were worth a paycheck. Moses, Ceasar, and Levi knew how the West was being won and what was being worn while winning it.

NINE

Two-Horse Power

You go fight a goddamn war, and the minute you get back and take off the uniform and put on a pair of Levi's and leather jackets, they call you an asshole!

—Willie Forkner, member of the Boozefighters Motorcycle Club, 1946

Denim jeans were cotton as God intended it to be used, a Midwest preacher might have reminded his Sunday congregation from the pulpit in the early decades of the twentieth century. Jeans were put on earth to get us through our chores, to be priced within reach of the farmer, miner, hod carrier, assembly-line worker, or any able-bodied ranch hand needing to rope a steer or obliged to crawl under the oil pan of a leaky pick-up. Cotton doesn't put on airs, and neither should we. Jeans prove it. Be thankful for your faded worn work pants, union-made and as American as strawberry shortcake. You can't buy another piece of clothing as right-headed as a pair of jeans. Amen.

However and wherever the sentiment was expressed, the message was always the same: denim jeans were cut to fit the virtues that helped make this country strong—humility and a devotion to the task at hand. No matter what your faith, as the twentieth century unfolded, work became the true religion of America, and cotton, particularly in the form of jeans, became the chosen fiber of a God-fearing national labor force. That set blue jeans on a curious path. They came to provide the media with a quick fix on the country's prevailing attitudes and values, especially among youth. Then youth came to dominate the entertainment media, shaping the aspirations of millions, and the distance between a star's navel and the waistband of her Versace low-rise jeans became an *Inside Edition* exclusive. By the late twentieth century, the same humble fluff that assembly-line workers wore with blue-collar pride now defined the zeitgeist of a culture that worshipped belly buttons and denim divas.

Jeans played an even more critical role for the American cotton industry. As the century progressed, it hit the skids, and by the 1970s synthetic fiber was threatening to obliterate cotton as the nation's favorite fabric. Denim alone resurrected the American cotton apparel industry.

Riveting Origins

None of those concerns were on the minds of a Bavarian and a Latvian who partnered in 1873 in a scheme to add rivets to waist overalls at stress points. These men were both thinking mostly about ten-penny nails, sharp tools, knobs, valves, hooks, braided wire, and assorted clumps of heavy rock-hard materials that ripped apart the pants pockets of common laborers. They were not thinking about making Calvin Klein a rich man, or staking Giorgio Armani to another Tuscan villa, nor were they looking to turn Marlon Brando and James Dean into bad-boy sex symbols. There were no such things as "new vintage jeans," artificially aged like thousand-year-old eggs. Diesel was the name of an engine yet to be invented, not a clothing company determined to make its fortune selling ten dollars' worth of indigo-dyed twill for twenty times its cost by strategically tearing the garment to create an illusion that the wearer was planning to lift something heavier than a double-shot decaf soy latte.

Levi Strauss and Jacob Davis didn't know from low-rise, Phat Farm, stone-wash, or the L.A. bootlegs. Fashion forward they weren't, but they knew what loggers and miners in the foothills of California and Nevada needed in their work clothes—the same things they looked for in their women back then: comfort, forgiveness, and a reasonable tolerance for neglect. Chopping down redwoods and crawling over boulders to pick-ax for ore popped seams and shredded pockets. By the time these two men joined forces, Levi had been peddling dry goods in San Francisco for twenty years. Dry goods in those days included just about anything you couldn't eat: underwear, collars, blankets, and coats, all delivered by ship from the east coast around Cape Horn. It was the Strauss family business, and Levi, then twenty-four, six years past arriving in New York from Bavaria, had been sent by his brothers Louis and Jonas in 1853 to open up the western territory, the same year he became a U.S. citizen.

By then he had anglicized his name from Loeb to Levi, or at least relocated it in his mind to where the Bavarian Alps met the High Sierra, but there was little chance of anyone mistaking him for a dinged and dented weather-beaten prospector. He was puffy and soft where they were creased and lean; his eyes also radiated a shrewd calculating intelligence more suited to the merchant trade than grub-staking. He was smart enough to seek his fortune outfitting men who sought their own in the foothills and almost never found it. Levi Strauss & Company—LS&Co.—a thriving wholesale business headquartered on Battery Street in San Francisco, made nothing it sold, and that was the beauty of it all—no raw material or manufacturing equipment costs, no burdensome overhead or worker payroll to drain the coffers. Dry goods arrived at one door by boat and left at the other by mule in covered wagons, and even the canvas covers of those wagons were sold by Levi. "The best job on the planet," he called his work. An unmarried man (and generous provider to relatives) who always left behind quality merchandise and a smile, he might have drifted into history beloved and forgotten.

But there was another European immigrant, a tailor up in Virginia City, Nevada, who approached Levi with a proposition that Strauss knew at once only a dummkopf could ignore. It came by letter along with a request for money. The tailor, Jacob Davis, also familiar with the many moods and uses of cotton, had run into a tough customer, he explained in fractured English to this man he'd never met. A woman who lived close to his shop wanted to buy new pants for her husband, a woodcutter. She paid Davis $3 to custom tailor a pair that would not wear out. Davis chose a light form of canvas, called duck, purchased from Levi's dry goods wholesale outfit. But from experience Davis saw that he needed more than tough threads to keep those pockets and seat seams intact. Finishing the pants, he caught glimpse of a pile of rivets on his workbench that he used for saddle straps. As an experiment he fastened a few at strategic points—in the pocket corners, at all places where bending and lifting and hauling usually decimated his good work. The pockets held, and so did the riveted seams. Word spread, and Davis soon became deluged with locals putting in orders for riveted work pants.

A transplant from the Baltic, Davis had an inventive talent that he'd put to use without much success trying to launch ideas such as a steam-powered canal boat. He didn't have the $68 he needed to patent his riveted pants, but he bet he knew who did—the supplier who sold him the duck. Many suspect Davis also had a hunch that since he and Strauss were both European Jews making their way through rugged untamed terrain, strangers in a strange land, they both spoke the same language of ambition and chutzpah. It was worth a stamp, no?

In his letter to Strauss, Davis claimed that his riveted pants were so much in demand "that I cannot make them up fast enough. My nabors are getting yealouse of these success and unless I secure it by patent papers . . . everybody will make them up." A classic pitch: act now or lose this golden opportunity. Davis sent along two samples of his riveted trousers, one in cotton duck, the other in denim, both supplied by Strauss, and he offered a 50 percent share of his invention if Strauss would front the patent money. Levi checked out the merchandise and wasted no time; scarcely a month later he had his lawyers file a joint patent application, granted in May 1874. That was the easy part. Metal-reinforced work pants were a great idea waiting to happen, Levi understood, but he had taken pains to stay away from manufacturing his own goods—costly and risky, even if potentially quite profitable. But Davis, a hands-on fabric snipper and sewer, could be both partner and production manager. Levi persuaded him to move to San Francisco to oversee the manufacturing. These two men with no history between them, one a kindly traveling *shmata* salesman, the other a sharp-featured clothes cutter, decided to trust each other's honesty and integrity on sight and set up shop as partners.

The pants they made and sold would become the most universal and yet most American piece of clothing in history. But far-flung fame would have to wait until they lined up their workforce. There was no factory. Instead, Jacob made the patterns and cut the cloth and almost certainly sent it out to seamstresses around San Francisco to sew and add rivets and buttons. Brown duck cloth was the most common fabric, initially, probably because it hid dirt. But demand for the blue version soon overlapped it—denim, made by the Amoskeag Mill in Manchester, New Hampshire, about forty miles up the Merrimack from Lowell.

Like hard-worn jeans, the name of the fabric itself, denim, took its fair share of abuse from nature—human, in this case. For some reason people go out of their way to scramble the foreign names of things as if it's a patriotic calling. Denim, in every sense a classic example, started out in the sixteenth century as a silk and wool blend, "serge de Nîmes," a twill fabric produced in Nîmes, France. It had nothing in common with the all-cotton fabric now known as denim except that serge too was woven on the diagonal. That might have been all the connection necessary for early English and American weavers, intent on improving the pedigree of their cotton cloth in order to boost its market value.

Gossypium hirsutum—upland cotton—in all its budding glory, from a nineteenth-century botanical. One cotton plant can produce twenty or more pods, or bolls.
(*Special Collections, National Agricultural Library*)

Detail from a Diego Rivera mural at the National Palace in Mexico portrays the various complex pre-Columbian processes in textile production.
(Courtesy of James B. Kiracofe)

Elegantly simple and efficient, James Hargreaves' multiple-spindle jenny almost cost him his life. Turning its handle, one operative could do the work of sixteen hand-spinners, endangering their livelihoods. *(The British Library)*

Arkwright's original Cromford cotton mill, at right, showing falls that powered the waterwheel. Before being transformed into a museum in the 1970s, it was being used to store toxic paint.
(Courtesy of Letterspace Origination)

Sir Richard Arkwright, posing with a waterframe component in the late eighteenth century. As gluttonous, self-satisfied, and surpassingly vain as he appears to be in this official portrait, he also displayed a rare generosity in providing for his mill workers.
(Courtesy of C.R.F. Arkwright and the Paul Mellon Centre Photo Archive)

Eli Whitney in a rare portrait as a younger man, a few years after he invented the cotton gin in 1793.

(The New Haven Colony Historical Society)

Below: A model of Eli Whitney's original 1793 cotton gin, at the Smithsonian Institute. Hooks pull cotton fiber through slots that dislodge it from attached seeds as the crank is turned and drum rotates.

(Smithsonian Institution)

Francis Cabot Lowell in silhouette. The only known likeness, probably drawn during his lifetime, hints at his delicate constitution. Lowell died young, at forty-two, in 1817. A few years later, the city named after him was founded by his brother-in-law and associates.
(Collection of the Boston Athenaeum)

Lowell's tidy, seminary-like boardinghouses line a road that leads to the Merrimack Mill.
(Lowell Historical Society)

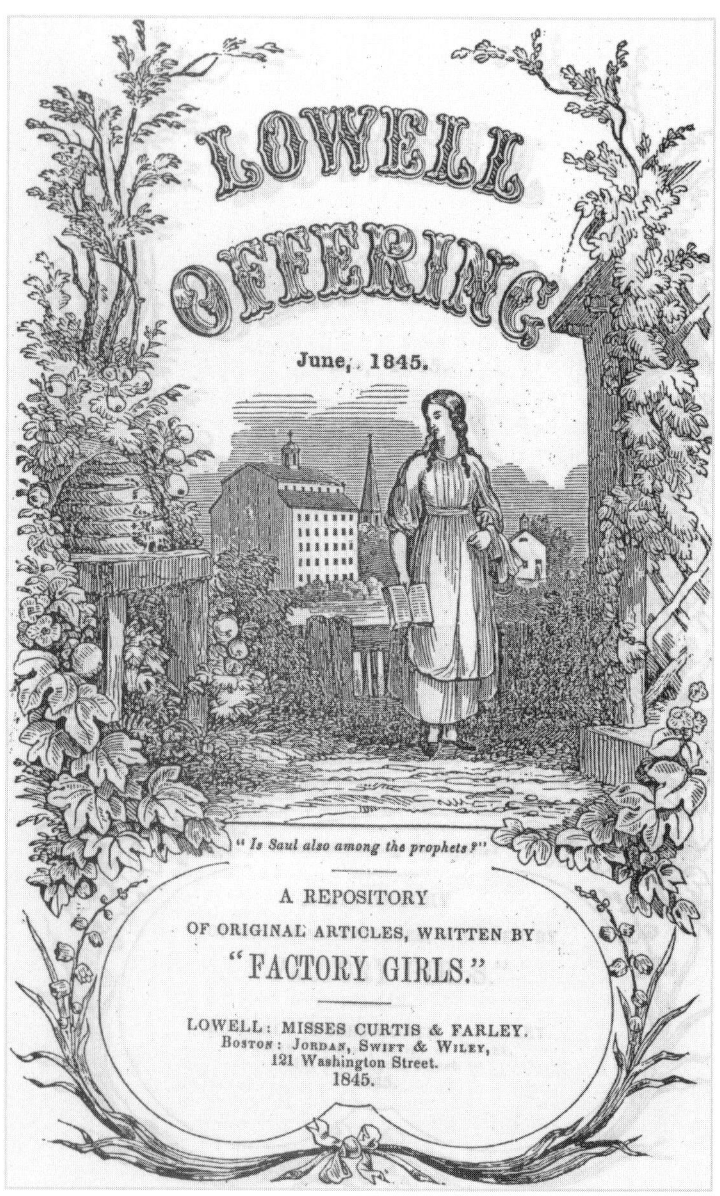

The working-girl magazine that infatuated Charles Dickens and brought international acclaim to Lowell's mill women, who wrote and produced it. *(Lowell Historical Society)*

These two immigrants, holding flying shuttles, were among thousands who found work for low wages in Massachusetts mills before the industry moved south.
(American Textile History Museum, Lowell)

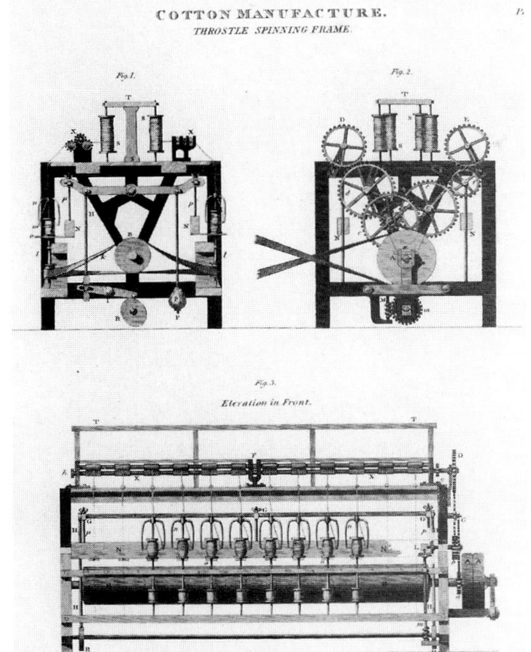

By 1850, complex machinery like this sophisticated water-powered spinning frame, seen in three views, engaged gears, pulleys, and cams to spin and twist cotton yarn at lightning speed. For all its improvements, the device owed its basic operation entirely to Arkwright and Kay's water frame of 1768.
(Kress Collection, Baker Library, Harvard Business School)

A Civil War engraving that deftly captures the dual nature of cotton: as pure as virgin snow in the field, and as satanic as the Devil himself in stirring up hostilities between North and South.

(Lowell Historical Society)

One steamboat could hold as many as 9,500 bales of cotton, weighing over four million pounds. Cotton-laden steamboats jammed the Mississippi River from Memphis to New Orleans before and after the Civil War.

(The Historic New Orleans Collection)

Men opening cotton bales at White Oak Mill, Greensboro, North Carolina. This punishing work was one of the few cotton mill jobs open to blacks.
(American Textile History Museum, Lowell)

The Tennessee Hagar Cotton Gin in the early 1900s. Cleaning and ginning cotton, blacks and whites by then routinely worked alongside one another—in marked contrast to the past. Horses or mules often supplied the power.

(Courtesy of the Albert Gore Research Center, Middle Tennessee State University, Walter King Hoover Collection)

Note the conflicting emotions and the wariness in the eyes of this Southern mill girl in Lewis Hine's haunting image from the early 1900s. Children often worked fifteen hours or more daily.
(The J. Paul Getty Museum, Los Angeles)

Black sharecroppers in Dallas cotton field, 1907. Living and working conditions for them were only marginally better than during slavery.
(Center for American History, University of Texas at Austin)

Above: The quarter-inch-long boll weevil, in all its evil glory. Note the long boring snout, used to drill holes in cotton bolls for egg-laying. Once infested, bolls produced no usable cotton fiber.
(Alton N. Sparks Jr.)

Left: If you can't beat it, salute it. The downtown Enterprise, Alabama, monument erected in 1919 that honors the boll weevil, so locally destructive it forced farmers to plant peanuts instead of cotton. Much to their surprise, they prospered.
(Special Collections, National Agricultural Library)

Mahatma Gandhi, as portrayed by Margaret Bourke-White in her classic photograph, "The Spinner." Home cotton-spinning for Gandhi was both a personal obsession and his political symbol for Indian independence.
(Margaret Bourke-White/Getty Images)

Levis come in from the range to clothe tenderfoots. In this 1930s ad, they've been taught city manners. But note the brand-savvy message, too: "The Rivets Are Still There."

(Levi Strauss & Co. Archives, San Francisco)

The USDA heavily promoted commercial cotton during the Depression as part of its directive to support impoverished cotton farmers.

(Special Collections, National Agricultural Library)

Surly but vulnerable, too, down deep. That was James Dean, circa 1955, the quintessential rebel in jeans. Note the thumbs curled into the front pockets, the cocked head. He made denim dangerous.
(Floyd McCarty/MPTV.net)

Brooke Shields for Calvin Klein: "Nothing comes between me and my Calvins." Sorry guys, that includes you, too. One of the most successful clothing ad campaigns ever devised. *(Calvin Klein, Inc.)*

A modern spindle cotton harvester at work. Mechanical harvesters span up to twelve narrow rows at a time. Their introduction after World War II drove masses of suddenly unemployed Southern black cotton-pickers to leave home and migrate north to find work.

(ARS Special Collections. Photographer: David Nance)

They had little use for the French, perhaps, but not for their fancy-sounding language, particularly when they could swipe it for free and charge more for its use. Crossing the English Channel, "serge" got tossed overboard, two words were compressed into one, and denim emerged as a stowaway from parts unknown.

Then along came jean. Since the seventeenth century the English had been importing boatloads of this durable cloth, a fustian blend of cotton and wool or linen, favored for its durability by the sailors of Genoa. The French referred to both the sailors and their trousers as "Génois," a backhanded acknowledgment of their Italian derivation, and decided to spell the fabric their way, jean. Heavily imported into England, it was woven all of the same color, while denim made its reputation as a weave of one dyed for every white thread. Most probably in the late eighteenth century, when Arkwright's spinning frame succeeded in creating a coarse but strong cotton warp, the mills of Lancashire first began turning out reams of 100 percent cotton jean, as well as cotton denim. The two coexisted but never met in the same fabric. Although jean cloth came in a variety of colors, it was usually dyed in indigo. From a distance, until you saw denim's variegated landscape of alternating blue and white pipelines, the two could easily be confused. Together they crossed the Atlantic to colonial Massachusetts. Shortly after becoming the nation's first president, George Washington paid a visit to a factory that produced both fabrics. He wasn't a tourist. Washington was trying to bolster an independent economy for a new nation that had chased away its foster parent and was now struggling to become self-sufficient. King George III, not the sharpest tack in the pack, had banned milling technology in the colonies. The *Salem Mercury* in 1789 reported that Washington was shown warp and weft spindles in action, and looms weaving "cotton denim, thickset, corduroys, [and] velveret."

As Alice Harris, an acute observer, notes, two centuries before Jimmy Carter christened jeans fit for the chief executive's wardrobe, the country's founding father gave them his vote: "In short," he wrote in his diary, "the whole seemed perfect, and the cotton stuffs which they turn out, excellent of their kind."

Reaching the United States, basic cotton denim went out into the bush as "waist overalls," in the wagons of peddlers like Levi Strauss, while jean traveled uptown in frockcoats and vests and trousers, often available from high-end manufacturers in chestnut, olive, or white as well as blue. The fabrics were not destined to go down in history literally joined at the hip—not until Davis came along with his promise that he and Strauss could "make a very large amount of money." Their fabric-and-metal brown duck pants quickly gave way to the much more popular blue denim variety, fitted with suspender and fly buttons. Competitors made and sold riveted bib overalls as soon as the Strauss-Davis patent expired in 1891, but none were accompanied by Levi's marketing savvy.

Quite soon in everyday conversation any technical distinction between denim and jean cloth became meaningless. Your denims were jeans, your jeans were blue, so you wore blue jeans and that was that. Strauss quickly realized that pants were more comfortable than overalls, also that people formed a curiously intimate relationship with their denims, much like the affection they developed for a pet. There was a reason. As blue jeans become more pliable through laundering, they caressed areas of the body they initially clutched. Of the millions of words lavished on denims since their appearance, the Italian philosopher Umberto Eco best summoned up their distinctive character, at least for males: "A garment that squeezes the testicles makes a man think differently."

He might have added that as the garment relaxes its grip, it dramatically increases your appreciation for it. Clearly "waist overalls" was a clunky name for such an intimate possession that often treated you more fondly than your spouse. The term soon disappeared, and the world still sighs in relief. Imagine Elton John wrapping his tonsils around "Waist overall baby / L.A. Lady /

Seamstress for the band." Or Brooke Shields purring for Calvin Klein, "Nothing gets between me and my waist overalls"? No.

Levi and Jacob did the hard work; the rest was left to the vagaries of common language usage. For reasons that remain uncertain, jean cloth eventually lent its name if not its character to denim pants, and waist overalls emerged in street lingo as blue jeans. At an intuitive level it all makes sense: waist overalls belong on brick layers who go to heaven, whereas blue jeans, like bad girls, go everywhere. It was in the destiny of these humble work pants a century after their birth to become the battle uniform of successive generations of beautiful losers, otherwise known as teenagers, as they mounted their desperate commando campaigns to scale the walls of adulthood and carried their crusades to the far corners of the globe. Other cool gear on store racks became ugly eyesores when hung off their knobby hip bones. Nothing fit exactly during those gangly years—except, of course, jeans. Of such forgiveness is eternal devotion born. But that was only their surface appeal; jeans could also express attitude in a flash with nary a word spoken. A thumb hooked through a belt loop could signal defiance at one extreme and a craving for shameless frolic at the other. Jeans created their own poetry of lust. You had only to be of a certain age with an unspecified hunger and yearning to recognize and respond to it. Denims confirmed the basic tenet of enlightenment, which is that we are all fundamentally connected, or at least that our fundamental equipment is connected by a universal fabric that packages it for the world to enjoy without regard to race, religion, nationality, or social status.

While it's anyone's guess what Strauss would have made of the long, strange trip of jeans from rock pile to rock concert, he clearly understood that astute marketing is the engine that drives the rag business. If the tipping point for blue jeans was still a few years off, the creative ingenuity of Strauss took form as early as 1886, when he devised the famous leather patch that displayed two horses trying in vain to pull apart a pair of "Levi Strauss & Co. Copper Riveted Every Garment Guaranteed" jeans. That rear waistband emblem on all of his 501s planted an indelible message in his customers' minds. Every time they slipped into their Levi's they caught sight of those rippling animals, both under whip and in midlunge, clearly so provoked they could easily split a giant oak yet held firmly in check by . . . denim work overalls just like these.

By 1900 Levi was pitching his 501s—named for the lot number assigned to the top of the line—as overalls that were "known the world over." That world extended from San Francisco barely past Sacramento, maybe another few hundred miles east to Reno at best, but no matter. A masterful entrepreneur, Strauss swatted away factual accuracy like a bothersome bug when it blurred his vision of world conquest.

Born to Work

At Levi's death in 1902 at the age of seventy-three, he left an estate valued at roughly $6 million in that era's dollars. By then a leading San Francisco philanthropist, he gave a sizeable portion of his estate to civic institutions. The majority went to his relatives, including four nephews who ran the company. A few years later the 1906 earthquake leveled the city and came close to shutting down their business. Fire engulfed the company's headquarters, destroying all of Levi's personal and business papers. That same year, Jacob Davis sold his shares back to the company. One of Strauss's nephews, Sigmund Stern, stepped in to take control and managed with considerable skill to keep it afloat during the hard times that followed the earthquake and fire, which required LS&Co. to rebuild from scratch. In 1919 Stern asked his son-in-law, Walter Haas, to join the resurrected firm—a shrewd move, as it turned out. Haas understood the power of advertising. Long before "branding" became a buzzword, he branded Levi's to set it apart from its competitors. Trading on the allure of the rugged lonesome cowboy, he linked the public's perception of his company and its products to the enduring romance of the Old West while competitors' denims were mostly targeted to the workplace in the form of bib overalls, jackets (called blouses), and utilitarian pants.

H. D. Lee in the Midwest and Wrangler, descended from Blue Bell Globe in the Southeast, expanded their territories rapidly—Levi's were sold only in eleven Western states until 1948—and they made basic improvements. In 1913 the Lee Company out of Salinas, Kansas, introduced the Lee Union-All. Legend has it that Lee, an oil millionaire, got tired of buying new clothes for his chauffeur and instead sewed together bits and pieces of dungarees into a one-piece chauffeur uniform that became the prototype for the Union-All. Another story makes the hero of the tale a mechanic who wrote to Lee's company suggesting that one-piece protective coveralls made more sense for many kinds of grimy work than separate pants and jackets. Whatever its genesis, "Union-All" was a brilliant name choice. It not only connected the product to the United Garment Workers of America who worked in Lee's factories, but it also helped create a broader identity for Lee's as the signature fabric of laborers everywhere. Lee's denims of any type became the uniform of the proletariat. Other companies picked up the theme, and "union-made" was soon de rigueur on all work wear. As the country prepared to enter World War I, the U.S. Army decided to adopt the Union-All for military use, and denim indigo transformed into the true blue of patriots. Lee's motto, "The jeans that built America," reflected that. About a decade later Lee made a different kind of contribution to a grateful nation, summarized in the company's official history:

> 1926—Three words: Jeans with zippers. To fully appreciate this development, speak with a man standing in an

Iowa cornfield in late December, his gloved hands desperately fumbling at his buttoned fly.

Soon after that, women, a new and burgeoning customer base for jeans, were offered side zippers. By now the large department store chains like Montgomery Ward and JCPenney were producing their own private-label brands. Ben Davis, Jacob's grandson, manufactured a popular line of blue-collar work clothes, "Union Made. Plenty Tough." Most consumers didn't know or care much that all of the denim they were buying came from the same few Southern mills, primarily Cone, once Amoskeag closed its doors on the last day of 1935. It was ring-spun cotton twill, tough and available in weights that ranged from 9 to 14 ounces—denim's heaviness is traditionally measured in ounces per yard at a 29-inch width, the span of an older loom——and dyed with synthetic indigo. During the '20s and '30s, as their popularity spread, blue jeans showed up regularly in the *Saturday Evening Post* and *Life* magazine, on aircraft carriers, navy crews, skyscraper workers—anywhere a he-man could sit down to eat a sandwich suspended on a beam thirty stories above ground or perch atop a massive tree trunk he'd just felled with his brawny arms and slender axe. The message was as clear as the eyes of any loyal citizen: do your part too by buying U.S. denim from U.S. cotton manufacturers, and put it to work to keep our country strong and safe from subversives.

Jeans made us proud. Wherever and whenever they were asked to perform, they never seemed to show stress or fail to rise to the challenge. In 1948, when hikers discovered a pair of Levi's 501s from 1890 inside the Calico Silver Mine in the Mojave Desert, they were still so sturdy that the woman who found them stitched them up and wore them for a time before selling them back for $25 to LS&Co. An earlier pair of Levi's from the 1880s was also bought back by the company in 2001—for $45,000 in an eBay auction—boxier in their cut, with a few tears and holes but otherwise suitable for wear. Many of these vintage jeans in fact were found in abandoned mines, because miners would sometimes stuff a pair into an air space as a makeshift wall plug. Often the pants outlasted the mines, and almost always the miners.

All that might have been expected. A garment originally designed to stand up to hard labor did its job as advertised. But less predictable were the rogue impulses that jeans were now stirring: they invited you to fantasize about the bodies they enclosed. Several decades into the twentieth century a nation turned its work-weary eyes to the silver screen, where jeans showed up on hunky movie stars and made cotton sexy for the first time in modern history. Good-bye preacher, hello sin!

From Ranch to Raunch

Gunslingers, snake-eyed varmints, low-down horse rustlers, and lily-livered scumsuckers bit the dust when John Wayne pulled out his six-shooter and started fanning that trigger. Teenaged boys and older men in the audience of those early Depression-era cowboy movies didn't stand much of a chance against him either. Once the film ended and the house lights came back up, their gals eyed them in unsettling ways: How come you're not six feet four inches tall with a jaw that juts out from here to the Rockies like Big John Wayne, and how come you don't stride across a room in two steps like he does? they seemed to be asking. How come you can't ride a horse at full gallop with the reins in your teeth and swoop a woman up like a sack of sweet potatoes with one arm? Who'm I with, some kinda namby-pamby?

There was disappointment in their eyes, and you knew it. Maybe in your whole life you never got closer to a horse than the local merry-go-round, but at least you could wear the same pants as Sheriff Wayne, even if you couldn't exactly fill them. Wayne, of course, wore jeans—just about all screen cowboy heroes wore jeans except the ones who looked like ballerinas in those tasseled outfits they sometimes were forced to parade around in. Maybe jeans weren't only meant for grease monkeys, then. Maybe they could even get you some hot action with the ladies if you shrank them to fit and wore them just right, kind of sloped at the hips, Wayne-style.

Or if Wayne was too rough-and-tumble for a cozy date, there were the Singing Cowboys, those guitar-strumming yodelers who also clad themselves in crisp jeans, turned up at least six inches at the cuff. It hardly mattered that Roy Rogers, Gene Autry, Rex Allen, and all the rest were about as sexless as linoleum, or that they saved their displays of true affection for their horses. They were all celebrating the passing of the real cowboy into legend, catching wisps of a vanishing wild frontier and the lawless libido that went with it. One company in particular—Levi Strauss—was strategically located to benefit from that connection. It was still headquartered in San Francisco, close to Hollywood. Production studios summoned LS&Co. to dress horse-riding heroes. Levi's showed up on movie screens all across the country, on Wayne and other sagebrush icons, long before the pants themselves crossed the Continental Divide. When they did, a decade later, they found an eager and ready young audience. By then Levi's jeans had become folklore celebrities, complete with their own fan clubs and adoring throngs.

In 1944 a man named Sam Neslin wrote to the company from Klamath Falls, Oregon, telling about a worker who was admitted a few nights earlier into a local hospital with a badly swollen and broken leg. The nurse told him she would cut off his trousers to expose the leg, not pull them off, in order to save him pain. "The man really blew his top," Neslin explained. "He said, 'Nothing doing, these pants are Levi's, they are too damned hard to get, you ain't cutting off this pair of Levi's, I don't care how much it hurts, pull them off.' " She did; he howled but saved his worn and dirty pants. Neslin decided "this man would rather suffer the additional pain than to have the pain in his heart that his Levi's were cut up and not usable. . . ."

One area of contention as the pants moved east was the damage done to furniture by the rivets and exposed buttons. Until they were covered in the late '30s, the company received a steady stream of written complaints about scratches and holes punctured in upholstery and horse saddles. "I want to thank you many many times for making them this new way," one women with a house full of marred chairs and shredded cushions wrote. Folks knew enough to cut Levi's some slack. It was a company that listened and responded: "Concealed copper rivets" proved it. Just in case anyone got nervous, the same pants tag announced, "The rivet's still there."

All except one, that is. Earlier, Walter Haas had made one important design change. In his eagerness to build pants tough enough to smother a bobcat, Jacob Davis had hammered rivets wherever he saw the possibility of damage through wear and tear, and that included the base of the fly. Cowpunchers sitting cross-legged and exposed around the campfire quickly learned that when metal heats up, it gets mighty hot in a hurry. But nothing changed until Haas himself went on a camping trip, moseyed on over to the fire, and took a seat. Soon enough, his burning crotch rivet sent him howling off into the night, and after that it disappeared from all models. The Haas family still owns the privately held firm of Levi Strauss.

Few knew or cared that two decades earlier cash-strapped LS&Co., pressured by competition, had almost tanked. Against conventional wisdom and at considerable expense, Walter Haas had continued to advertise heavily even as the company sputtered downward. He picked up on Levi's original marketing concept, which was to differentiate his denims as the best-made brand. Haas decided to attach a red Levi's tab to every pair, the first outside label to appear on American apparel. The stitched seagull wing on the rear pockets—technically, an arcuate—marked them as singular and special. The implied message was that any other blue jeans were inauthentic. Quality was Haas's selling point, but in truth John Wayne and Gary Cooper were his greatest unpaid salesmen. Together they turned around a failing enterprise.

Women showed up wearing them on screen, too, as leisurewear in urbane '30s comedies. When Clare Boothe Luce's wickedly catty play *The Women* was adapted for the screen in 1939, audiences discovered a group of sophisticated East Coast divorcees-in-waiting holed up at a ranch in Nevada (a state where quick divorces were granted), all decked out in narrow-waisted, broad-hipped "Lady Levi's," western denim chic. Their jeans and jackets never looked more at home than in the company of a silver martini shaker. By then dude ranches had become a common source for denim wear. New Yorkers and other city folk would head to Nevada or Arizona for vacation and come back loaded up with Levi's and H. D. Lees for themselves, family, and friends. Being hard to come by, they were all the more desirable. Even better, they demanded special attention and required rituals. If you took the time to step into a bathtub or jump into a body of water wearing a new pair a size or more too big and then let them dry on you, they'd return the favor by customizing their shape to flatter your womanly curves or hug your masculine thighs. You probably were never told, but that really wasn't necessary after 1933, when a chemist named Sanford Cluett invented a process to preshrink cotton fabrics, including jeans, completely and permanently. That's why preshrunk fabrics have been sanforized (easier on the tongue than *cluettized*) before you buy them. Somehow jeans from all the big makers ignored that technical advance for years, until the '60s, when prewashing also became common and reduced much of the stiffness associated with new jeans.

What Hollywood started, Airstreams finished. As America's roads filled with people on the move in the '30s, whether by motor home, car, or truck, regional brands of jeans migrated into new territories. They were fast becoming democracy's showcase garment, at home where upper-crust swells gathered to sip cocktails at a clambake or where down-home folk danced the hoochie-koochie around a barbeque pit. They went to battle again in World War II, and this time denim coveralls clothed Rosie the Riveter at home as well—all those women whose factory jobs helped support the troops. Under orders from the War Production Board, Levi Strauss removed the stitched-wing arcuate from rear pockets to save wasteful threads. A pretend-stitch wing was painted on instead. Still Haas refused to bow to pressure and would not reduce the denim weight of his 501s. Cosmetic changes were one thing, maintaining quality was another.

For their minor sacrifices, blue jeans companies were awarded with one of the greatest public relations coups of the century. Since they clothed the military out of uniform, blue jeans accompanied American soldiers wherever they traveled—and in a World War, that's just about everywhere. No one had to explain their cool quotient to anyone under twenty-five. In Asia, Russia, and all of Europe, they stirred the deepest yearnings for apparel that expressed attitude. Your blue jeans, soldier, what you want for them? I trade you my family's hand-knotted Bokhara heirloom rug, my putt-putt motorbike, a grandmother's antique silver necklace perfect for your girl back home, a case of Polish buffalo grass vodka, very rare, anything you see, price no object, soldier. . . .

Jeans became a prime currency of the postwar black-market barter. Levi's ruled; they had movie-star status and a reputation for surviving hard livin' with style. They roared out of the west across the Mississippi after the war ended, following an advance party of fiercely loyal evangelizing consumers. The G.I Bill enabled all WWII veterans to attend college; farm boys in their jeans now sat in the same classrooms as suburban sons and daughters, and their curious wardrobe—jeans and overalls—became the newest fashion trends. Middle-class youth is always eager to adopt the dress of the underclass, sociologists assure us. In this instance they got it right. Soon Levi's owned the nation's young and restless. A waist-overall born to work in was quickly growing up to become the rites-of-passage pants to leave home in.

First, though, it had to define itself as outlaw garb, the renegade duds of the ultra-cool sociopath. Beat writer Jack Kerouac took them on the road and into the unexplored wilderness of improvised jazz and midnight poetry readings. Soon Marlon Brando and James Dean stepped up to add a crucial seasoning of danger to the stew. Their jeans did most of the talking; often the pants were easier to understand than the tongue-challenged method actors inside them. Brando showed up in 1954 in *The Wild One* on a bad-ass Moto Guzzi motorcycle, in a wing collar leather jacket with zippers everywhere and "Johnny" scrolled in bold script on one shoulder in case he forgot his own name. He wore a busman's cap tilted on a rakish angle and button-front Levi's turned up at the bottom. To teenaged girls he looked like a ducktail dream, off-limits and irresistible. To boys he held out the possibility that juvenile delinquency was a reasonable career choice.

Then arrived James Dean in 1955 in *Rebel Without a Cause*, as a mumbling misfit breaking out with a bad case of angst. Like Brando he was handsome beyond reason yet with an attitude that didn't alienate or threaten guys. In *Rebel* he wears the black trousers of a dorky student until his first brawl with the high school toughs who taunt him. After that he shows up in jeans, a tee-shirt and a red windbreaker, primed to kick butt. James Dean sulked and smoked and fought his way through three movies dressed only in denim and then totaled his Porsche before he was asked to change into something more appropriate for company. He became an instant icon and left behind a dress code for raging adolescents of all ages: blue jeans were the thing you wore to live life in the fast lane.

And to make sure female teens had their very own off-the-shoulder role model, Marilyn Monroe showed up in *River of No*

Return in 1954 as a blowsy saloon singer wearing a camisole and blue jeans and little else. Girls got the message: in the same jeans you could be a 4-H Club farm girl by day and a wholesome tart by night. A short time later The King arrived from Tupolo, Mississippi, and set jeans on fire. Elvis wriggled his hips and girls screeched, which prompted Ed Sullivan to censor his writhing crotch on national television. Still, it was too late; we were all wild as a bug. "There's only one cure for this body of mine / That's to have that girl that I love so fine!" In *Jailhouse Rock* Presley grabbed Judy Tyler and planted a big one on her lips. "How dare you think such tactics would work with me!" she protests. "They ain't tactics, honey," he tells her with a curled-lip smile. "It's just the beast in me." Most of us guys didn't know we *had* a beast until Elvis and the gang dropped by in blue jeans to show us how to satisfy its appetite at feeding time.

From Elvis forward the future of blue jeans in America and in the world at large looked so bright you had to wear shades. "About 90 percent of American youths wear jeans everywhere except in bed and in church," a 1958 syndicated newspaper column announced. By now Wrangler had targeted and won over working cowboys as the functional jeans, cut just right.

John Kelly, who's been raising horses most of his life in northwest Colorado, explained in 2003: "Rodeo cowboys only wear one kind of jeans, and they're Wranglers. They don't wear Levi Strauss. Levi's are more of a fashion statement. They're also a little bit too tight in some of the wrong places, and they break down. You'll see kids walking down the street in San Francisco and their butt will be patched up and it'll be frayed and the knees will be frayed. Those are Levi's. Cowboys don't like having holes in the knees and a hole in your butt. They're very particular about the way they look. Wranglers are just much tougher."

Kelly wears jeans "365 days a year," he says. "You don't have to think about 'What am I going to wear today.' That's one of the best things about them. It's really simplicity . . . like your Mao suit—remember when the Chinese all dressed the same way? I just dress the same way, and all my friends just know that's what you wear. There hasn't been anything that's been produced that's as good an idea. And it's cotton, so it's really comfortable."

Fierce loyalties to individual brands or styles occasionally led to barroom brawls. That was to be expected. You had to be ready to defend your pants with your life. Behind the Iron Curtain that took on a different, more literal meaning. Jeans, rare and expensive—a month's salary on the black market—came to represent Western capitalist decadence to the Kremlin Politburo, which banned their import. To Soviet bloc youths, they were a symbol of status and liberation; also, they fit, which was not a feature of clothing made in Russian factories. They ignited a lust for raucous pulsating rock music and the squeal of fast cars. Long live decadence! Jeans were the beginning of the end for the old hard-line regime. You could probably graph the rise of their popularity in Communist countries with a parallel descent of politburo support. Ronald Reagan called the Soviet Union an evil empire. Almost: to a new generation of Communist youth it was an evilly dressed bunch of old coots in suits cut like cardboard boxes from the empire of bad tailoring.

Ivan, a teenager in southern Russia in the '80s, remembers: "It was hard to get. Out of ten kids, only two can buy jeans. But whatever they do over here they want to match over there. They thought America was the best country in the world. Was every kid's dream to come live in America." Communist youth refused to choke on any more toxic waste emanating from the Kremlin's gasbags. Reagan and the last Soviet president, Mikhail Gorbachev, took justifiable credit for toppling the Berlin Wall in 1989, but wise observers also understood that jeans outfitted and emboldened the frontline commandos. There they perch in news photos ecstatically celebrating victory on top of the crumbling Wall, denim-clad young men and women wrapped in a symbol of freedom as powerful as any national flag.

Thanks, Ben

There was another revolution going on during this era, much closer to home and much more intimate. Skin-close, in fact. By the mid-'70s, polyester miracle fiber, child of '50s, had gained significant market share against cotton in the United States. Wrinkle-free double-knits were the proud test-tube babies of fabrics made from the by-products of crude oil. Polyesters were huge. They fit the needs of the Me Generation, which had no time to iron the belongings of Significant Others. Ironing chores kept us from going with the flow, sharing, and giving our mellow selves the space to come from our fully self-actualized potential. The Maharishi endowed us with our own personal mantra; our public one came from DuPont: "Better Things for Better Living Through Chemistry."

Into this wash-and-wear, permanent-press universe denim jeans wandered like a missionary sent to save the soul of a fallen outcast. Initially the nation's large cotton producers had reacted to synthetics by frantically scrambling to reassure fleeing consumers that cotton was "the fiber you can trust." They discovered the only thing about it consumers trusted was its tendency to shrink and wrinkle. Between 1960 and 1971 cotton's share of the American apparel market slipped from 66 percent to 34 percent, and for the next few years it continued to fall at a fast clip.

In time a new marketing arm funded by growers, Cotton Inc., began an aggressive media campaign that encouraged wayward consumers to "come home to cotton"—an inspired effort to remind them that as opposed to polyester, cotton breathed and did not reek like a sweaty sneaker. But the grower's true allies turned out to be General Westmoreland, LBJ, and the Pentagon. They promised a quick end to the Vietnam War and cooked the books to support their delusions. Outraged American youth took to the streets to protest against dishonesty in all forms. Natural cotton jeans—honest, genuine—quickly became the battle fatigues of the counterculture. The flower-power children of the late '60s and early '70s adopted them as their unofficial uniform of dissent. During this period the film *The Graduate* opened and became a smash hit. In its most quoted scene, young Ben Braddock (played by Dustin Hoffman), befuddled recent college grad, is pulled aside by a pudgy friend of his father who advises him with solemn authority that the future is in "plastics." That did it. In one word the film captured an entire generation's repulsion for everything it considered to be superficial, bogus, hypocritical, and contrary to its generation's interests and concerns. "Plastics" became shorthand street lingo for what you were barricading the dean's office to protest against. You could spot the enemy, those deceitful political warmongers, by their wrinkle-free, drip-dry threads as surely as the Minutemen of Concord spotted Redcoats.

Cotton Inc.'s advertising agency, Ogilvy & Mather, jumped on this opportunity and made the most of it. In one TV ad a hip disc jockey tells us, "You know, in a plastic world some things should never change—like blue jeans." You could almost hear his Patchouli-scented pass-the-roach audience nodding: "Right on! Outa sight!" By the early 1970s jeans were rapidly becoming a medium for personalized statements: walking down Haight Street in San Francisco or anywhere in New York's East Village you came across flared bell bottoms with Tibetan bells dangling from the seams, floral embroidery, astrological signs, and winking moon faces needle-worked onto rear pockets, metal eyelets and studs, attached feathers, carved scrimshaw hung from belt loops, multicolored peace signs, and even antiwar graffiti.

As for the jeans themselves, they were now beginning to show up in new styles, moderately adjusting to hippie trends but still a few years away from the next great revolution. Walking these same streets in 1969, an enterprising young retailing couple in San Francisco thought it glimpsed a retail opportunity. Donald and Doris Fisher came up with a name that seemed to perfectly reflect the yawning schism between anyone under thirty and the universe of adults. Journalists were calling it a generation gap. The Fishers simply shortened the name of their first small store on Ocean Avenue to Gap, filled it wall to wall with Levi's jeans, vinyl records, and eight-track cassettes, and with only a few employees, they opened for business. Donald and Doris were playing a hunch that today's protesters would become tomorrow's mainstream consumers and would remain loyal to a company that shared their values and offered crash-pad clothing at affordable prices. A good guess and an inspired business decision.

Don guessed right again about cotton a decade later, in the early '80s. Inspired by the romance of colonial Britannica conjured up by Rudyard Kipling, a young California couple, writer Mel and designer-artist Patricia Ziegler, had combined talents a few years earlier to create natural fiber clothing and travel gear that harked back to the look and feel of the far-flung British Empire. In their first cubbyhole retail outlet in Mill Valley, California, decorated with elephant tusks, pith helmets, and lush, towering tropical foliage, they sold surplus WWII cotton goods from warehouses in East London that were cleverly updated by Patricia. The Zieglers filled their store with primitive artifacts and jungle sounds and decided to call the company Banana Republic in tribute to their renegade spirits. The wittily written, handsomely illustrated apparel catalogues they produced soon became collectors' items. In them, celebrities like Doonsbury's Gary Trudeau and power-lawyer Melvin Belli reviewed the company's clothes. "Khaki," Mel declared, "is the denim of the eighties." Banished were all synthetic fibers. Cotton ruled. Customers used to sterile, predictable retail environments flocked to BR as if to a theme park. Fisher took note and soon became hopelessly smitten himself with the epaulets, Bushman's shirts, and many-pocketed photographers' travel vests that were now Banana Republic's signature styles. He muscled its acquisition through his board of directors, installed his two sons in key management positions, and faster than a speeding water buffalo, Banana Republic became a national phenomenon. Although the tusks, foliage, jungle calls, and playful whimsy soon disappeared, the khaki remained. By 2003, Gap Inc. had become the nation's leading apparel chain whose 4,250 stores grossed more than $13.6 billion. Chinos? Sure. But denim still rules.

Few of these customers, then as now, cared too much about the subtleties of jeans construction or the processes and materials that went into their choice of garments. Sales clerks were usually no help. The received wisdom was—and still is—that such details are of interest mainly to people in the trade. Fit is everything. Yet the enduring appeal of jeans probably owes as much or more to the way they are dyed and woven; also, no self-respecting ex-hippie from that era would take a manufacturer's word for anything without getting the real dope beyond the hype. The truth about jeans, like life itself, lies within.

Inside Jeans

Most all-cotton blue jeans age with us the way we'd all like to—gracefully, with style. "If we were to use a human term to describe a textile we might say that denim is an honest fabric—substantial, forthright, and unpretentious," *American Fabrics* magazine commented years ago. It is a fabric, too, that seems to mimic they way we prefer to see ourselves: tough enough for action on the outside, but gentle and welcoming within. There are reasons for that, having mostly to do with the complex interaction between indigo dye and cellulose cotton fiber. Like love, sooner or later it all comes down to the compatibility of molecules, or chemistry. Waxy cotton, which rejects practically all other dyes and needs to be strenuously manipulated to accept them, has a molecular affinity for indigo. That dye doesn't penetrate to the core of the fiber. Instead, its microscopic particles cluster around the core much like magnetized beads. With each successive laundering minute specks of indigo dye chip off and abrade. That gives the blue jeans a softer hand; as the density of the dye dissipates they also fade in color to light blue. You get to wear them in as you wear them out.

Indigo is a dye with a long and rich history all its own, one that profoundly affected the economic, religious, and social institutions of numerous countries including the United States. Indigo, which gets its name from the Greek word *Indikon,* meaning "from India," was originally extracted from the leaves of an Asian bush, *Indigofera tinctoria,* that grows profusely in Bengal and elsewhere. Ancient Egyptians, Guatemalans, and others reserved indigo blue for royalty, but it became significant in Europe only after Da Gama opened up trade routes. Its technical superiority—what gave it great commercial value—is its ability to dye both animal and plant fibers, and to remain colorfast.

The last thing you would expect looking at an indigo plant is that beneath its frilly green vegetation there lies a midnight-blue soul. To penetrate into the heart of that darkness, you need only two simple elements: air and water. In ancient India, the intricate process began when the plants were cut, bundled, and dropped into huge vats of water to start fermentation, producing a powerful telltale stench. Once the plants' leaves were saturated, the vats were continually stirred to introduce oxygen, the key element in transforming the colorless liquid to its deep blue hue. For several days, fermented scum was removed from the tops of the vats while a specialist regularly tasted the brew for a level of sweetness that indicated the fermentation was at an optimal stage. At that point the plants were removed, and the liquid drawn off to a clean tank, where girls and women stirred with sticks and paddled for hours, introducing more air until it turned yellow-brown and the blue particles that began to appear joined together and precipitated out to the bottom of the tank. Lime or soda was sometimes added to speed up the process. Finally, the indigo dye-stuff on the bottom was collected, dried for several months, and packaged into cakes, each bearing the factory's stamp, to be reactivated by dyers in water.

Marco Polo noted this process in his thirteenth-century journals. Five centuries later, when Levi Strauss was buying denim for his first waist-overalls, it had not changed in any significant way. A weaker blue dye, woad, was extensively grown and processed in Europe until sea trade opened with the East. Superior to woad in fastness and intensity, indigo became a valued import and triggered angry protests from local woad growers. They called it "Devil's Dye."

Inevitably, as the nineteenth century progressed, laboratory scientists—a new breed—were working away at trying to analyze the structure of this curious indigo plant so that its dye could be synthetically reproduced. One experiment built on another, and by the mid-nineteenth century it became clear that a coal-tar compound called aniline was also an oil produced in indigo: that proved to be the break-through discovery. Not long afterward the renowned German chemist Alfred von Bayer used aniline to create the first synthetic indigo, with properties identical to the prized by-product of the plant. It began to be produced by BASF in Germany in 1897 only four years after the Strauss-Davis patent was awarded. Within two decades, Indian natural indigo exports fell from 18,700 to a mere 1,000 tons. Today, practically all indigo in industrialized countries is synthetically manufactured.

But whether the dye is natural or not, indigo and jeans form a perfect union. Rarely if ever have a fabric—cotton—and dyestuff been more surely made for each other. To manufacture authentic denim, some of the cotton yarn that will be woven into it—the warp—is first dyed. About 300 to 400 strands are bathed in a series of indigo vats, then woven with the undyed horizontal weft, or filling yarn. The blue threads appear on the outside of the jeans, the natural strands—the whitish yarn you see when you roll up a cuff—appear on the inside.

Until the late 1960s these threads were created by a process called ring-spinning, a traditional time-intensive technique that delivers strong, soft threads with a "true twist" that give them a more distinctive look and feel. (Levi's continued to be ring-spun until the early '80s.) Denim aficionados—and there are many—refer to the unique surface characteristics of ring-spun fabric as slubs. Slubs are imperfections in texture, most commonly a slight endearing lumpiness that mark any pair of jeans as real and authentic. Ring-spun denims were reintroduced in the 1990s as "new vintage" jeans became popular. The alternative, open-end spinning, was adopted by textile mills in the 1970s as a cost-cutting method that eliminated a few middle stages in the process. Open-end fibers are "mock-twisted"—blown together—and look and feel bulkier; their weave is coarser and they wear out more quickly. While some are spun with "faux ring" attachments that create more surface character, open-end is to ring-spun denim as military music is to Mozart.

Then, too, there are various denim weaves to consider: right-hand, left-hand, and broken twill. Right-hand twill is the traditional weave that Levi Strauss first bought and sold. The diagonal yarns descend from right to left like the spine of a "Z" and are twisted in a way that tightens them and makes them more compact; most jeans are still woven this way. Left-hand twill (favored by H. D. Lee) opens the yarn and produces softer denim after washing. Broken twill, whose diagonal line does not run straight from top to bottom, was developed to accommodate the needs of cowboys; in the early days of crudely woven twill the legs of their jeans would often spiral out of shape, and Cone Mills solved the problem by staggering the line. Once aware of these differences, a jean's twill pattern is easy to spot.

Those considerations and style issues aside, there is not a substantial difference between the quality of $24 Wal-Mart and $240 Diesel jeans—or any comparable competitors at either end. All are made from inexpensive varieties of medium (and short) staple cotton. Stefano Aldighieri, former director of fabric and finishing at Levi Strauss, estimates the quality difference in fabric and dyes to be "about 25 percent or so, at most." An independent testing laboratory in England in 2000 performed a variety of tests to measure the strength, durability, and quality of chain-brand jeans against Guess, Diesel, Levi Strauss, Next, and other top labels. There was only one notable weakness among chain-brand jeans: they were more likely to fade after twenty washings. In all other respects they outperformed at least one of the expensive boutique design products. Some high-end designer jeans—Tommy Hilfiger and Calvin Klein among them—make jeans with inferior open-end spun fabric. Even so, there has been much concern voiced recently about whether Levi's will compromise its standards to meet the low wholesale price demanded by Wal-Mart, now that the company has agreed to sell its jeans through that powerful retail outlet.

Andrew Olah, who has been selling denim from mills around the world to Gap and other manufacturers for more than twenty years, knows twill. "Don't think fabric alone," he advises. "Think small details of the finished product. Think wine. Think hamburger. You can buy a McDonald's or a delicious $15 New York restaurant version. They won't be prepared the same." Specialized finishing techniques—sanding, whiskering, bleaching, and so forth, add cost. So too do denims manufactured by top factories like Kurabo in Japan, generally considered to be the world's best denim fabric maker. Olah explains that a textile factory's yarn blends of raw cotton from different countries, each with its own characteristics, are their most closely guarded secrets. He also knows that in the United States denim garment buyers "get the same stuff from the same factories. I know where they all do their business." He won't argue that the retail price truly reflects quality differences, but there's still *something* about a pair of high-end jeans—intangible maybe, but worth more. Olah seems to be torn between heart and mind, which is only appropriate—ever since the mid-'70s, when jeans jumped the barricades of social protest and landed in the designer showrooms of Calvin Klein, their appeal has triggered impulses and desires not easily explained away by pure reason: Andrew's, yours, or mine.

Fashioning the Future

Designer jeans were either a brilliant or absurd idea waiting to happen, depending on whether you chase cattle for a living or drop beats as a hip-hop DJ. Possibly they were both—absurdly brilliant—and unquestionably synched to the social and political rhythms of the times. Jeans had breathed oxygen into the collapsed carcass of the cotton apparel industry, but in 1975 the Vietnam War ended and with it the common cause that united America's antiwar youth. Cotton's market share was hovering around 33 percent, a historic low. Hordes of disenfranchised suburban and urban twentysomethings on their way to becoming middle-aged Baby Boomers were caught in midtransit. Suddenly you were too old to walk around in the same jeans you crashed in for a week at the commune's pad, and not about to don wash-and-wear. Also because you were perpetually horny you looked for jeans that advertised your wares; that was now easier to do as their cut subtly changed to accentuate the sensual for both genders. They'd gained hot-celebrity status when Dennis Hopper and Peter Fonda dressed themselves in 501s and slimline Truckee jackets and drove their hogs cross-country in *Easy Rider*. They'd made the cover of the Rolling Stones' *Sticky Fingers* in real fabric complete with a real fly zipper. All that was needed now were jeans that smoothed your reentry into mainstream America but still identified you still marginally dangerous and equipped for action. Enter Ralph, Calvin, and fast on their heels, Gloria Vanderbilt. Rebellion? Forget it. Get a job? Maybe. Show off your cool? Always. "Levi's," said Bill Blass, "are the single best item of apparel ever designed."

Saturday Night Live's Gilda Radner had a slightly different take. When Vanderbilt's jeans appeared in the late '70s, the comedienne observed, "She's taken her good family name and put it on the asses of America." That proved to be smart business. The funky '60s were quickly giving way to the svelte '70s; stretch denims appeared for women as if ordained by an act of God. Suddenly your jeans moved in the same direction as your body. Overnight they went uptown in stone and acid washes; they blossomed into a swirl of styles and colors, with waists that hugged your navel one month and dropped to your bikini line in the next.

"Jeans are about sex," said Calvin Klein. Good guess. After Brooke Shields revealed that nothing got between her and her Kleins, his company's sales climbed to $180 million from $25 million the previous year. By 1977, 7,489 million square yards were being produced, a tenfold increase over the mid-'60s. Within another decade, cotton would soar back to reclaim more than 60 percent of the U.S. apparel market. By 1994, more than one billion square yards of denim a year were being manufactured in the United States.

Jeans alone had the range to attract high-end designer boutique shoppers looking for maximum fit and status and yet remain true to their blue-collar Asbury Park roots, born in the USA right alongside Bruce Springsteen. Long before that they'd been adopted by the gay community as well. When Andy Warhol was invited to a White House dinner he kept his jeans on under his tuxedo trousers for comfort and security.

In 1981, Americans bought 502 million pairs of jeans, more than ever before. The innocent homespun "jeans art" of the '60s, crude and endearing, soon became a calculated corporate onslaught of embroidery, feathers, lace, decals, sequins, exotic washes, and fur trim traveling down the catwalks of the world to boost sales for Versace, Dolce & Gabbana, and later Diesel, Paper Denim, and Cloth, Seven, Guess—a vast and highly competitive army of producers, ever-changing, all frantically fighting to stay ahead of the curve and deploying their troops to knock off one another with no ethical compunctions, at the drop of a stitch.

Levi and Jacob were long gone, and maybe it was just as well. Whatever they stood for probably didn't include artificially manufactured authenticity. A man in Henderson, Kentucky, named Bart Sights has made a lucrative career since the late 1990s as the industry's leading counterfeiter. Sights uses sandpaper, rasps, hole-cutters, a workbench full of abrasives, and puncturing tools to recreate the old jeans he loves and collects. "Dressing up for me is wearing something that is 75 years old," he told the *New York Times Magazine* for its December 1, 2002 issue. With his sandblasting and stonewashing machines and his sharp eye for authentic detail, Sights replicates classics like the 1890's double-X Levi's for companies that sell his "new vintage" jeans at a premium. A "well-weathered" pair starts at about $150. Tom Ford's line of distressed, ripped, and beaded jeans sold out at $3,500 each. Sights understands that real jeans got weathered by people doing back-breaking work. That may breed a certain nostalgia among the young and the aimless for manual labor, but these replicas also touch a deeper longing for stability and simplicity. As the twenty-first century unfolds in a series of cascading tragedies beginning with 9/11, we seem to take comfort literally and figuratively in clothing that harks back to a more placid and reassuring epoch.

Probably nothing illuminates the zone where style and statement overlap better than hip-hop. Damien Lemon, twenty-five, artist-coordinator at *Vibe* magazine, explains the origins of baggy jeans on rappers: "You go to jail you gotta take out your belt, you know what I mean? Because people would hang themselves with that. So your jeans would be hanging off your ass and that became a hot little thing, and then everybody started doing that on the outside too." But fashion alone won't ever substitute for quality and authenticity: "The hip-hop consumer is, if not the most, one of the most savvy customers in the game. We'll buy something impulsively if it's the hot new shit, but if it can't stand up to the quality or what we need it for—if it's not durable—it's going to fall off quick, you know what I mean? That's one of the thing that Akademiks is all dope with. Their quality is bananas while it's stylistically dope, too. You know what I'm saying? You definitely want your form. It's all about quality, it's all about fit. Jeans has always been there."

Indeed they have—on the range, on the assembly line and cat-walks, wherever clothing can be totally ignored or worn as a statement, dressed up or down, dirtied or prettified. Jeans remain cotton's greatest contribution to twentieth-century popular culture, even up against the tee-shirt; they transcended fashion to shape values that shook the youth of the world out of its complacent acceptance of authority.

That revolutionary act aside, the century just past was a progressive struggle for cotton in America—in the field, in shut-down textile factories, wherever the industry put food on family tables. The shelves of Gap and its rivals today tell one part of that story: miles and miles of cotton from denim to flannel, barely a yard spindled in the United States and less of it grown here than ever. We've become dependent on the once-poor nations we help support to furnish our dry goods at the expense of our own economy. Cotton has had no persuasive answers. It says it is the fabric of our lives but what it's not saying is that its own life is in serious jeopardy as a premiere American crop and all but defunct as an American-made product.

Anyone stepping back in time a hundred years or so might have seen the clouds forming. Just as the South was regaining its self-esteem through industry, its cotton fields fell victim to a tiny beetle that would create $22 billion in crop damage over the next sixty or so years, and decimate millions of acres of farmland. In hindsight the bug was a harbinger of hard days ahead. But with all that there would be twentieth-century triumphs to celebrate as well, and the birth of a Delta music that owed its heart and soul to cotton. The first notes resonated one day a long time back when a black Mississippi sharecropper fixed cotton baling wire to a slab of wood, built himself a crude guitar, and began to sing and shout the boll weevil blues.

TEN

Boll Weevil Blues

The Bo Weevil say to the farmer,
"You can ride that Ford machine
But when I get through with your cotton
You can't buy no gasoline.
You won't have no home, won't have no home . . ."

—From traditional blues song of the 1920s

Driving down Main Street in Enterprise, Alabama, a small, piney-woods town in southeastern Coffee County near the Georgia line, you come to a circular fountain that surrounds a towering white sculpture of a woman wearing a flowing toga. A floral wreath decorates her hair. Arms raised high above her head, in her hands she holds aloft a swollen black creature about three feet across with a jagged, protruding snout that at first glance appears to be the muzzle of a weapon. The creature looks at once both prehistoric and intergalactic, menacing and yet curiously vulnerable. It is a boll weevil, the legendary invasive cotton beetle, many times enlarged.

The question that springs immediately to mind is not so much what is it, but why is it being memorialized in a farming area that the insect all but ruined? The answer lies in the fields surrounding Enterprise. In 1911 the boll weevil had arrived in Coffee County on its remorseless eastern migration from Texas, where it crossed the Rio Grande from Mexico in 1892. The tenant farmers and landowners who grew cotton had been struggling for decades against erratic market prices and a lien system that produced more sweat than profit, but at least provided for basic necessities. About the time they were finally close to breaking even, as if visited by a vengeful god with an unlimited capacity for cruelty, the boll weevil appeared.

Barely larger than a toddler's fingernail, the female beetle used its snout to drill a hole into the plant's flower bud, or square, which was about the size of a pencil eraser. Into each square she deposited a single egg. Alternately, she laid her egg in the young seed pod, or boll, left behind when the flower fell off after three days—between 100 and 300 eggs in all. Each plant could sprout twenty or more pods. As hatching larvae destroyed the tissue of squares or new bolls, the vegetation fell off. Rows of dead squares littering the ground overnight became a common, dreaded sight: they signaled boll weevil infestation.

A Georgia county neighboring Enterprise produced 13,862 bales of cotton the year before the weevil arrived. Four years later it produced a scant 333 bales, 2 percent of the former yield. There were no effective methods for eradicating the weevil, although a dozen or more had been tried. Facing bankruptcy, local farmers prepared to move—but where? The scourge of the weevil was sure to soon follow. One man in Enterprise, H. M. Sessions, took an alternate approach. "Why fight the beetle?" he asked. "Why not learn from it to diversify our crops, grow something impervious to its destructive nature, and see if we can rise up from the dust?" Or, he asked, "Do we throw up our arms in surrender to a common bug?"

The locals listened. They might be destitute but they had their pride. "Pull for Enterprise or Pull Out" was the town's motto. Meetings were held. Think peanuts, said Sessions. Peanuts? Yes, legumes! There's this Negro man named George Washington Carver, Sessions explained, and he's come up with three-hundred-plus uses for the lowly goober. They're reported to be easy to cultivate, they put back in the soil nutrients your cotton has sucked away, and they fetch a fair market price. One farmer, G. W. Baston, took Sessions up on his idea. The men made a pact. Sessions supplied the peanuts for planting and a picker for harvest, and Baston agreed to sell his peanut crop, if any, for $1 a bushel and split all profits. To the surprise of both men, peanuts flourished in the piney woods soil. In his first year Baston produced 8,000 bushels and got out of debt. Local farmers bought his peanuts as seeds for their own crops. One year later, in 1917, Coffee County was producing more peanuts than any other county in the nation.

They all knew what little critter they had to thank for their liberating good fortune. In 1919 about 5,000 people—most of the town's population at the time—attended the dedication ceremony. An inscription at the base of the towering downtown monument, still standing today, reads: "A profound appreciation to the Boll Weevil and what it has done as the herald of prosperity."

Intractable Foe

When the beetle first appeared we were a nation of solutions, not problems. In those dawning years of the new century we had unlimited natural resources from lumber to coal and all of the food and cash crops any country could hope to possess. Also, we had spunk. We saw ourselves as ready and able to lead Western civilization into a piston-driven, electric-wired, steam-powered future. Optimism radiated out in all directions. No quarter-inch bug could obstruct us from our appointed rounds. But the years of infestation dragged out into decades, and decades stretched into reams of time that engulfed entire generations. As each new attempt to subdue the boll weevil at first succeeded only to ultimately fail, these accumulated missteps at the very least raised questions about the potency of science to subdue nature.

From that perspective the war against the weevil tracked the nation's slow but steady descent from the Age of Abundance to the Age of Anxiety. Self-confidence gradually eroded into uncertainty. In the second half of the century we lost jobs and industries overseas, sent 50,000 American men off to die needlessly in a humiliating, losing war effort, and discovered we were befouling our own streams and contaminating wildlife in the name of progress. The boll weevil for its part served to remind us that the forces of nature can lay waste to the most ambitious designs of man. In the time it took to conquer this tiny bug, America abandoned horses for automobiles, invented the airplane, fought in two world wars, developed the nuclear bomb, reached Mars and the moon, created computer and satellite technology, and learned to replace human heart valves with artificial ventricular assist pumps.

Cotton's Blues

It might have been sheer coincidence that a new melancholy form of popular American music emerged during these years—the blues. Did it owe its dejected mood to a sense of discontentment that seemed to be enveloping the country? Probably not—it rarely traveled past the aching lovesick human heart. Even so, its melancholic chords captured a pervasive social malaise. Soon enough, like a musical messenger from the dark side of the American dream, the blues cast its dim shadow over the nation's sunny optimism. It had to rise up from the cotton fields of the South, and from slavery: Where else was that dream denied to so many for so long?

The first blues notes were shouted, not sung. Before the Civil War, overseers often forbade slaves to talk to one another, fearing insurrection. The slave gang's leader, himself a slave, would instead holler a musical phrase—a sentence, a fragment—and the workers would respond, as in gospel. Because the rhythmic call-and-response chanting helped coordinate physical labors, overseers usually allowed it; unknown to most of them, that field holler turned into a vehicle for sending encoded messages. If the gang leader learned that an abusive master was on his way to the fields, he'd alert the other hoe hands, especially the women, with his yell. In many cases a female slave had two choices: she could submit to a master or kill herself. That warning might give her a chance to hide.

Then in time came the boll weevil, heartless and cruel beyond reason. In no time it found its way into the blues, too, like these lyrics inspired by Charley Patton's "Mississippi Bo Weevil Blues":

> Well, the Merchant got half the cotton,
> The Boll Weevils got the rest.
> Didn't leave the poor Farmer's wife
> But one old cotton dress,
> And it's full of holes, all full of holes.

From the moment W. C. Handy first heard a slide guitarist in 1903 running his tablespoon up and down the steel strings to make that instrument talk, every blues song that wasn't about a two-timing woman seemed to be about a single-minded bug. The boll weevil in these early blues lyrics rides on a Memphis train, sits in the farmer's rocking chair, puts on his overcoat, strips a man right down to his bare bones, cleans out his pantry, and moves on. That insect embodied uncharted reaches of evil that few mortals could hope to fathom. The blues and boll weevil might have each existed in the absence of the other, but with nothing like the earthy profundity, depth of desperation, or resigned acceptance of fate that the ruinous pest bestowed on the music. There was, however, one crucial difference. While the weevil stole food off the sharecropper's plate, for a lucky few

Delta blacks with a musical gift, the blues provided a ticket out of town as far away from those cotton fields as a train's tracks would take them.

Cotton Man's Blues

Growing up in a sharecropper's cabin with no electricity on a Delta cotton plantation outside of Greenwood, Mississippi, the young boy listened to his great-grandmother, a former slave, talk about the beginning of the blues. Singing, she said, mattered just as much as breathing, sometimes maybe even more.

In time the young boy's great-grandmother died, and so did the most wonderful woman in his life, his mother, while she was still in her twenties and he not yet ten. His father had run off long before. Orphaned, the boy refused to move in with nearby relatives and instead lived alone as a child in that isolated shack out past the fields. He took some of his meals with the family of the white plantation owner, Flake Cartledge, a kind and fair-minded man he greatly admired. One day the soft, gentle voice of a black singer and musician, Lonnie Johnson by name, came floating over the speaker of his Aunt Mima's crank-up Victrola. Then she put on another 78 rpm disc—this one rough and raw, the shouts of Blind Lemon Jefferson. They were bluesmen both, but so different from each other, and they both electrified the boy's soul. He knew right then he'd found his life's work.

His first real guitar, a red Stella acoustic guitar, was soon stolen from the shack. Before that he'd tried to imitate the sound of a vibrating string using broom wire. He saw some boys who nailed cotton baling wire to the side of their house, then slipped a rock or can under one end as a bridge and called that single-string guitar a diddley bow. Once you've got that string attached, as one of them explained, your whole house becomes the resonating box on the guitar. Eventually he was able to replace the Stella with another for $15 and began to copy Johnson, Jefferson, then Robert Johnson, hard-drinking Charley Patton, and his mother's first cousin, Bukka White—any bluesman he heard—until he could move from raw pain to a silk caress within the same lyric line and bend notes all the way around a corner. By then he'd dropped his first name, Riley, and settled instead on B.B. The "B.B." stood for nothing or anything he chose. The music he played and sang stood for everything, as it does to this day. It saved B.B. King's life.

"I kept loneliness close to my heart, like a secret or a shame," he recalled as an older man. The blues were his only outlet, a form of music that followed naturally from his church gospel-singing boyhood. The music traveled on guitar strings like an angel of salvation from the Delta's cotton wage hands and croppers to loggers and levee-builders in their work camps. All three occupations along the 300-mile Delta basin required the sweat and toil of black men—and sometimes women—who quarreled, laughed, made love, and broke each other's hearts while working for the Boss Man.

Born in cotton fields and in gospel, the blues became the language of raw emotion among blacks at a time and in a place when fearful whites permitted no other expression. Rage and joy were both too dangerous, too unpredictable. "Plantation bosses were absolute rulers over their own kingdoms," King remembered. "It gave you a double feeling—you felt protected, but you also felt small, as if you couldn't fend in the world for yourself." Will Dockery's farm, made famous in an early Charley Patton song, functioned like many others as a self-contained community with its own post office, commissary store, churches, school, and graveyard. Laws might have changed in the seventy or so years between the end of slavery and King's Mississippi boyhood, but sharecropping, King knew, was but a slight variation on that traditional master-slave connection. It ran as deep as the alluvial soil beneath his feet.

Like other black Southerners who lived and worked on plantations, King viewed cotton as more than a cash crop. "In the

Mississippi Delta of my childhood, cotton was a force of nature. Like the sun and the moon and the stars above, it surrounded my life and invaded my dreams. I saw it, felt it, dealt with it every day in a thousand ways. It's how I beat back the wolf. Cotton turned me from a boy into a man, testing my energy and giving me what I needed—a means to survive. But I did more than cope with the crop. I actually loved it. It was beautiful to live through the seasons . . . there's poetry to it, a feeling that I belonged and mattered. Hoeing and growing cotton has a steady rhythm; it's a study in patience and perseverance."

Nobody, not King or the thousands of others who got close enough to cotton in the fields to witness its struggles as a seedling—and if it survived, its blossoming—could ignore the deeply human feelings it evoked. Cotton was like a woman, some said—demanding, plain one moment, and beautiful enough to take your breath away in the next, never entirely allowing you to relax but making you aware of the dangers of getting overly comfortable, too. There was a lesson there for any man.

Cotton taught King to take nothing for granted, and it moved him from one part of his life to another. Not yet a teenager, King went to work as a cotton picker between "can" and "can't": between the break of day and nightfall, when it's said that God turns off the lights in the fields, forcing you to quit. Moving in at length with another aunt and uncle eight miles outside Indianola, King learned from his landowner how to plow and chop the rows. He spent four years behind his mule in the cotton fields, six months a year, six days a week, twelve hours a day. "The mule will shit, piss, and fart in my face—he'll do it all—and I'll keep holding him steady by the ropes. . . . Me and the mule are in this thing together. I feel for him, and I have a feeling he feels for me."

Picking up to 500 pounds of cotton a day during the fall harvest, King earned $1.75 for that bale. He threw himself into the work partially to bury the painful loss of his mother, and learned that it required a maternal instinct of its own. "Cotton, you see, is fragile. I'd been protecting the crop and praying it'd come up good. . . . The cotton is in the blossoms, and the cotton is the victory, the money, the success of the cycle."

Most bluesmen had less affection for the crop. They hated picking and chopping, but then, too, their music generally showed less capacity for warmth and compassion than King's. Another man dealt the same hand—orphaned, impoverished—might have carried his bitterness with him. King learned a few life lessons from cotton that served him well. It could kill you or save you. The choice was yours.

As a boy he knew that a sharecropper got paid a lump sum once a year after harvest if lucky, maybe an advance of a few dollars a month plus provisions if the weevil hadn't eaten up profits. Wage hands, though, got paid by the week. A blues guitarist on a hopping Saturday night in the town of Clarksdale or down Highway 61 at Indianola could play for coins and bills. The crowd at the juke joint was all black, rowdy and poor, but at least wage hands had leftover pay in their pockets, and King, like other bluesmen, learned they'd fork it over to hear and dance to music that might tell their story and move a lady to cuddle close. A teenager during the late '30s and too young to be admitted, King stood outside Johnny Jones's Night Spot in Indianola on Saturday night, week after week, peeking in through the cracks in the side wall. He had an eight-mile walk into town and back, but it didn't matter. On the other side of that wall he heard Count Basie, the boogie-woogie pianist Pete Johnson, the blues singer Walker Brown, Charlie Parker, Lester Young, blues harpist Sonny Boy Williamson—a pantheon of blues and jazz artists in every style and tempo imaginable. That's where the foxy women were, too, B.B. noticed with a practiced eye. That's where B.B. himself would one day be. On the other side of that wall.

The next step in that direction for King coincided with a major breakthrough in cotton cultivation. At long last, tractors came within economic reach of the moderately successful farmer; they began replacing mules. King knew tractor drivers to be well paid and highly sought after by the ladies. Both were equally important to him. Trained by his field boss, he learned to cultivate and plant cotton with his tractor and do a score of other chores with it as well; he earned $22.50 a week, a small fortune. "Where dozens and dozens of mules had been used before, it now took only nine drivers to plow a big plantation. To be one of the nine drivers was like being in an elite bomber squadron," he remembered. What he loved most was turning the brittle land into something soft and receptive. He liked the fat worms, the crawling bugs uncovered by his churning discs, and the smell of the earth. At the time he was playing his acoustic and singing gospel songs on the curb near Jones's Night Spot early on Saturday evenings to make money. He got praise, but no coins dropped in his box. What he needed to bring to his act was exactly what he loved about his tractor driving—its contact with the earthiness of nature. There is nothing more visceral than cotton in the field. It gets picked, not thrashed, which for most of its time throughout history meant that someone, not a machine, had to reach into an open boll and pull out the contents by hand—as millions of poor foreign farmers continue to do. It was ugly one day, glamorous the next, and then stripped clean and left to die.

King, like many others, never could fold the lessons from all of that into his music so long as the music he performed remained choral, directed up into the heavens, and above the fray. But then, necessity spoke. Singing the praises of God on the streets of Indianola brought in not a penny, just compliments like "You doin' good," so King changed his tune—literally. One evening he began to improvise on a Sonny Boy Williamson harp (harmonica) blues, adding his own lyrics and guitar riffs. Something about a woman that done left him—what else? And wasn't he feeling down? Oh yeah, 'bout down as a man can be. And wasn't this one lonesome town? Oh yeah, weren't no more lonesome town nowhere. This momma, she loved B.B. real good.

Now he's the laughingstock of the neighborhood. Twelve-bar blues, as simple and predictable as popping one cotton seed into every hole on the row, and just as likely to produce, King soon discovered. A listener who'd earlier passed by on the sidewalk when he was singing gospel wandered back with a friend now that King had come down from the heavenly firmament to sing about lust, love, and loss. When he'd finished, the man reached in his pocket. "Keep singing, son," he said, and handed King the first dime he ever made from his music.

As B.B. knew, cash in a Delta bluesman's pocket meant freedom; if he didn't drink it all away he could start saving for a ticket to ride. One of King's idols, Big Bill Broonzy, put it this way:

> I got the key to the highway;
> I'm booked out and bound to go.

Got to get out in a hurry, Broonzy continued. Walking just won't do, that's way too slow. In that he echoed the sentiments of another traditional blues song:

> Ain't but the one thing that I done wrong
> Stayed in Mississippi just a day too long.

Delta-born blues singer Muddy Waters from Coahoma County had his own take:

> If I feel tomorrow like I do today,
> I'm gonna pack my bags and make my getaway.

No one had to tell any of these men—or Howlin' Wolf, John Lee Hooker, or Sonny Boy Williamson, Tampa Red, Willie Dixon and the rest—that the blues were their ticket upriver to the clubs of Memphis. Ever since Reconstruction their ancestors had been free to mingle with whites yet barred from joining mainstream society. So close yet so far, and nothing had changed. No one had to press these musicians, either, to coax a cry and moan from their instruments, to wrap their plaintive tales around alienation, weariness, and oppression. Picking cotton from sunup to sundown and living with ten others in a two-room shotgun or dogtrot cabin took care of that. The woman in their songs represented a lot more than the memory of a lady with sassy hips, a come-hither smile, and a treacherous purr. She stood in for everything unattainable, unaffordable, or untrustworthy—the bane of a poor black man's existence in a world where he wasn't valued or respected. And the boll weevil? That insect proved how bad things could get even when you'd swear up and down there's no way things could get any worse.

Beat the Devil

Some, like Nate Shaw, knew otherwise. All God's dangers ain't a white man, he determined. As a black tenant farmer in east-central Alabama who grew cotton and raised a family during the first decades of the twentieth century, Shaw (whose real name, Ned Cobb, was changed by his biographer to protect his anonymity) had seen his share of shotgun-armed mobs set on torching a sharecropper's row house to the ground, as well as every form of racial terrorizing that could follow a black man deep into his sleep and slip through the cracks to soil his dreams. Shaw never in his lifetime expected to dread a thing worse than a riled-up, liquored-up backwater white. Then in about 1915 the boll weevil came along. This danger had six legs, not two, and a snout that could bore through your life savings in an eye-blink. He and his children watched as the tilled ground between his cotton plants turned to a sea of mangled fallen squares overnight, all the work of that bug. They would fill buckets and sacks with those dead cotton flower buds, then baskets, and set them on fire to destroy the weevil inside. Shaw gave his children pennies and nickels for picking up squares. In the end it all come to nothing. The ones you missed would winter over in the nearby woods and come spring they'd be back, armies of them scheming to eat you out of house and home. More than a few whites put the onus for the coming of the beetle into the country squarely on the black man. Didn't matter they were talking trash; give 'em enough pulls on a bourbon bottle and they convinced themselves four ways to Sunday. Black or white, that weevil, he'd kill you.

"I was scared of him to an extent," Shaw confided late in life to an earnest young listener from the North, Theodore Rosengarten, who got down the man's first-person history in a book he called *All God's Dangers*. Shaw's story, said the *New York Times*, is one of only a handful of American biographies of surpassing greatness. It is all that and more. Nate Shaw delivers his epochal portrait of the early-twentieth-century South with lyrical, spellbinding power in language as rich and fertile as Black Belt soil.

Once the weevil lays its eggs, he continued, "it don't take 'em but a short time to raise up enough out there in your field . . .— in a few days, one weevil's got a court of young uns hatchin." Taint long, taint long, says Shaw, until the young one cuts a hole in the square to exit, little sneakin devil, you look at him, he's green and sappy. And when he comes out of there he turns into a rascal with gray wings like ash. "You can't thoroughly understand the nature of a boll weevil. . . . He's a very short fellow . . . And he'll stick that bill into a cotton pod, then he'll shoot his tail back around and deposit an egg—that's the way he runs his business. And he's a very creepin fellow, he gets about, too; he'll ruin a stalk of cotton in a night's time."

What's most bad about the weevil, Shaw continues, is the way he gets to where he wants to go. A born naturalist, Shaw stood in his cotton fields for hours at a time observing the smallest detail of the weevil. "And you watch him, just watch him, don't say a word. And he'll get up—he ain't quick about it, but he'll get up from there and fly off, you looking at him. Common sense teaches a man—how did he get in your cotton farm out there? He got wings, he flies."

Shaw had it right. The weevil flew east inches at a time, sixty miles or so a year, with the relentless determination of death itself, until it reached the Atlantic seaboard twenty-four years after it showed up in Brownsville, Texas; from there it worked its way north another few hundred miles to infest both Carolinas by 1921. It never destroyed the South's entire crop; the region continued to supply raw material to three-fourths of the world's mills, including Lancashire. One writer in that era called America's cotton the only natural monopoly of a worldwide necessity. The fabric now clothed close to 75 percent of the global population. India, our closest competitor, supplied a mere 11 percent of the fiber; and Egypt, 6.5 percent. In 1916 Southern mills alone consumed about 2,300,000 bales. But those figures deceive. A few select areas like the Delta, relatively unharmed by infestation due to soil and climatic conditions, produced most of that cotton while the Old South—including Georgia, Alabama, and the Carolinas—suffered huge losses. The weevil never spread to California, where cotton had only recently been planted, but it reached as far west as Arizona and profoundly affected the course of American agriculture by testing our capacity to contain and conquer an elusive, relentless natural enemy.

Chapter and Verse

Some tenant growers like Shaw who watched the weevil survive one insecticide campaign after another didn't begrudge the weevil its tenacity: "Everything, every creature in God's world, understands how to protect itself . . . ," Shaw decided. "And everything God created He created for a purpose and everything drops to its callin, and most of the things obeys His rulins better than man do."

It was nothing like a fair contest. At every turn the pest outflanked the predator. Thousands of acres were going barren. The $22 billion in losses over a century from crop devastation and control costs first reached into the pantries of starving families from Mississippi to Savannah; many could afford only chicory or dandelion root coffee to serve as the family's food for two meals a day. More than 2,100 gins in Louisiana and Mississippi as well as half the cottonseed oil mills had already shut down in the earliest years of the infestation. Banks went bust. Cotton growers in Concordia Parish and elsewhere ripped apart their gins and converted to rice. As early as 1902, Seaman Knapp from the U.S. Department of Agriculture showed up with Booker T. Washington to demonstrate containment methods to blacks and whites alike, and agents from a federal extension service after 1914 fanned out to educate growers.

Cotton had divided the nation; cotton's enemy now reunified it. Before the arrival of the weevil, lawmakers up on Capitol Hill were generally regarded by Southern farmers as shifty-eyed pols with the morals of a goat in heat. Now, in the face of an evil, alien force of nature—known to scientists as *Anthonomus grandis* Bohemian, and to Delta blues singers and bereft farmers alike as Mister Bo Weevil—these growers sought any and all help, federal or otherwise. In spirit the rural South reattached itself to the Union through a bond of trust born out of desperation for the first time since the Civil War. Having established a foothold, the federal government would continue to play a pivotal role and come to the aid of farmers everywhere to assist them through a series of subsequent disasters like the calamitous Great Mississippi River Flood of 1927, which washed out levees, transformed the Delta into a three-hundred-mile lake, ruined crops and homes, and precipitated a massive federal engineering project. In return, as New Deal farm programs took effect, the government began to exert increasing control over regional agricultural practices. Today, cotton in America would not exist without federal support in the form of subsidies.

Any creature that withstands an all-out assault over such a long span inevitably becomes mythic. Because the crop it chose as its host happened to be cotton from the evangelical South, the weevil plague took on a biblical dimension as well. The entire saga of America's cotton by now contained so many recurring incidents of punishment, redemption, catastrophic loss, and reversals of fortune that at times it seemed torn from pages of the Old Testament. Prosperity lingered never more than a fiber's width away from calamity. At any moment the heavens threatened to open up to reveal a wrathful Almighty, pointing in fury to a burning cotton bush, thundering, "I warned you!"

Nate Shaw and thousands of other blacks knew down to their bones that the weevil might look like a bug to all the world but in its wicked little heart it had to be part redeemer, too: maybe one of God's messengers like Gabriel. You just fix on the way it laid low the white man and ask yourself if you're not reminded of what the white man did to the black man. That weevil gave him a good dose of how it feels to be reduced to crawling, then to get thrashed to the ground just as you rise to your knees.

Black tenant farmers had to tip their caps to the weevil for that—also for giving the black and white farmer a foe in common. You might be leading two different lives under the Jim Crow laws, but there were no segregated restrooms at the wrong end of a dead square. That gave blacks some leverage. In infested areas cotton became a hit-or-miss proposition, and landlords who took a chance on planting a new crop became more dependent than ever on wage hands for part-time work when normal operations collapsed.

Planters tried to convince black laborers in their communities to stay put; they needed their help. The more civic-minded threw picnics for local blacks that included lectures on "the advantages of the Delta as a home for Negroes." Still, they were undermined in their efforts by the most extreme racists among them. Ever since the Civil War, and particularly in areas like the Delta where blacks predominated, frightened, violent whites throughout the South frequently resorted to a reign of terror. Lynchings became even more common in the first decades of the twentieth century as a perverse show of strength. No black felt completely safe in the company of white strangers, and not always secure among familiar faces. Any perceived insult or alleged crime could lead to mayhem. In 1903, in Greenville, Mississippi, white town citizens lynched a "miserable negro beast" suspected of attacking a telephone operator. Strung up in full view of an applauding crowd that included Greenville's civic leaders, the lynching became a social occasion. A journalist reported that as soon as the corpse was cut down, the crowd moved off to see a baseball game. Before another lynching, the fingers of the victims, a husband and wife, were chopped off while the couple was still alive and passed around as souvenirs.

There were 188 lynchings over three decades in the Delta alone, one about every five months, and numerous shootings of blacks, often for the slightest infractions. The law stayed far away. Economic decline only exacerbated tensions. Where the weevil migrated, alarmed white landowners held back on breaking the ground. Their black tenants and croppers, fearful as always and now destitute as well, abandoned hope of planting their own cotton with any certainty of success, and could not be sure to find work as wage hands. As conditions worsened, rumors circulated about jobs up north. The Illinois Central Railroad now connected the Delta to northern midwestern cities like Chicago. As one observer pointed out, the Delta became a kind of staging area for Southern blacks seeking to escape to make a new life for themselves. Arriving in Clarksdale or another Delta town, they encountered fierce white hostility, meager wages, and no promise of a better tomorrow. Desperate families packed up and left in vast numbers to seek jobs in the factories of New York, Detroit, Cleveland, Youngstown, and Chicago. One migration motivated another.

The black writer Richard Wright understood why: "Cotton," he said, "is a drug, and for three hundred years we have taken it to kill the pain of hunger, but it does not ease our suffering . . . ; we travel down the plantation road with debt holding our left hand, with credit holding our right, and ahead of us looms the grave, the final and simple end." The only other choice was escape. "We look out at the wide green fields which our eyes saw when we first came into the world and we feel full of regret, but we are leaving."

Between 1916 and 1928, more than 1,200,000 Southern blacks left the South for America's urban areas—70,000 moved to Chicago alone—all but emptying out many rural communities. Another mass migration would follow after World War II, when mechanical cotton harvesters first appeared, eliminating the need for hand-pickers. (One machine could harvest 190,000 pounds in a day.) Harlem's black population doubled. Never, said Wright, has a more utterly unprepared folk wanted to go to the city. Chased by poverty and the weevil, these tenant farmers and sharecroppers had to learn for the first time to form relationships with things, the items manufactured in factories, when before their relationships had been exclusively with people and animals. Northern whites spoke a new clipped language they could barely understand. The accents of Poles, Italians, and Eastern Europeans made their conversations undecipherable. Slum tenement housing crammed five or more family members into a single steamy room. But there was something else missing, too—signs that read For Whites Only above public restrooms, and signs that ordered blacks to the rear of the bus. You might be trapped, as Wright said, by the brutal logic of jobs, but you could sit down anywhere you chose, eat where you wanted, and never have to step into the gutter when a white woman passed.

Still, there were rich memories of soil warmed by the sun, swollen streams, and swaying fields of white lint and potato pies and meats smoked outdoors over open fires. There were memories of other Southern ways, too, that you never expected you'd miss but did—a rhythm to life that left space for slow-cooking pleasures. What you had as a reminder might not be all that much, but it helped: music that carried your ache. The blues, born and bred in cotton and always ready to hop a midnight train, followed you up the Mississippi or over the Ozarks to the tenements, front stoops, and clubs of Chicago or New York or wherever a cousin or an uncle had put down roots. Singer and songwriter Lizzie Miles came down with a bad case in 1923:

> Look at me. Look at me.
> And you see a gal,
> With a heart bogged down with woe.
> Because I'm all alone,
> Far from my Southern home.
> Dixie Dan. That's the man.
> Took me from the Land of Cotton
> To that cold, cold minded North . . .
> Just cause I trusted. I'm broke and disgusted,
> I got the Cotton Belt Blues. . . .

No, you didn't need to see your crop stripped bare by that evil weevil to sing the blues. You had only to step inside the empty shadows left behind by the sadness of an abandoned home or a lost love. That would do it, every time.

Agents of Change

Although tightly intertwined in our popular culture and history, with cotton a common ancestor, the boll weevil and the blues would part company forever as the twentieth century progressed. One would grow up to transform American music; the other would exert a powerful influence over the business and chemistry of agriculture. The revolutions they inspired had nothing to do with industry and commerce in the tradition of Arkwright's England, the Boston Associates' factory empire, or the Southern mill migration.

Cotton textiles, in fact, were stagnating. Despite the success of apparel retailers and department stores like Macy's and Marshall Fields in home furnishings, between 1923 and 1940 the cumulative net income of domestic mills barely equaled their losses, and assets declined by $580 million. "Only wars have bailed the industry out" wrote *Fortune* in 1947, when cotton still retained 70 percent of the country's total textile market, before foreign competition from Japan and underdeveloped countries kicked in. The nation's largest cotton manufacturers had strapped themselves with archaic business practices. They remained fiercely independent, as *Fortune* explained, and refused to share information within the trade; as a result they operated largely by guesswork rather than verifiable knowledge and built up huge, costly inventories that exceeded demand, particularly in unbleached, unfinished "gray goods." Between 1925 and 1939, one-third of America's cotton mills shut down.

No, cotton was making its bones elsewhere. As the progenitor of the blues, it was indirectly responsible for knocking down racial barriers in popular music. Its reach extended to touch a young generation of whites in rigidly conservative families and to shake-rattle-and-roll 'em loose from their parents' segregationist values.

Following WWII and the introduction of mechanical harvesters into the South's cotton fields, a second wave of unemployed black cotton hand-pickers migrated north. Some settled in Memphis en route, and the city's blues players would soon cross the color line long before judges and politicians got around to making democracy safe for racial equality. At Sun Studio in Memphis, producer Sam Phillips recorded the most commercially viable ones for a racially mixed audience. As bluesmen, they formed a community that quickly became the country's center of soul. "If you were a black man on Beale Street on Saturday night," musician Rufus Thomas told a friend, "you'd never want to be white again." And if you were white and able to cry, moan, and wail on key, you might well be accepted as a kindred spirit. In a photograph from the early '50s, young Elvis Presley wraps a friendly arm around the shoulder of a black man in his mid-twenties, B.B. King; that image captured the men's relaxed camaraderie at a time when Southern blacks and whites risked life and limb to show physical affection for one another before a camera; segregation in Memphis was literally built into the city's architecture in the form of separate drinking fountains and facilities. Racism was so entrenched in its social order that King Biscuit Flour sponsored a wildly popular blues show on WDIA, the city's black radio station, reasoning that the black domestics who listened bought all the provisions for their white employers. They were right; sales soared.

In time the blues clung tight to its mournful roots while its rightful heirs, R&B and rock and roll, surged ahead, accelerating the tempo and adding a hefty quotient of funk and sin, as in "Sixty Minute Man." That replaced despair for the young and restless in a swift hurry. Joints jumped, Stax horns blared. Otis and Sam & Dave and Mick and Keith and Jimi put a match under those languorous bent blues notes and lit up the night. Some of the musicians, like Eric Clapton, reached out to introduce relatively obscure chitlin'-circuit bluesmen—most notably, B.B. King—to a new audience of young, predominately white listeners. After his first triumphant Fillmore appearance in the late '60s, overwhelmed with joy, King cried backstage. In twenty years he had never before played before a packed auditorium of cheering whites.

As for the boll weevil, by midcentury it still had farmers and government agents singing the blues to the tune of $10 billion in estimated losses and 600,000 square miles of infested fields. By then it was doing a little too much damage for its own good. Too many government officials with large budgets and dogged determination were now hiring the nation's top entomologists to bring it down, and private chemical companies like DuPont and Monsanto saw a little too much profit in devising powerful insecticides to help farmers as they helped themselves. An elaborate USDA-funded Boll Weevil Research Laboratory opened in 1960 at Mississippi State University. This was all-out war. "It was clear," said North Carolina's agricultural commissioner at that time, "if we did not eradicate the boll weevil, the boll weevil would eradicate the cotton farmer." He spoke for growers in eleven states. A single female beetle could give rise to six generations all within one year. If only 10 percent of those generations survived, the sixth generation would produce 729 million weevils.

That was the enemy, impossible to miss and easy to hate. But what of the chemical weapons we used against it? They at first represented the contribution of our best and brightest. Yet as time wore on and we learned more details about the unintended effects of these toxins on the health of humans, wildlife, and vegetation, a more subtle and elusive adversary began to emerge: our own methods of destruction. Like Agent Orange in Vietnam, the arsenal itself became a self-inflicted wound. By the time we acknowledged that we were killing scores of birds and fish, contaminating our air and groundwater, and spreading cancer across the Cotton Belt, we'd developed three generations of potent pesticides that we had been applying to cotton and numerous other crops in torrential quantities for close to fifty years. Planters and governmental counterparts weren't guilty of being cavalier so much as ill-informed, complacent, and defensive in the face of mounting evidence. During that period American agriculture moved away from natural resources and into the chemistry lab to fertilize and protect its output, and it has never left. For cotton growers, prime movers in that revolutionary shift, the choice was simple. Nature brought devastation in the form of the weevil, boll worm, and other pests. Science brought hope. Looking back, it all made sense except to a few visionaries like Freud's disciple Ernest Jones. "The control man has secured over nature," he said, "has far outrun his control over himself."

Armed and Dangerous

Nate Shaw observed the pesticide-weevil war at close range as one of its earliest foot soldiers. Beginning in 1919, shortly after arsenic, the first insecticide, came into widespread use, Shaw spread it by hand over his cotton crops. The poison arrived in two popular compounds, calcium arsenate and highly toxic copper acetoarsenite, commonly known as Paris green. Farmers applied three million pounds of those powders in that year. "I carried poison to the cotton fields maybe four or five rows and . . . shake that sack over the cotton and when I'd look back, heap of times . . . that cotton would be white with dust, behind me," Shaw recalled late in life, "and I'd wear a mask over my mouth—still that poison would get in my lungs and bother me." At first the powder seemed to work wonders. But nature was on the weevil's side, not the grower's. "Old weevil, he can't stand that [arsenate], he goin' to hit it out from there; maybe, in time, he'll take a notion to come back. . . ." Sure enough, he did. You might scare him out of your fields but not out of your life, Shaw soon discovered. Farmers found they could dump as much arsenic as they chose on their emerging plants; they could grow as much corn and grain for safety crops, too; they could convert to cattle. In the long run, it hardly mattered. Cotton was all these farmers ever really cared about and the weevil seemed to know it. The growers would come back to the crop, and when they did, in time they'd find the wily weevil ready to move in. Like the man sang:

The Farmer took little Weevil
And put him in Paris green
The Weevil said to the Farmer,
It's the best I've ever seen.
I'm going to have a home, a happy home.

Although the names and formulas of the insecticides would change as the century progressed and as science produced more sophisticated weapons, the pattern was set. First came the weevil or many another natural predators, then the toxin, and then, after a significant decrease that seemed to signal its eradication, back came a tougher, hardier weevil, and so on. Call it a roundelay or a dance of death and resurrection. Darwinians know it by another name: natural selection. The few survivors of lethal, nerve-paralyzing insecticides usually carry a gene that produces an enzyme that neutralizes the spray or powder, or one that provides them with a protective shield that blocks out the deadly molecules. To date, more than 500 insects have built up resistance to insecticides. Nature is infinitely dexterous in such matters—man, less so. An early user's typical response to reinfestation was to dump as much calcium arsenate as possible on his cotton crop, sometimes mixed with molasses to attract the insect. That greatly increased the odds of genetic resistance among the weevil and other pests, since more random survivors had more time and opportunity to breed resistant offspring. These same toxins also wiped out many of their natural enemies.

Up in the northeast corner of North Carolina, another farm hand was learning many of the same lessons as Nate Shaw, ones he'd eventually be able to use to help eradicate the insect. Marshall Grant and the boll weevil arrived on his family's 800-acre cotton farm on the Roanoke River in the same year, 1924, so he considered boy-versus-bug to be a fair fight. As a youngster Grant was given the job of pulling a mule duster to apply calcium arsenate over infested plants. He soon learned to wear long pants to protect his skin against the stinging powder. He also took showers day and night. The papers were full of stories about children on farms forgetting to wash up, becoming violently ill from the dust, and occasionally dying. Rains carried the insecticide into nearby streams. After a few years, most of the rabbits and small game that Grant (and Shaw, in Alabama) loved to hunt began to disappear. Grant didn't have to be told why; local doctors held the weevil powder responsible for the sudden rise in illnesses they were seeing on farms even if they didn't have positive proof. Drift—wind-blown toxins from crop dusters—was now a problem as well, ever since aerial spraying had begun two years earlier.

Over the next few decades North Carolina's cotton acreage declined 97 percent, from 2 million to 45 thousand acres, and Georgia's cotton yield tumbled almost tenfold from 1.5 million to 190,000 acres. Clearly, arsenic wasn't working as well as expected, although in some areas, like Louisiana, it substantially increased yield.

Recognizing the dimensions of the crisis, the federal government under Franklin Roosevelt set up programs during the Great Depression to funnel money to destitute growers, inaugurating subsidies, called farm payments. It also widened extension-agency assistance for America's cotton farmers. In unaffected areas, where there might be too much inventory, some were paid not to produce a portion of their crops. Still, no new compounds were coming out of the nation's agricultural science labs to defeat the beetle.

Far away from the decimated fields, cotton factories also suffered. As the Depression worsened, they continued to shut down in the North, leaving behind thousands of unemployed workers and towns on the brink of collapse; in the South, the descendents of the first mountain whites who worked for beggars' wages took to the streets to demand fair pay and the right to unionize. At the height of the General Textile Strike of 1934, directed by the United Textile Workers of America, nearly 500,000 workers in twenty-one states from Maine to Alabama walked off the job. That strike remains the largest in American history. Georgia's flamboyant governor, Eugene Talmadge, promised the state's cotton workers, "I will never use the troops to break up a strike." Once elected, accepting a bribe from the state's leading mill owners, he immediately called out Georgia's entire 4,000-man

National Guard. The troops beat the oppressed mill workers, threw many into detention camps, and killed at least one man. Most of the striking workers were not rehired.

Cotton's financial and social problems in its fields and factories mirrored the country's. Once, we'd lived the life of a millionaire, spending our money without a care, but nobody, it seemed, wanted us now that we were down and out of spare change. Impoverished white sharecroppers were quickly becoming America's national social embarrassment. Much of the federal money made available by the USDA to help support them went instead to their wealthy landlords and to large, prosperous cotton growers. The plight of these destitute white tenant farmers, really indentured servants who suffered the same threadbare miseries as their black neighbors, came to the attention of writer James Agee and photographer Walker Evans. During the Great Depression in 1936 the two men lived with three white Alabama sharecroppers and produced *Let Us Now Praise Famous Men*, one of the most painfully ironic book titles in American literature. These were peasants so unfamous as to be all but invisible, too poor to afford shoes, the nation's anonymous discards. A century of progress had bypassed them without pause. "Gudger has no home, no land, no mule; none of the important farming implements," Agee wrote. All furnish, or supplies, came from the landlord, at the cost of a half-share in his crops—cotton, cottonseed, and corn. Nothing had changed since Reconstruction. Gudger's neighbors, the Woods and the Ricketts, also owned no home and no land; they did own mules, but owed the landlord "the price of two thirds of their cotton fertilizer and three fourths of their corn fertilizer, plus interest."

Agee created poetry out of the dust of destitution, and found beauty in the commonness and integrity of these people. "It seems to me . . . that a tenant can feel, toward the crop, toward each plant in it . . . a quiet, apathetic, and inarticulate yet deeply vindictive hatred and yet at the same time utter hopelessness and the deepest of their anxieties and their hopes: as if the plant stood enormous in the unsteady sky fastened above them in all they do like the eyes of an overseer."

That book jolted America's decision makers awake but too late to matter much. The damage had been done for decades to whites and blacks alike. When sharecropping began to die out a few years later, it happened only as a result of our entrance into World War II. The South's rickety social and economic institutions collapsed and with them a culture that had lingered on far beyond its rightful life span. As C. Vann Woodward remarked about the racism, antiquated mule-farming techniques, corrupt politics, and brutal social tyranny fostered by the Old South, "It would take a blind sentimentalist to mourn their passing."

Nothing in the lives of these sharecroppers might have changed except for the worse, but in 1939 Hitler invaded Poland and suddenly Europe came knocking on our door for help. Two years later bombs fell on Pearl Harbor, and we were thrust into World War II. It was a terrible and terrifying period for the country at large, and a short-term boon to many of our agricultural, fuel, and materials industries, including coal, steel, and cotton. Military demands created an urgent need for an endless array of textiles—for clothing, coverings, for linters used in smokeless gunpowder, and for a host of other war-related items. Once hostilities ceased, few American industries returned to business as usual. A vibrant postwar economy created strong consumer demand for new products; technological advances in production methods, brought about by the exigencies of war, now streamlined operations. Cotton temporarily benefited as well in its battle with the weevil—but at an ecological cost no one could foresee.

Death Spiral

If we could drop an A-bomb on Hiroshima, surely we could annihilate a lowly snouted beetle in our own backyard. For help, our agricultural scientists, funded in part or whole by the government, turned to recent insecticides developed through organic chemistry—"organic" as in synthetic carbon-based, not as used today to mean natural and chemical-free. They came out of wartime experiments and killed a wide variety of insects on contact, while arsenic had to be ingested. In 1939 a Swiss chemist, Paul Müller, discovered that an organic chlorine originally synthesized in the nineteenth century killed insects with remarkable efficiency. That poison with an unpronounceable name quickly became known by its initials, DDT. From the moment of its appearance, DDT was hailed as the twentieth century's miracle disease-eradicator, and by 1943 it was solely responsible for keeping thousands of GIs alive. By then, more American soldiers were dying from insect-borne diseases in the tropical Pacific than from combat. As many as 50 million civilians had died from malaria and typhus during the previous decade. DDT powder quickly began to reverse that carnage by eliminating the insects responsible, along with houseflies, bedbugs, fleas, hornflies, and lice. Once the war ended and supplies became more widely available, DuPont and a few other chemical companies pumped out over 100 million pounds of DDT per year, most of it targeted toward domestic agriculture. In combination with another new insecticide, BHC (benzene hexachloride), DDT at the outset proved effective in controlling a variety of cotton bugs, including the weevil. It seemed that the weevil would disappear within months. In some areas it did—only to return yet again. Within a decade the weevil would build up immunity to BHC; without it, DDT was impotent against the beetle.

Initially no one thought there might be a downside. No one was focusing attention on damage to wildlife; also, there were no signs of imminent danger to human health. To question the wisdom of heavy DDT spraying was to challenge the power of science to rid us of our natural enemies. It also suggested you were underestimating the ability of agribusiness lobbyists to twist arms in the halls of the nation's capital by pairing patriotism with pesticides.

Posters displayed delighted children leaping through plumes of liquefied DDT as if from a gushing summer water hydrant as a testimony to its safety. An advertisement in *Time* magazine featured livestock, vegetables, and a farmer's wife cavorting together and singing, "DDT is good for me-e-e-e-!" "After WWII, synthetic chemicals . . . came to be the gospel of how you farm. . . . ," says Southern agricultural and social historian Pete Daniel, curator in the History of Technology division at the Smithsonian Institution. "There is an ideology that believes in chemicals the same way some people believe in God."

By 1950, a pro-pesticide network made up of USDA members, agricultural scientists, and growers dominated congressional hearings on farm policy. They left no doubt that God and toxic insecticides were on our side. Disbelievers were un-American. "In the decades after World War II, people in the agricultural research service would lie to Congress and lie to their constituents," Daniel argues. "They would deny wildlife kills and tell anyone in the fields they'd be fired if they objected to the relentless spray campaigns." That mindset, he adds, was directly responsible for starting and maintaining the pesticide treadmill that continues to turn fields of green into slabs of dead dirt across the nation.

But if congressional committees cooperated, nature didn't. Relentless spraying brought about the first wave of environmental wreckage—dead fish stinking up streams, red-breasted robins disappearing from front lawns. Throughout the 1950s, articles began appearing in the popular press with headlines like "DDT, Miracle or Boomerang?" They were dismissed by the industry as hysterical and misinformed. There was simply too much money, and too many careers, invested in the success of DDT. That conspiracy of silence ended abruptly and permanently in 1962 with the publication of Rachel Carson's *Silent Spring*. Cotton's fate would change with it.

An accomplished marine biologist as well as a respected author and firebrand, Carson startled the country with her exposé on the hazards of wanton pesticide use, focusing much of her attention on the careless and irresponsible application of DDT during the previous two decades. She leveled her artillery against the federal government's efforts to keep incriminating information from public view. A primary cause for alarm, Carson explained, was the cumulative effect of DDT. Instead of flushing through the digestive systems of mammals and birds, nonlethal doses remained stored in fatty tissue, where they accumulated. DDT was shown to have a half-life of eight years. Over time these residues reached harmful levels; in many species, including humans, the poison was released in mother's milk. DDT, as was later proven, also interfered with the formation of calcium and compromised the strength of eggshells. That threatened the extinction of bird species like the brown pelican. Its diet consisted of small fish, many contaminated with DDT. The female pelican was soon producing thinned, brittle shells, easily crushed by incubating parents. In 1970, on Anacapa Island in California, only one fledgling was born in an already stressed population of 552 nesting pairs.

"The 'control of nature' is a phrase conceived in arrogance, born of the Neanderthal Age of biology and philosophy, when it was supposed that nature exists for the convenience of man," Carson wrote. "It is our alarming misfortune that so primitive a science has armed itself with the most modern and terrible weapons, and that in turning them against the insects has also turned them against the Earth."

Carson's chapter headings capture her righteous indignation: "Needless Havoc," "And No Birds Sing," "Rivers of Death," "Beyond the Dreams of the Borgias." You might expect the attack to be too strident to be readable. Instead, she pulls us in by the lapels and holds us transfixed with well-researched stories about animals and humans in peril. Her book became a best-seller within weeks. Suddenly, for government officials, there was nowhere to hide. For years the Food and Drug Administration had warned that excessive use of chlorinated insecticides represented a definite health hazard, only to find its concerns detoured or dismissed by the USDA, under heavy pressure. Carson quotes from the Department of Agriculture's own 1952 yearbook on ensuring insect containment: "More applications or greater quantities of the insecticide are needed than for adequate control." That, of course, is the perfect recipe for creating more resistant insects and ecological disaster. Carson charged that we'd put people in charge of the future of our country's agriculture who either were ignorant about the health and environmental consequences of their actions, or worse, knew and chose not to pay attention. After scores of congressional meetings, newspaper editorials, fact-finding commissions, and graphic television documentaries, DDT was finally banned in 1973.

Where was cotton in all this? Pretty much out of the limelight, but if anyone had cared to take a closer look, it was a poster child for chemical pollution. At the time Carson was writing, cotton farmers were applying more than 41 percent of all pesticides in agricultural use in the United States. The South accounted for two-thirds of them.

There was one dim ray of hope. A chemical cousin of DDT had also come along after WWII, an organophosphate called malathion that, in limited usage, clobbered weevils without jeopardizing every other living thing within sight or building up to dangerous levels in fatty tissues. It was developed out of war-related experiments in nerve gases conducted by German scientists; that was hardly a recommendation for its safety, but unlike DDT, the liver of humans and other mammals detoxified it in small amounts. Unfortunately, cotton growers began repeatedly to spray malathion, pyrethroids, and carbamates in torrential sheets at considerable cost and in concentrations that potentially caused harm to humans. If a way could be found to target and hit only specific infested areas within a large plantation with malathion applications, costs would go down and unnecessary damage to flora and fauna would be reduced. What field cotton needed was a hero, someone who could outmaneuver the insidious weevil at its own game and help farmers predict in advance where it would strike.

Enter James Tumlinson III. He proved to be an insect maven with the patience of a saint, the devious mind of a con artist, and the attention to detail of an obsessive—everything you'd look for in a man who figured out after years of arduous research that sex, not poison, was the ultimate weapon. Mae West, of course, could have saved Tumlinson endless hours of exhaustive effort. "What took you so long, big boy?" she might have chided.

Scent Sleuth

Until he'd recently met and married a woman who fully understood his work, Jim Tumlinson had had an ongoing issue: trying to explain what he was up to. Fresh out of the Marines in 1964, Tumlinson decided to go for his master's degree in agricultural chemisty at Mississippi State; he soon joined his graduate-school colleagues in the university's federally funded Boll Weevil Research Laboratory. Fellow researchers knew all about his primary area of interest, but try telling the guy on the next barstool that you specialized in insect chemical communication and chemical ecology, including pheromones and other semiochemicals that mediate insect-insect and plant-insect interactions, or that you were particularly fascinated in the biosynthesis of pheromones and plant chemical signals and insect behavior that included learning, mediated by those same semiochemicals.

On the other hand, you could make an end dash around those gnarly details and run for daylight with a grandiose claim that you and your lab buddies were on the very brink of solving one of the South's most devastating problems, the eradication of the boll weevil. That sometimes drew a crowd. Then came the fun part: if there was a female in the listening audience, you could gently remind her that she and a quarter-inch beetle had a lot more in common than she might care to consider. Procreating, for example. You had to be real careful here, of course; she might be hanging on some guy's arm. Pheromones, you continued to explain real fast before her date dropped you with a left hook, were those chemical messages that sea urchins and boll weevils and homo sapiens emit as sexual scents of attraction. Male weevils send them out and sure enough—wouldn't you just know it—female weevils flock. The trick in the lab was to duplicate and synthesize that male pheromone just as Jonas Salk did with polio vaccine. Once you could manufacture it in a test tube and put it in traps, you could use those traps to accurately monitor weevil infestation in specific locations and greatly improve the odds for targeted eradication. That's why he and his lab buddies were on the brink of rescuing millions of acres of wasted cotton fields along with the agricultural economy of a dozen or so Southern states. Better yet, those same traps would also drastically reduce repeated blanket crop spraying and help protect wildlife and drinking water from toxic drift.

A major breakthrough came a few months before when researchers noticed a huge congregation of wild weevils gathered in the shrubbery outside the window of the lab's rearing room where they bred the beetles. They had attributed that at first to odors from cotton oils in the lab being blown out by a nearby exhaust fan. To test that theory, the lab contracted with Mississippi's notorious lock-up, Parchman Farm, to organize prisoners to pick squares from cotton grown on the penitentiary's plantation so that researchers could study the constituents of the essential plant oils. Tumlinson liked to say that for weevils, cotton was a smorgasbord; there must be some component in the plant they found irresistible. He and his research partner, Jim Minyard, distilled 3.5 tons of squares to produce about three ounces of essential oil—all in vain. To their dismay, female weevils weren't turned on. I'm sorry, they said in effect; I have a headache tonight. That's when Tumlinson and company began to suspect that it wasn't the cotton itself that drew the ladies to the shrubbery, but rather the plume of pheromone produced by the thousands of male weevils in the rearing colony. If true, they now needed to isolate and synthesize a quantity of those pheromones to test their theory in the field.

As hard as it was, that was the easy part. Before they could move forward they had to determine what exactly was in the pheromones that excited the female, and that's where Jim Tumlinson stepped up. He was young enough to take on any challenge and tenacious enough to plod through the seemingly endless processes required to collect those pheromones and analyze their components; he also showed himself to be extraordinarily talented in determining through trial and error which were active and in what particular combination they performed most effectively. He began by grinding up sixty-two thousand quarter-inch male weevils, and then extracted an infinitesimal amount of secretion by drawing air over them and passing that air through activated charcoal with chloroform. After countless, exhausting hours, Tumlinson made another glum discovery—there was next to no pheromone available for extraction. What to do? He had an idea. In the section of the Boll Weevil Lab responsible for rearing millions of weevils, he knew they deposited a huge amount of feces—technically known as frass—under their cages. Why not sweep up the frass and search out its pheromone content? If the secretion existed he'd have himself an ample, ready supply. Analyzing the first batch of swept-up frass, Tumlinson made a happy discovery: it contained substantial quantities of pheromone—enough to examine, even to break down into identifiable components. He went to work. With a colleague he extracted and steam-distilled 135 pounds of boll weevil frass, hoping to deduce the structure of its essential ingredients.

That involved a complex, delicate chromatography separation process and years of painstaking work; for openers he and his team had to feed, determine the sex of, and monitor the ages of these weevils, then successfully bioassay their excretions. That was where art met science. Many researchers performed the tasks properly yet failed to get results. But Tumlinson found a lab technician, Nevie Wilson, who had a magical touch, and by then a coresearcher, Dick Hardee, had come up with a term to describe the combination of the four terpinoid compounds that made up the pheromone. He called it "grandlure," and the name stuck. Tumlinson and Hardee also figured out how to measure the relative strength of each one of these four pheromone components in attracting females. None by itself excited them, but in specific blends they set off the appropriate bells and whistles as predictably as any live male weevil hunk.

By 1969, after five years of intense, often wearisome, experimentation, Tumlinson—now Dr. Tumlinson, Ph.D.—had for the first time isolated, identified, and synthesized the structural elements of grandlure. His work held up in lab bioassays and preliminary field tests. That enabled chemists to create a synthetic version that mimicked the properties of the naturally produced pheromone.

Grandlure didn't travel immediately to the cotton fields of America, but its formulation set the stage for an ambitious USDA boll weevil eradication program that began in earnest in the early 1980s and systematically filtered west from the eastern seaboard. A pilot program in 1977 had successfully used improved pheromone traps to detect weevil populations when insecticides are most efficient, just before the beetle enters diapause, or winter hibernation. Malathion was the chemical weapon of choice; while no friend to wildlife or humans in excessive quantities, it was proving effective against the weevil. With the introduction of accurate grandlure monitoring traps, minimal targeted treatments replaced fifteen to twenty typical widespread applications of organophosphates each season, saving money and helping to restore ecological balance. From California to North Carolina, bright green fluorescent grandlure monitoring traps soon sprouted up as standard fixtures in the nation's cotton fields. Since 1987, more than 20 million have been in continual use. They do not kill off the weevil—there are too many complicating factors to produce a trap that also eradicates—but they provide such a close reading on weevil populations that they act like a highly sensitive weather balloon precisely pinpointing hurricane activity.

In practice, systematic eradication fell to growers like Marshall Grant, who grew up working his family's cotton plantation in North Carolina. They formed committees and cajoled reluctant Southern state agencies and stubborn farmers into initiating an ambitious and uniform boll weevil program. "Ain't no damn government man coming on my farm telling me what to do," Grant heard again and again. But he'd been raised behind a mule himself. Small growers couldn't dismiss him as a suede-shoe salesman or briefcase bureaucrat. Besides which, they were hurting. Georgia's cotton yield had tumbled almost tenfold by the early '80s. They knew what Nate Shaw had learned half a century earlier: the weevil left, only to regroup and return again. The program promulgated by Grant and others in partnership with the USDA required farmers and state environmental agencies to agree on a blanket insecticide spraying during the initial season, and then selective applications based on trap monitoring data. Nothing happened overnight, or without protest. One eagle's nest in a cotton field held up the spraying of a Georgia plantation for weeks, until all parties agreed to spray only when the eagle had left on its hunting rounds. An airborne scout reported by walkie-talkie on the eagle's daily flight pattern to coordinate timing with the crop-duster.

Another twenty-five years would pass by before anyone was willing to go on record claiming certain victory against the weevil. It took that long, even with the advent of global positioning satellites and a host of related technological advances, to evict the insect from its cotton-boll home, or at least to chase it back to Mexico.

By then, in 2003, most cotton industry people had long forgotten Jim Tumlinson's contribution. Many hadn't yet been born when he identified the pheromone's active ingredients. They were reminded, along with the rest of America, by a *New York Times* editorial, "Good Boll Weevil News," that praised the eradication program's advocates, and one man in particular, for getting the job done. The tiny, adaptable weevil, said the *Times*, "has clearly met its match in human competitors like James Tumlinson." He and his group "outweeviled the weevil . . . to find the precise biological perfumes that few boll weevils can resist."

Inducted into the National Academy of Sciences in 1997, and now a professor at Penn State University, Tumlinson has moved up and on. So has cotton. Thanks to his work, production in Georgia is up from 190,000 acres annually to 1.5 million acres; animals have reappeared. Pheromone traps have cut back eighteen or more blanket applications of malathion a year to a few strategic, isolated treatments. All the same, Tumlinson's experience taught him never to make presumptions about the woody, leafy world that surrounds us. "Nature," he says, "has a way of jumping up when you least expect it and swatting you down." Nate Shaw, long dead, would have been the first to agree.

ELEVEN

The Shirt on Your Back

Man can hardly even recognize the devils of his own creation.

—Albert Schweitzer

So much for things past. In a world currently plagued with apocalyptic nightmares like random urban bombings, bodies sprawled on sidewalks, and terrifying threats to our homeland, many of us would settle for a problem as easily definable as weevil infestation. Barring that, as global citizens we'd be grateful for a quick and handy way to grasp the geopolitical, social, economic, cultural, and moral dilemmas that confront us daily. Cotton provides it. It's called the shirt on your back. From seed to store shelf, your cotton shirt wends its way through a welter of current controversies and perfectly encapsulates how America operates in a global economy. It also sheds light on the coming deluge of biotech crops and the conflicting emotions they kindle; on the heated debate over whether America's self-absorption is currently sabotaging its own war on terrorism; on the battle between sustainable agriculture and agribusiness; and on the collapse of core American industries and the possibilities for their rebirth. Cotton, a crop at one moment, a textile at another, and a food product at yet another, also by far the most popular fabric on earth, figures prominently in each of these scenarios. It's about as ubiquitous and inescapable as the air we breathe, also crucial to the livelihood of millions around the globe. Unraveled strand by strand, the shirt on your back offers insights into those primary forces that shape the way we live at present. It has even more to say about things to come.

The shirt got there quite possibly by cheating some poor West African ox-plow cotton farmers out of a fair price for their meager harvest and in the process creating tensions that collapsed the World Trade Organization conference in Cancun, Mexico, in 2003. It was probably spun from a blend that included genetically modified fibers, embroiling itself in debates about health, safety, and bioethics. It was almost certainly not made within several thousand miles of America's borders, and even so it might well contain ginned American cotton that this country's textile factories can no longer afford to convert into fabric on native soil.

There's no need to look further for a quick take on the beleaguered state of American manufacturing either. China, now the world's largest cotton manufacturer and grower, might well have her fingerprints all over that shirt at some stage. That fact alone sets teeth to gnashing in Washington, D.C., and pits various departments of the federal bureaucracy against one another in a furious debate over priorities.

Woven into the shirt, too, is a definitive manual for special interest groups on how to get what you want from people who matter in the halls of Congress, authored by the National Cotton Council. No one twists arms holding bags full of subsidy money more expertly than the American cotton industry. And bringing a barely imaginable future into the very real present tense, cotton has become a primary player in the world of nanotechnology, as close at hand as a J. Crew store near you. The concept of structures built from infinitely small atoms may be too difficult for our relatively underutilized brain cells to comprehend, but when spilled ketchup slides off a pair of new khaki cargo pants without a trace, that's nanotech at work, using cotton as its agent of change. Someday soon these same embedded nanofibers will create lightweight fabric seventeen times tougher than the Kevlar used in bulletproof vests. You get the idea. Your cotton shirt has places to go, stories to tell, and gossip to share.

Seeds of Conflict

The cotton item you buy at a Gap store or any other soft goods retailer begins, of course, as nothing more than a kernel with a tough exterior and a tender heart. With the addition of water at a proper temperature in loose soil, moisture seeps slowly through its seed coat and swells the interior embryo until it bursts; a root appears, then the first cotyledon leaves, and a seedling is born. All that has been understood, analyzed, and fully documented for centuries, from India to Indianola; from ancient Peru to modern Uzbekistan; and from the pre-Columbian tribes of the Southwest to the contemporary agricultural scientists who develop new varieties of that same Hopi pima cotton today in the same regions of Arizona. For thousands of years intervening humans have manipulated the DNA of cotton and other crops by selective breeding. For the first time in history, that DNA might now also carry the genes of unrelated species.

For about six thousand years since *Gossypium hirsutum* was first domesticated, a cotton plant was just a plant. Then, in 1996, thanks to the enterprising mischief of scientists at Monsanto—a ferociously aggressive, ambitious agrichemical company headquartered in St. Louis—it became a living laboratory, a genetically modified organism, or GMO. The head of Monsanto's agricultural division at that time, Robert Shapiro, had fallen head over heels for biotechnology, and with the fervor of a religious convert he and his colleagues were busily committing all of the company's resources, including many millions of dollars, to it. Cotton, a Motel 6 for every visiting insect and wanton weed, was ideally suited to take advantage of this breakthrough technology, Shapiro reasoned. As such it joined corn, canola, and soybeans as the test-tube babies of the new age of American agriculture. No longer would growers be at the mercy of nature's whims. The seeds they sowed would instead be encoded to protect their plants as well as their financial investment against the vagaries of weather, insect, and weed. Chemical costs would plummet; productivity would skyrocket. All would profit handsomely—Monsanto, first and foremost.

By 2002, six short years later, over 70 percent of field cotton in the United States was being grown with genes inserted into each seed from an outside source, and close to 80 percent by 2003. Around the world, farmers were planting more than 131 million acres of transgenic crops. By 2010, and probably long before that, all American cotton will be grown exclusively with biotech genes except for a tiny fraction produced by a dedicated core of organic cotton growers; in China and elsewhere transgenic cotton is quickly becoming the fabric of people's lives, and Monsanto has branded itself as synonymous with the controlling technology. Only the poorest countries in Africa and Asia, not able to afford the more expensive genetically altered seeds, will lag behind.

You can barely utter the words "genetic modification" in any gathering without stirring up a blizzard of controversy, particularly in European countries where the Green Party has developed significant political muscle, and specifically on the subject of GM fruits, vegetables, and other foods. Transgenic crops have given birth to a contentious group of antitechnology activists whose suspicions are regularly festooned across the pages of tabloid Fleet Street newspapers with supporting photographs of sudden hair loss and decayed molars. Fear has always been a marketable commodity. In this instance, it may one day prove to be justified, but at present there is little or no hard evidence to back up claims of harm to mind, body, or land due to gene alteration. Still, the European Union has instituted a three-year moratorium on commercialized GM crops.

Then, too, there's the threat of economic dependence on a global biotech oligarchy that includes Monsanto, Astra/Zeneca, and DuPont, which purchased the world's largest seed company, Pioneer Hi-Bred, in 1999. Where corporate money leads politicians are quick to follow. At its core, the debate in the European Union, as well as in Asia, centers on choice and trust—on who gets to choose whether GMO products are deemed safe for consumers, who gets to decide if they should be clearly labeled, and what choice of alternative non-GMO items, if any, will be available in the coming decades from the produce bin or clothes rack if genetically modified foods and textiles proliferate. In this power struggle between common consent and governmental regulations, emotions run high.

Legitimate mistrust in authority has been spurred by recent memories of the mad cow disease epidemic. During that outbreak the British government misled consumers about the extent of the risk to human health and continued to sell animal-based feed to other European countries while slaughtering thousands of the nation's own cattle. Despite repeated assurances that the feed posed no health risks, its contamination apparently led to occurrences of mad cow disease in Ireland, France, Germany, Denmark, Belgium, and Italy. Once lied to, twice wary. Environmentally conscious activists have blocked the import of biotech food crops into any European Union countries since 1998, despite protests from the U.S. government. In the view of one seasoned English GMO protestor, Zoe Elford, "When you release genetically altered plants up and down this country, those are sites of living pollution, and that pollution will replicate itself. Once it's out there, you can't get it back." Monsanto and other multinationals have launched a "manic, myopic scheme," but curiously, Elford has no problem eating foods that might be genetically modified. "I'm worried more about the big picture, really."

Tomatoes grab the headlines here, not transgenic cotton bolls. The general sentiment among most consumers is that if you can't eat it, it can't hurt you. Never mind that GM cottonseed oil happens to show up in thousands of processed American supermarket foods from salad dressings to marinades to Pepperidge Farm cookies, or that one bale produces enough cottonseed oil to cook 6,000 bags of potato chips. "It hardly matters that it's from biotech crops," says Monsanto spokesperson Karen Marshall. "There's no modified-gene component in the oil because it contains no proteins, and that's what gets encoded." Still, many consumers are left to wonder when a biotech food is not a food? The accepted official answer: When its edible seeds and oil derive from a plant categorized as an industrial textile. At present, Monsanto's cotton biotechnology operates in relative obscurity in the United States. Its GM soybeans, canola, and corn receive almost all of the press. That enables transgenic cotton—regulated by the USDA and EPA—to continue along its path free from meddlesome public scrutiny, and it allows interested observers to view it as a test case.

The majority of America's cotton growers now germinate GM seeds with two traits introduced through Monsanto's bioengineering: herbicide tolerance and insect resistance. Both resulted from experimental research that first garnered national attention in the early 1980s. While those initial successes appeared to signal a sea change in the direction of American agriculture, in reality the alliance between cotton and biochemistry by then was nearly as old as the blues.

Brothers in Arms

Enlisted to join the lengthy, hard-fought campaign against the pernicious boll weevil, chemical companies like DuPont, Dow, Monsanto, and American Cyanamid took up permanent residence in the country's fertile fields. Once entrenched, they increased productivity by manufacturing synthetic nitrogen-based fertilizers, herbicides for killing weeds, defoliants for stripping away excess vegetation that hindered mechanical harvesting, and broad-spectrum insect killers that helped ensure the investment of bankers and growers in farming machinery, seed, equipment, and land. There was a high price to pay in the loss of flavor in produce and in the welfare of the land itself, but growers traded these off for reliable yields as family farms gave way to giant agribusinesses with vast acreage.

The high cost of treatments infuriated and alienated many growers. As damaging evidence mounted, environmentalists and concerned farmers also began to weigh the consequences of repeated applications on the welfare of the farmland, farm workers, and their surroundings. They discovered these synthetic pesticides killed off numerous beneficial insects as well as the soil's crucial microorganisms. Where insecticide dosage was heaviest the dead earth itself had to be kept alive by artificial resuscitation in the form of inorganic fertilizers. Taken off life support, it could no longer function. The only temporary fix appeared to be more of the same synthetics and growth hormones in ever greater amounts—a junkie's diet that characterizes the vast majority of our agricultural land even today. America's cotton farmers soon were applying more than 41 percent of all insecticides in use—and that percentage rose to one-half the total amount by the 1970s, creating a phenomenon commonly known as the pesticide treadmill.

America's cotton planters and chemistry companies had long since become so interconnected as to be inseparable. Like Latin lovers, they continue to quarrel, make amends, fight, badmouth one another, smooch, separate, and renew their vows, only to start the cycle all over again. The world of consumers, largely ignorant of that bond, continues to approach cotton as the last unmolested textile on earth, as natural and pure as clean mountain air. Growers themselves continue to approach the crop as a weed-choked, water-guzzling, persnickety plant prone to attracting and throwing a banquet for the latest destructive worm, fly, beetle, or all of the above. Until the mid-'90s, they applied insecticides and herbicides using methods that were simply sophisticated versions of the sacks of poison that Nate Shaw broadcast by hand behind a mule: load up an aerial or groundspraying device, take aim, and fire away. Most food crops received similar saturation treatments.

Monsanto's biotechnology revolutionized the industry. It was now possible to replace the neurotoxins sprayed on cotton that poisoned a broad range of birds, mammals, and insects with *Bacillus thuringiensis,* or Bt, a natural insecticide harmful only to certain species of bugs. Biotech wizards had finally figured out how to insert Bt into the genetic blueprint of every plant. That new approach was invisible to the eye and pretty much incomprehensible to the average intellect no matter how confidently you spoke about recombinant DNA, plasmids, and the like. Overnight, all pliers and wrenches down in your workshop became useless for swapping parts in and out. A genetically modified cotton seed looked identical to the unleaded variety, but within its hull it contained the ability to tolerate Monsanto's weed-killing herbicide Roundup as well as the ability to ward off the menacing tobacco budworm, bollworm, and pink bollworm, all of whom had been feasting on cotton for more than a century. The weevil wasn't affected, but by then trap monitoring and malathion were limiting losses.

Biotech was God's way of reminding cotton farmers that the world of genetic science was about to pass them by in a nanosecond unless they caught this midnight express to the future. They did, and they quickly discovered you did not need a degree in molecular chemistry to make excellent use of its rewards or to grasp the governing principle: cut into a chain of genes from one organism, cut out a specialized one and pop it into the chain of genes of another organism. Like a guided missile with a mission, that inserted gene will perform its function in its new home. As the plant's DNA continues to replicate itself, it will continue to reproduce these new specialized properties and pass them down from one generation to the next.

Although Monsanto did not commercialize GM cotton seeds

until 1996, the company's scientists had been tinkering with transgenics—organisms created by bioengineering—for close to twenty years. These researchers at an early stage had isolated a microbe capable of splicing its own genes into plant cells. Once they had built this bacterial Trojan horse and helped it smuggle its contents behind the host cell's protective fortress walls (by attaching a bodyguard gene that prevented the cell from rejecting the foreign invader), they spent countless hours attempting to manipulate the microbe, a bacteria, to carry desired traits such as insect resistance and herbicide tolerance.

The world awoke to these brave new experiments in 1983 when a young Monsanto researcher, Rob Horsch, announced at a conference that he and his team had grown the first genetically altered petunias. A *Wall Street Journal* reporter picked up the story, and Monsanto's breakthrough made front-page news. Genetic manipulation, until then an esoteric pursuit, suddenly jumped from the lab to the media to the boardroom, all in one leap; Monsanto's corporate honchos looked at Horsch's puny petunia and saw before their eyes the lush hanging gardens of Babylon, now a majestic cascade of leafy budding greenbacks. A few other chem companies like DuPont and Pfizer made halfhearted attempts at gene-splicing, but only Monsanto committed the money, effort, and ingenuity needed to take out the patents that would give it a virtual monopoly on biotechnology. At least part of the necessary funding was generated by Monsanto's agrichemical superstar, Roundup, the trade name for glyphosate, a "broad spectrum" (read "kill-everything-in-sight") herbicide first introduced in 1976, and an immediate best-seller.

Home gardeners and farmers liked Roundup's relatively low toxicity for animals and humans and the fact that sunshine and rain quickly decomposed it, reducing its environmental impact. Quick minds at Monsanto were already scheming. What if soybean, cotton, and other seeds could be altered to carry a gene resistant to Roundup, making them immune to the weed-killer?

Cotton growers and food-crop farmers who spent a fair portion of their work days during growing season arduously removing weeds would now be able to spray an entire field, thousands of acres at once, with a blanket dose of the herbicide. The Roundup-ready plants would survive intact; the weeds wouldn't. Monsanto could not only sell sizeable quantities of Roundup but at the same time sell or license patented seed technology engineered to resist the herbicide. Whatever they called that at Harvard Business School, Monsanto's execs delighted in cooking up their version of a tasty double-revenue burger.

At Monsanto such visions of world dominance vanquished any lingering ethical conflicts between pure science and pure profit. Cell by altered cell, the company was intent on building itself into a multinational biotech corporate empire, and it was designing the potential magic-bullet GM seeds to lead the foray. Robb Fraley, appointed team leader of the project, arrived with a scientist's scorn for crass commercialism: if the whole point of biotechnology was to sell more herbicide, he announced, "we shouldn't be in this business." Fraley eventually backed off when he discovered that Roundup-tolerant seeds were supporting his biotech research at Monsanto.

By then his fellow researchers under Horsch's direction had isolated and identified the transgenic key to the kingdom. That turned out to be the glyphosate-sensitive enzyme that produced the essential amino acids all plants, weeds included, need in order to grow. Monsanto's researchers now understood how Roundup destroyed vegetation over a period of weeks. It deactivated this enzyme and by doing so deprived the plant of its amino acid nutrients. Once the scientists located the specific enzyme gene, they were able to alter it to make it immune to Roundup in cotton, soy, corn, and canola seeds—Monsanto's targeted crops. By keeping the enzyme active, they could keep the plant itself alive and well while killing surrounding weeds—in theory, that is. In practice, most of the 1980s would come and go before these bioscientists solved a series of exasperating problems and finally achieved their goal. In the meantime, farmers continued to rely on heavy external herbicide and pesticide applications.

Leasing the Future

While all that was transpiring in one lab complex at Monsanto, a radically different plant gene-splicing effort was underway as well at the company and in the labs of its European rivals. Since the turn of the twentieth century, scientists had known that the Bt microbe caused caterpillar worms to shrink and die by puncturing their digestive membranes. For forty years, both organic and conventional farmers had been using Bt in liquid or powder form to control the pest. The caterpillars, unlike the weevil, showed up in cotton fields around the world, boring into pods and squares and creating massive damage. Most bugs, like humans, like to nestle, and cotton bolls, with their soft lint centers and tough protective walls, provide cushy quarters and security for the borers. All that's missing for newborn larvae is a crib with a bubbly bright butterfly mobile playing Brahm's *Lullaby.* Once inside, the destructive caterpillars have a much better chance of escaping the effects of the externally applied Bt. But if that bacterial poison could be genetically encoded into each plant, there would be nowhere for the worms to hide. Although Monsanto had assembled a group of talents as adept at slipping new genes into plants as any in existence, they were immersed in their herbicide-tolerance labors and cranky about taking on new challenges. In certain respects research scientists and farmers were a perfect match—two of the more obstinate, willful groups of unbending individuals on God's green acre, and damned proud of their intransigence.

Still, Monsanto's executives pushed hard. Finally, even as the Roundup-ready experiments continued, a new insect-impervious

(Bt) cotton seed emerged. There was one problem. Monsanto was not in the business of selling seeds. Rather, it provided biotechnology for use by another outfit that bred and distributed a delivery system for Monsanto's biotech—a company that sold the actual seeds. After a complex set of negotiations, the firm teamed up with Delta and Pine Land (D&PL), located in tiny Scott, Mississippi, a town it owned outright. D&PL's chairman, Roger Malkin, knew that his firm controlled 70 percent of the cottonseed market. He knew that under newly promoted CEO Robert Shapiro, Monsanto kept the lab lights burning until 1 A.M. One determined master of the arable universe recognized another. As Daniel Charles reports in his entertaining and incisive chronicle of biotechnology, *Lords of the Harvest*, Malkin also knew that seeds were cheap for cotton growers—$8 an acre. Insecticides cost a bundle—often $150 an acre, a huge expense. If D&PL sold Bt seeds that drastically reduced insecticide costs, it could raise its prices and still keep the planters happy. Malkin could charge as much as $32 an acre for D&PL seeds, split the proceeds with Monsanto, and please everyone.

Transgenic cotton seeds encoded to tolerate Roundup and to repel insects quickly became objects of wonder and desire for cotton growers across America. Some were "stacked" to do both at once. There was, however, one catch. Those seeds were not for sale—at least in any traditional sense. To obtain a Roundup-ready or Bt or stacked seed, each grower had to sign a contract with D&PL prohibiting him from saving and replanting GM seeds from one harvest to the next, or face a civil lawsuit. That seed-saving practice had been a way of life for farmers for at least 10,000 years. No longer. A farmer now legally leased these seeds; violating the agreement not to replant could make him liable for damages, "which will be based on 120 times the applicable Technology Fee," or seed cost. D&PL and Monsanto meant to make sure that farmers using any of their GM seeds from soy to cotton were obliged to purchase a new supply each season.

The nation's cotton growers, among others, at first raised hell.

By temperament, no one who rides a tractor for a living likes to be told what he can and can't do. That's why he'd rather tromp through muck in a reeking hog sty before sunup for a nickel than sit in an office cubicle and answer to a dimwit department manager for a dime. But ironically, nature itself stepped in to help seal the deal. By 1996 a class of chemicals called synthetic pyrethroids were losing a battle against the devastating tobacco budworm—cotton's most potent enemy after the weevil. Growers in the Delta floodplain and elsewhere were spending up to $140 an acre to ward off infestation—and watching helplessly as squares fell and the crops failed. By then most tenant farmers in the Delta had long disappeared, giving way to an elite corps of Mississippi plantation owners with large holdings and impressive political power in Washington, D.C. They were about to tear into Monsanto and defy its edict when the budworm struck across the Cotton Belt with the force of a demonic plague. "It was the most horrible feeling you ever had in your life," cotton farmer and entomologist Jay Mahaffey remembers. In that crisis mode, the promise offered by Monsanto–D&PL's new Bollgard, the trade name for its Bt seed, outweighed all other considerations. The cotton fields of America were dying. Within a year 75 percent of those fields in Alabama were planted with Bollgard seed, 60 percent in Arkansas and Georgia, and well over 90 percent in the Delta. As Charles remarks, "Bt didn't just clobber the tobacco budworm, it also delivered a staggering blow to the companies that sell insecticides to cotton farmers. Before 1996 [Delta grower] Frank Mitchener was spending up to $140 per acre fighting insects. After 1996 he spent an average of $90 an acre, and a third of that was the 'technology fee' he paid for the right to use Monsanto's Bt gene."

Green activists soon began trading one worry for another. They noted the positive effects of encoded Bt. Now that it reduced infestation by the budworm and pink bollworm, even if less effective against the bollworm—it mitigated the unintended damage caused by synthetic insecticides. Crop-dusters that had saturated cotton fields across the South with toxins ten or fifteen times a season were in contrast able to spray three to four times in the hill regions and floodplain, an astounding decrease that Rachel Carson, among others, might well have applauded.

But to evolutionary biologists like Fred Gould at North Carolina State University, there are no permanent solutions to ridding crops of the tobacco budworm or any other insect. There are only temporary victories in the eternal struggle for dominance between man and nature, never a winner and no endgame. DDT, Gould liked to point out, had already lost its potency against numerous insects while it was still contaminating humans. To Gould and others in his field, anyone who assumes that genetic manipulation eliminates the process of natural selection knows nothing about either. Bt might well be a glorious gift. If the past is prologue, using it indiscriminately almost certainly assures that someday a few chance worms will develop resistance, breed, and circumvent man's transgenic hanky-panky, an event sure to jeopardize millions of acres of cotton and other crops. Then what?

Fearing just such an event, the USDA and EPA instituted a series of safety precautions. Bollgard farmers are now required to set aside at least 4 percent of their land as a "refuge"—that is, to keep it Bt- and insecticide-free in order to welcome the invasion of tobacco budworms and their relatives. The idea is that any potential Bt-resistant worms will mate with nonresistant neighbors and prolong the likelihood of producing immune offspring. Cotton growers at first protested vigorously. They were being asked to transform a chunk of their acreage into an experimental insect-control laboratory without compensation. On the other hand, they had little choice if they wanted to outwit these boring insects and take advantage of the newest technology.

Ironically, man's most sophisticated manipulation of things that grow since the dawn of history may have created the likeliest opportunity for nature to once again go about its own business without interference. Beneficial insects such as spiders and wasps that feed on destructive pests have begun to return to GM cotton fields. Five years from now we may be praising genetically modified cotton and food crops for their help in restoring a natural balance—or damning them for creating unforeseen hazards in the process.

Skeptics like environmental science consultant Dr. Charles Benbrook, however, are quick to point out that we're dancing on the lip of an active volcano. If Bt-encoded plants become compromised, whether corn or cotton, Benbrook argues that it "would be analogous to the loss of antibiotics in the ongoing struggle against human infectious diseases." As a former executive project director for the House Committee on Agriculture, Benbrook is a critic with credentials who gets quoted frequently in the popular media. He has no faith at all in long-term control of insects by Bt crops.

He thinks the EPA should require larger refuges, closer monitoring—that, in short, the regulators need to be regulated. Other scientists, including University of Arizona entomologist Bruce Tabashnik, are wary but less anxious. "Within the next five years, insects are bound to adapt," Tabashnik says. "That doesn't mean we shouldn't be pursuing genetic modification. We just have to recognize that transgenic crops are one tool to be combined with other control methods; they're not a permanent solution." He made news in 1990 when he identified the only insect thus far, the diamondback moth, to build up resistance to externally sprayed Bt. A 2003 study conducted by Tabashnik and colleagues showed no evidence that insect pests had evolved resistance to Bt transgenic crops. Monsanto, taking no chances, was already marketing Bollgard II by then as a "better two-gene Bt product."

There are a host of "what-ifs" in these biotech scenarios, and few certainties. One is that most of us walking around in a genetically modified cotton shirt have no idea if we are benefiting from human ingenuity or funding future chaos. We don't have the scientific training to assess whether we're poster children for technological progress or lab mice in a risky microcellular experiment of the sort that goes awry in apocalyptic science fiction. Welcome to the twenty-first century. When every giant step forward for technology may or may not be an eventual leap backward for humanity or an exquisite pirouette into the abyss, you know you're in cotton country.

Foreign Invasion

For those of us in the dark about the future of biotechnology—and that's all of us, by the way—there's at least one reason for optimism. That would be the second half of the word. Technology is a pursuit the United States continues to excel at and market successfully to the developing nations no matter what the consequences. Cotton was there at the beginning—in fact it *was* the beginning.

America's first infatuation with technological ingenuity occurred, of course, when Francis Cabot Lowell introduced mechanically spun and woven cotton fabric to the young nation. Gears meshed, belts rotated, shuttles flew, and a plant's dewy hairs miraculously became a shirt. Biotech is a legitimate stepchild of that same industrial revolution, as far removed as it may seem. Sooner or later some of our cleverest inventive minds were destined to turn their attention back to the plant itself. Using instruments that seem as baffling to us as the spinning mule once did to our colonial ancestors, they are retooling that plant's DNA to improve yield and reduce costs. (It can be only a matter of time until biotechnicians also insert genes that make cotton plants perform more efficiently for conversion to fabric. When the 500,000 or so fibers inside the bolls no longer stick to their seeds and arrive at the factory uniform in length, thickness, and hue of whiteness, each bale identical to any other, many steps in the production process will be eliminated or facilitated.)

That same technical expertise that marks us as a pivotal

player in today's global community also explains why America's true ambassadors to many developing agricultural nations are no longer the diplomats who maintain our embassies in dusty backwaters. Monsanto and its determined competitors have been handed the job. As they introduce genetically modified seeds to farmers in countries dependent on food and cash crops like cotton for survival, they also broadcast and promote our values and priorities. It's an odd dynamic. Futuristic, cutting-edge biotechnology arrives in remote villages lucky to be wired for electricity where men still plow their few exhausted acres behind oxen. The companies' invading forces show up in seed sacks, not in amphibian assault craft. Still, surrender by the natives remains a key objective.

Cotton is leading this onslaught in China and India and in underdeveloped areas of South Africa and Mexico. The Philippines and a host of smaller Asian and African countries will soon be cajoled to adopt GM seeds as well. In India, the third largest grower of the world's cotton behind China and the United States, almost all fiber comes from small family farms, is ginned in the growing region, and moves along a convoluted distribution chain to be eventually blended by Indian or neighboring textile mills into fabric purchased by buyers for Gap Inc. or scores of other international retailers. Some observers close to the action, like Washington University anthropologist Glenn Stone, who travels extensively throughout India's farmlands, note that GM seeds pop up in places where efforts to enforce control of any new technology quickly become subordinated to the law of the cash-and-carry jungle. "Bollgard was supposed to be sold only by authorized dealers who had taken a day-long training course so they could answer farmer questions, fend off criticisms, and arm-twist the farmers into planting refuges," Stone reports. "But a secondary market developed immediately, as I demonstrated by buying a box from an unauthorized dealer who didn't know diddly about it. GM cotton worsens the process of Indian farmers losing control of agricultural information and farm management strategies. The GMO information vacuum is a deep problem."

In India, where more than 50 percent of all pesticides are used to destroy cotton pests, black markets flourish and unlicensed sellers routinely ignore environmental or health safety. Few if any effective controls exist. A Gujarat seed company, Navbharat, with no genetic engineering facilities, simply crossbred Monsanto Bt cotton from open field trials with conventional cotton, and in 2001 produced an unlicensed hybrid to resist the bollworm that was plaguing that region and Punjab.

In the previous year the pest had wiped out 90 percent of Punjab's cotton. Farmers using the improvised Bt seeds were elated—until the government stepped in under pressure from Monsanto and its Indian bio-seed partner, Mahyco, and ordered them to burn all of their 11,000 acres of "unapproved" crops. To a great extent Navbharat's ad hoc solution defines the way that daily commerce operates in India, and has for millennia. Expediency rules among the millions who live on the brink of starvation. Concern for unwanted consequences is a luxury few can afford or even contemplate. Unapproved Bt hybrid seeds quickly became a hot underground commodity. Farmers planted them to grow varieties of GM cotton that failed or that proved ineffectual against numerous pests. The Bollgard initiative broke down even as it began.

One of India's most vocal anti-GM activists, Vandana Shiva, points to the dismal events of the recent past in the Warangal District of the state of Andhra Pradesh as proof that in India's chaotic, unsupervised black market, there are no safeguards in place to prevent agriscience from quickly turning to a man-made disaster. In 1998, cotton farmers there were sold a new chemical spray that ruthless dealer-moneylenders promised would rid them of a destructive bollworm species certain to wipe out their young crops. As Stone observed, while Indian farmers have a long history of ingenuity in cultivating local varieties, they have problems protecting crops planted with New World cotton species that are vulnerable to many Indian insects. That sometimes leads to desperate measures. These growers invested all of their meager savings in a pesticide that proved to be worthless.

Narsoji, a grower in Kadavendi Village, lost his entire crop and owed the moneylenders about $3,300—or two-and-a-half years' earnings. Having sold his two oxen, Narsoji had nothing left to his name beyond a few containers of the useless poison. Facing ruin, he drank one down, convulsed, and died in his open-air kitchen. In a nearby village, another desperate grower named S. Sailman squirted the contents of his pesticide sprayer into his mouth and died, leaving behind his illiterate pregnant widow with two small children. "Sell your oldest son to a landlord," a man in her village advised, pointing to her six-year-old. In that year, bondage paid eighty dollars a year. Narsoji and Sailman were two of the more than 600 bankrupted cotton growers in Andhra Pradesh who committed suicide by swallowing the ineffectual pesticide. Monsanto seized on the tragedy, which received scant attention from the Western media, as proof that Indian cotton farmers needed its Bt seeds to combat those pesticide-resistant bollworms. The company's India marketing director claimed at the time that its Bollgard could have prevented the suicides—an arrogant, erroneous assessment, since Monsanto's transgenic seed technology did not effectively combat the bollworm species that caused the destruction in Andhra Pradesh.

To Vandana Shiva, vehemently opposed to GM, unregulated biotech seeds will someday lead to even more horrific consequences by promoting unstable monocultures. Monsanto's position, says its communications director Karen Marshall, is that "it is absolutely wrong to deny poor, penniless growers this technology." Perhaps—at least in theory. But a constant stream of reports from the fields of India contains stories of confidence tricks played by seed dealers on farmers like Venkat Reddy, who switched over to Bt cotton and found it grew inferior short-staple crops he couldn't sell. "It is clear that the Bt cotton has failed on all counts and . . . has neither improved the yield nor reduced the pesticide usage," one farm scientist team concluded after a field visit.

"Seed-saving is so fundamental to Indian rural society that any threat to the practice is a threat to the society itself," the Christian Aid Organization in Great Britain argues in an overview of Bt it calls "Selling Suicide."

No one escapes this controversy untainted. The antitechnology faction appears to be made up of a few too many rigid extremists who would rather fight the future than embrace it with reasonable caution. They rarely if ever present satisfactory economic alternatives. Proponents like Monsanto appear to want to impose their solutions with a perilous disregard for local customs and conditions. Cotton's lesson to the world in all of this is that to be safe and effective, bioengineering needs to be accompanied by a practical set of regulations that can be realistically implemented. That rules in America and at least for the moment rules out India and most Third World nations where the need exists but where survival is at such a premium and enforcement so lax that genetic modification remains a treacherous gift at best, an invitation to disaster at worst.

There is bound to be considerable media attention focused instead on China's adoption of GM cotton. According to *Business Week*, by 2001 more than one million of that country's farmers were already using modified seeds. Currently, Bt cotton in China represents the world's most widespread use of transgenic seeds. They've reduced pesticide use by an average of thirteen sprayings a year and increased production. The future looks bright, but few Asian or African countries are able to control and monitor their agricultural practices as closely as China. In a messy universe where greed regularly trumps wisdom, GM may come to be seen in the short term as a purveyor of grief rather than a savior of the downtrodden. Several generations from now, however, today's bug-infested fields might well be overflowing with healthy gene-manipulated produce and cash crops.

Eco-Cotton

Maybe it's that fear of unknowable consequences or simply a determination to decide for yourself what to put in and on your body that currently drives huge numbers of conscientious American consumers to scurry in the opposite direction about as far as an organic carrot can sustain them. Whatever their motivation, this rapidly expanding group has turned its attention to eating fruit and produce grown without chemicals, to making environmentally responsible purchases, to avoiding processed foods, and to supporting local farmers' markets whenever possible. Way back when, the parents of these folks may have crashed in communes on the funky side of town. Now their offspring shop Whole Foods, compost, and drive a Prius. They're members of an all-volunteer, earth-friendly army that recycles and subscribes to the tenets of LOHAS—Lifestyles of Health and Sustainability.

But what to wear to war? Hemp, that all-purpose politically correct fabric does everything but provide creature comfort; most of the currently available selection is too knobby and coarse against the skin. As for cotton—you might treat it like family, but by now you've probably come to recognize that shirt on your back doesn't quite qualify as purebred and eco-cool. If its genes haven't been tinkered with, its fibers have been laced with pesticides, and as a fabric it has no doubt been doused with harsh, environmentally hazardous finishing chemicals such as optical whiteners, and heavy-metal dyes.

In a better world, you might argue, there would be an acceptable alternative—and there is, barely: organic cotton. Its current status offers an insight of what happens when products that address larger concerns like the welfare of the planet are pitted against the realities of the marketplace. Information is the wild card. You first have to know that organic, nonchemical cotton exists in order to consider buying a bedsheet, baby blanket, or garment made from it; you then need to find a retail or e-commerce outlet that sells it, and so on. Large soft goods chains are no help. In fact, it's not in the interest of retailers to have customers clamoring for a product they don't produce in large enough quantities to meet their margin profitability goals, which partially explains why organic cotton gets almost no mainstream publicity. As a consumer making purchases that reflect your socially responsible values, you've go to do some detective work to find a source, most probably by searching the Internet. There may be a Whole Foods nearby, but no Whole Cloth. How come?

"You don't stuff your shirt into your mouth," Patagonia's director of environmental analysis, Jill Vlahos, explains, pretty much nailing it. She and others in the trade estimate that public demand for organic fiber is lagging ten to fifteen years behind food. Most of us, LOHASians included, care more about what's in us than on us. When we do, we look for products that reflect our values without sacrificing taste or style. Avoiding synthetics—petrochemical fibers—is about as close as we can get. The few name companies that use and promote organic cotton have taken pains to create appealing designs, whether manufacturing apparel or home furnishings. Patagonia, a leading upscale outdoor outfitter, switched over to 100 percent organic cotton in 1996 when founder Yves Chouinard, after considerable research, concluded that "the most damaging fiber used to make our clothing may actually be conventionally grown, 100 percent 'pure' cotton." Discovering how environmentally destructive cotton is, he added that "it would be unconscionable for us to do anything less." "We've only got one planet," says Vlahos, "and our customers think it's worth saving." Patagonia shares its organic sources with competitors.

Gaiam (a contraction of "Gaia," the Minoan word for "Mother Earth" and "I am") is the country's largest retailer of organic cotton—almost exclusively to women, and primarily in the context of yoga wear, linens, and items for infants. One of the obstacles for organic cotton has been that consumers have to pay its premium price—up to 30 percent more, but Gaiam keeps that within reason. "We don't want to sell eco-products only to eco-heads," says textile product developer Bill Giebler. "I want to get my mom on board, too." It can be a tough sell even to Gaiam's aware customer. "She won't eat a tomato filled with crap but textiles have no direct link to health, so they're down on her list. Until she has a baby. "Then," Giebler says, "only organic cotton can touch its newborn skin."

There may be no precaution too extreme for a young mother, but for infants through seniors, allergic skin reactions to conventional cotton are rare. Industry spokespeople point out that just about all of the harmful chlorine bleaches, caustic soda, formaldehyde, and heavy metals used extensively in processing and finishing cotton cloth get rinsed and leached out by the time these products reach an apparel or home linens retailer; so, too, do residues of harmful pesticides. Environmental activists maintain that can be true and still be beside the point. "When planes still sweep down and aerial spray a field in order to kill a predator insect with pesticides, we are in the Dark Ages of commerce," sustainable agriculture advocate Paul Hawken contends. "Maybe one-thousandth of this pesticide actually prevents the infestation. The balance goes to the leaves, into the soil, into all forms of wildlife, into ourselves. What is good for the balance sheet is wasteful of resources and harmful to life."

Nike, Timberland, and Norm Thompson's Early Winters outdoor catalogue are among recent converts. "From this day on, sustainability is part of your job," Norm Thompson's CEO told his employees in 1999. It wasn't an option. "This means you share a common mind-set, earn recognition and contribute to the health of the earth." Or take a hike. The company currently uses some organic cotton in 35 percent of its Early Winters apparel, which will be all-organic by 2006. One such sweater is described in its catalogue as providing the wearer with "a soft, easy, chemical-free layer."

And so it goes—but why, exactly? What's the rap against good old cotton? Organic cotton's advocates need little prodding to unleash a torrent of startling facts and figures. One-third of a pound of chemicals is used to produce the cotton in every tee-shirt. Three-quarters of a pound of chemicals goes into every pair of jeans, and about one and a quarter pound into every set of queen-sized sheets. Five of the nine pesticides used in U.S. cotton are classified as carcinogenic by the EPA, including dangerous cyanazine, finally phased out because of its proven link to breast cancer. "The pesticides used on cotton, whether in the U.S. or overseas, are some of the most hazardous available today," Doug Murray, a pesticide expert at Colorado State University, comments on one of the organic community's numerous Web sites. If supermarket tabloids ever picked up the story, they'd likely headline it, "Cotton's Dirty Little Secrets!"

In West Texas alone, every year cotton accounts for 13.8 million pounds of agricultural pollutants dumped into the soil, water, and air. There are 300,000 pesticide-related farmworker illnesses annually in the United States. The USDA reports that 84 million pounds of pesticides were sprayed on U.S. cotton as recently as 2000, making it the second most heavily sprayed crop behind corn.

Around the world cotton is currently grown on about seventy-seven million acres in eighty countries, or 3 percent of the earth's arable land. It represents a small fraction of all crops in production, yet it uses 25 percent of the world's insecticides and more than 10 percent of all herbicides and defoliants. In India 91 percent of male workers in pesticide-laden cotton fields experience health disorders. The toxicity of those chemicals made headlines when a leak at the Union Carbide pesticide plant in Bhopal, India, in 1984 caused forty tons of lethal gas to spill out, killing more than 3,500 people in the first three days, and more than 10,000 overall. Another estimated 120,000 survivors experienced severe respiratory, liver, kidney, and related ailments. (Union Carbide paid $793 to each family of the deceased.) Organophosphates like malathion and carbamates primarily used for cotton were a major component of the deadly gas.

Conventional cotton may be safe to wear and may feel as comfortable and welcoming as a soft hand on the cheek, yet it can endanger our grandchildren, the inheritors of tainted streams and tap water filtering runoff chemicals potent enough to peel the chrome off a trail hitch—or so organic cotton's supporters argue.

The mind reels. This is not the kind of news most of us want to wake up to. Cotton, after all, is about the only thing we can cover our skin with to go dig a ditch and keep in our medicine cabinet to clean off the scrapes we get while digging it. There's some relief to be gained from industry observers who assert, correctly, that the great majority of the synthetics used to grow the cotton in jeans and tee-shirts actually comes from fertilizers, not toxic pesticides. There's more comfort still in the significant recent drop in pesticide use in this country as a result of the widespread introduction of Bt seeds. But that alternative doesn't placate many of organic cotton's advocates. Genetic modification has been rejected, says the Pesticide Action Network, "because of the potential for genetic contamination and its continued reliance on artificial chemical inputs." Unwanted cross-pollination and fouled-up insect management strategies are frequently mentioned problems.

If size alone mattered, no one would be paying much attention to any of this. Organic cotton accounts for less than one-third of 1 percent of worldwide production—most of it originating in Turkey, Peru, and Uganda. Only one-tenth of 1 percent of domestic cotton is grown organically. In 2002 that amounted to fewer than 9,100 acres out of millions in the United States, a drop of 22 percent from the previous year. Those numbers, according to a spokesman for the National Cotton Council, "round out to zero."

But like any shrewd guerrilla movement, the organic cotton industry marshals its meager resources to strike where it can be most effective. In California, the Sustainable Cotton Project, founded by organic farmer and activist Will Allen, has arranged bus tours of the San Joaquin Valley's vast cotton farms each fall since the early '90s for representatives from the largest corporate cotton-users—Ikea, Levi's, Gap, Nike, to name but a few. To attending company employees, who for the most part dwell and work in urban environments, these field excursions leave a lasting and vivid impression. They may see a crop-duster swing down low from the skies like a bird of prey to blitz crops with sheets of toxins at close range. Too close. Even at a safe distance an acrid scent fills visitors' nostrils, and just about the time they come up for air they arrive at a nearby dairy farm, where they find themselves face to face with mountainous pyramids of cotton trash discarded during the ginning process; the trash looks like lumpy gray mattress stuffing. Heavy-equipment movers deposit it into cramped cattle bedding areas that sit beneath the corrugated roofs of buildings easily mistaken for minimum-security prison camps. There is no grass in sight. Cattle shove their heads through narrow openings between bars in order to eat cottonseed meal. They live in mud and trash, dragging one to the other until all distinctions between wet and dry, vegetation and muck become blurred. The stench is overwhelming. Dairy owners buy gin trash from nearby cotton farms for the cows to sleep on, as well as hulled meal favored for its high protein content, for feed. In turn, cotton planters find an eager, local market for their gin waste and for the seeds that are not sold to oil manufacturers. Dairy cows routinely eat the pesticide-laden cotton trash they sleep on.

"I couldn't swallow food for a week, I was so sick to my stomach," Ikea representative June Deboehmler recalls. She had no idea, she says, that chemically treated cotton ends up indirectly in our milk and in our meat. Ever since, Deboehmler has been lobbying Ikea, an environmentally conscious company, to blend organic cotton into its fabrics.

Theoretically, consumers need not worry about malathion or a dozen or so pesticides currently sprayed on cotton. The FDA monitors for unacceptable levels in food products. The conventional wisdom is that these pesticides generally break down before harvest or get removed during processing and present no health risk. But reports from the front sometimes tell another story.

"If cotton were a crop that we ate instead of one that we wore, the EPA and FDA wouldn't allow us to spray it with some of the things we use," a Department of Agriculture extension service agent, Jerry Williams, told the *New Yorker* in 1991, almost twenty years after DDT had been banned. Beef and dairy cattle consume three million tons of cotton meal a year.

Still, you are hard-pressed to find cotton growing organically anywhere in abundance in California, second only to Texas in production of the conventional fiber. In a state where the fields of one nonorganic grower, J. G. Boswell, cover more square miles than Rhode Island, maverick farmer Pete Cornaggia, living near Chowchilla, plants 108 acres of organic cotton a year and wonders why he's still at it. "I got a mess," he says, adding up the higher cost of using turkey manure and compost and beneficial insects to bring in his crop—"about $40 an acre extra." To Cornaggia, "It don't pencil out. There's not that many people out there who's gonna pay the price." Also, there's the cost of conversion to organic, and the wait—three years to gain certification. "The state charged me $3,200 to be certified, then $300 more a year to clean up the environment and all this crap and I don't understand it. Plus, lately I can't sell my bales. I keep hoping it will turn around."

In 2003 it did. Suddenly demand for organic cotton exceeded supply, and conventional cotton shot up to about one dollar a pound as crops failed in China and Pakistan due to inclement weather, flattening out the difference in market price. Cornaggia emptied his warehouse. That's farming, a crapshoot at best and fool's wager at worst, and everyone who does it knows it. "I got into organic on account of it was something different," Cornaggia says. "I got laughed at. A few neighbors tried it; most have gone bankrupt. You don't get twenty to forty cents more per pound for organic, you're dead in the water. If organic can't support itself there's no reason for it, is there?" Pressed, he confesses to a weakness for nonchemical farming as if admitting to an embarrassing character flaw. It won't bring wealth any time soon. Maybe peace of mind, though. "Try puttin' a price tag on that."

Preaching Purity

Cornaggia is one of fewer than a hundred cotton farmers out of about 25,000 in America who plant organically. They typically work small spreads since labor and materials costs rise substantially above 400 acres. Their fiber, hand-picked rather than machine-harvested, arrives at the gin cleaner than conventional cotton but in most other ways it closely resembles it. On the high plains of West Texas in O'Donnell, near Lubbock, organic cotton growers LaRhea and Terry Pepper have become the uncrowned reigning sovereigns of the movement. LaRhea, a large and forceful woman who operates with the kinetic energy of a turbo-engine, helped cobble together a loose confederation of thirty like-minded farming families in the area who formed the Texas Organic Cotton Marketing Co-op. It now produces about one-third of the country's organic cotton.

"I'm preachy and I know it," LaRhea explains. "My question, the one I ask myself and other farmers hereabouts, is this: Can you look in a mirror tonight and say you did the best with the resources you have?" LaRhea is not the sort of person you readily say "no" to, or, for that matter, "Let me think about it and get back to you." If she's sure, you're sure. Her father, Jack, puts his daughter's passion in perspective: Within a few miles of their farm, he says, every day he passes the homes of eight cotton widows, women whose husbands died prematurely in their pesticide-laden cotton fields. He encouraged Terry and LaRhea to go organic in the late '80s, damn the expense, for a simple reason: he wanted West Texas cotton farmers to live longer, especially his own kin. Terry didn't need much convincing: his father, who used "whatever chemicals the agricultural schools were pushing," had died of acute leukemia at fifty-seven. In Terry's opinion, there has to be a strong connection between those chemicals and his dad's early death.

The Peppers quickly discovered that unloading their harvest was as much of a challenge as growing it. "The mills don't buy on a *maybe,*" LaRhea explains. You can farm with the best intentions and still go broke growing organic cotton unless you deliver an agreed-upon number of bales by a specific date to a mill with a firm purchase order for it. Since as an organic grower you don't chemically defoliate your crop with paraquat, a restricted-use pesticide, or use chemical growth regulators, desiccants, and boll-openers to boost your yield per acre and prepare the crop for harvest, that's always a crapshoot. You're dependent on a hard freeze to drop leaves. The first chilling frost arrives on nature's schedule, not yours, and it can lay waste to the best-laid plans. If you get your crop in on time, you still might not find a mill to buy it. In their first full year, 1991, LaRhea and Terry were caught holding 400 bales they couldn't move.

A textile factory floor manager knows that running organic cotton will cost him up to eight hours of downtime while every piece of equipment from carding machines to looms is first cleaned to remove leftover conventional lint. That's a required part of the drill; the expense is passed on, and it's a primary reason that organic cotton products can cost twice as much as their conventional counterpart and, as a result, often meet consumer resistance. That in turn creates problems that ripple all the way back out along the supply chain to the grower.

Another time, braving the winds of fate, the Peppers ambitiously picked up the manufacturing tab for 4,000 yards of the Texas co-op's organic denim without knowing if they could wholesale it. No one wanted the denim—not until LaRhea got on the phone and starting making calls, many calls, most of which went something like this: "Here's what we got for sale and here's why you need to offer it to your customers even if you don't know organic from Adam and couldn't care less about the welfare of our planet. Because you should, we all should, besides which, organic cotton is the *only* fabric for people with skin sensitivities, and whether or not you know it, you've got customers who break out from resin residues like formaldehyde in regular cotton and can't find any other choice; they will stay loyal to you to their last dying breath if they can trust you to supply them with truly pure cotton, you hear? How much have I got? Not enough to go around, but I tell you what, you lock in your order with me here and now and I'll guarantee your allotment."

That was then, back in the early days. More than a decade later, in 2002, LaRhea Pepper sits in a booth at the Natural Products Expo West in the Anaheim Convention Center surrounded by sample merchandise manufactured by Cotton Plus and Organic Essentials, the companies she started with the organic cotton co-op. Cotton Plus is a comprehensive source of organic fabric from chambray to knit, while Organic Essentials, which sells feminine hygiene products, cotton balls, and personal care products to a niche clientele, is growing at a rate of 40 percent a year. The Peppers now sell their cotton to a Mexican factory that makes organic tee-shirts for Patagonia.

Many of the 20,000 or so Expo daily attendees saunter up and down adjacent aisles stuffing an endless array of product samples, from soy-cheddar bunny munchies to organic hemp protein powder, into their bags. The Expo attracts an eclectic crowd, from U.S. Senator Orrin Hatch to reggae singer Ziggy Marley. From her perch, LaRhea takes it all in. By now she's been on a sales mission to Japan, corralled organic dairies in Colorado to use her uncontaminated cottonseed for feed, and buttonholed congressmen in the halls of the nation's Capitol Building to help bolster the fortunes of her farm and her vision. She seems to know everyone, as well as everything on earth about the benefits of organic. "Cotton fibers spiral, and in their natural state that spiral doesn't shrivel or get scrunched, like it does from chemical growth regulators and Bt and Roundup-ready genes. What that means is, organic cotton's the softest going. It drapes better, too. Also, because no resins are used in finishing it, it's best for skin sensitivities."

The country's leading organic cotton broker, Chris Hancock, drops by her booth for a chat. Although Hancock specializes in long-staple cotton grown in Peru, Turkey, Uganda, and Pakistan, and the Peppers grow a shorter variety, they share industry news with the casual intimacy of comrades in arms. Hancock lives in New Jersey, but his broker's heart is in Okinawa. "These days," he says, "Japanese consumers account for about 90 percent of the organic cotton market. The environment is part of every aspect of people's lives. They don't have enough open space to be careless about what's in their air and water. Even the slightest amount of runoff from low-impact dyes is considered impure." A compact wiry man with slicked back hair and a thin moustache, Hancock puts to rest any Birkenstock images that attach themselves to environmental crusaders. He speaks the lingua franca of his trade mostly out of the side of his mouth with the rasp of a gumshoe in a '40s *film noir;* it serves as a refreshing antidote to New Age smarminess. "I'm running Tangus in Peru and Izmar in Turkey," he confides. He could be talking about running guns from Uzbekistan.

Hancock is a man who can sigh in at least ten different ways. Brokering organic cotton will do that to you. There's either too much or not enough to go around. It's either spoken for or rotting away in some obscure warehouse when you desperately need it. Timing is everything. In 2003 Hancock was frantically searching for raw material to meet a sudden boom in demand only to discover that Nike had gobbled up most of it for its new organic cotton activewear blend. Between the migraine headache of matching niche buyer and seller, and being dependent on the vagaries of commodity trading, there's no persuasive business rationale for his efforts. But Hancock grew up in cotton mills. Although he loved fabric, he hated what went into it. He saw and smelled a few too many vats of the acrid, corrosive chemicals used to bleach cotton, to give it a satin sheen, or to dye it white—as in a white cotton shirt—or with all the colors of the rainbow, and finally it provoked him to follow his own path. "If I wasn't involved in organic," he says, "I'd be out of the textile business."

High-Tech Couture

Almost certainly the shirt on your back contains no organically grown cotton. If it's made to confront the elements, it's probably a synthetic like CoolMax, Nature Tex, Icetec, Omni-Tech, or Diaplex Laminate, all derived from petroleum by-products. "The outdoors is the gym of the '90s," said Beth Gillespie, a spokeswoman for Portland, Oregon–based Columbia Sportswear, about a decade ago. But nothing stays in one place very long these days, including the outdoors. It's moved inside, fiber by microfiber, into clubs and sports bars and a wide range of venues where casual clothes rule. In their reincarnated form, today's technical fabrics bear little resemblance to the clingy, stinky polyesters of the '70s. They breathe, wick away moisture, and can be treated to protect against just about everything from insect bites to sunburn to body odor.

In the film *The African Queen*, Humphrey Bogart's gin-soaked Mr. Alnutt is scolded about his drinking by Katharine Hepburn's prim Miss Sayer. "Geez, missy," he groans in self-defense, "it's only human nature." Hepburn wants none of it. "Nature, Mr. Alnutt," she snaps back, "is what we are put into this world to rise above." In a single retort Hepburn encapsulates the philosophy that drives the world's leading textile science and technology firms.

That would seem to leave humdrum natural cotton far behind, dangling from the clearance rack under the rear eaves at JCPenney. As its history demonstrates, however, cotton doesn't always conform to expectations. Its acolytes are not about to let the fact that it naturally wrinkles and mats to the skin when wet interfere with its welfare. "Cotton is dead," says top Hollywood costume designer (*Men in Black*) Mary Vogt. As fashion, perhaps, but "dead again" may be probably closer to the mark. In *The Botany of Desire*, Michael Pollan looked at the way some plants like the tulip and potato use their sweetness, beauty, and so forth to make us fall in love with them. By benefiting us, they get to take advantage of our best efforts to benefit them, and as a result they prosper and endure. That's cotton. It's nature's way of reminding us that by comparison we know next to nothing about how to seduce and bewitch over the span of a lifetime. Faltering here, it rallies there. Consider cotton's current liaison with nanotechnology. Synthetics may have passed spun cotton by with barely a nod, but nanotechnology, so cutting-edge it makes CoolMax look like a monk's medieval fustian, has eagerly embraced it.

Nanotech operates by manipulating individual atoms to form desired structures in a molecular universe viewable only through a scanning electron microscope, where each nanometer is one-billionth of a meter, and eighty thousand equal the width of one human hair. Whoa! This world of self-replicating consumer goods and nanorobotized industrial fabrications remains primarily in the science-fiction experimental stage, at least a generation away from practical application. But one exception is fabric. Eddie Bauer, Levi Strauss, and Gap Inc., among others, have begun to weave nanofibers into their apparel.

Cotton, they discovered, is the ideal vehicle to deliver this new technology to consumers. Rolls of it are immersed in vats containing nanofibers; resembling whiskers at the microscopic level, trillions of these fibers bond with cotton and permeate the fabric. Each has a unique property—stain-resistance or moisture-wicking or flame-retardation, for instance. Nanosponge fibers soak up hydrocarbons that create unpleasant odors. Oven-drying the permeated cotton permanently binds these nanofibers to the threads, producing garments and home furnishings that shrug off spilled wine and never crease or work up a sweat throughout their life span.

In the foreseeable future these fibers will provide daily doses of vitamins as well, monitor heart rates, and eventually even power electronic devices. One leading player is Nano-Tex, a subsidiary of Burlington Industries. It provides the technology that makes Eddie Bauer's Nano-Care cotton khakis or Levi's Stain-Defender Dockers impervious to our messy habits and grimy labors. But do we really want clothing that walks through life unsoiled by dirt and immune to human frailty? Isn't there some required penance attached to mopping up the red wine we spill on a white couch, some crucial need for self-humiliation we satisfy by dousing the stain with club soda and emptying a box of salt on it while party guests scrutinize our pathetic efforts with a mixture of pity and disdain? Are we tampering with evolutionary instincts here? Hard to say, but cotton's role is less uncertain. It has nudged its way into high-tech couture simply by being so expedient and adaptable, exactly the attributes that induced our ancestors to domesticate it more than 6,000 years ago.

Forecasting a sunny future for cotton would be a fitting way to close out one of history's great love stories between a fiber that promised to do it all and a human race that surpassed itself in resourcefulness and ingenuity to bring that promise to fruition.

But being made for each other isn't quite the same as living happily ever after. From field to factory, America's cotton industry today is in more trouble than it's been in since the arrival of the boll weevil. That's hardly news to any of the country's 200,000 textile workers who have lost their jobs since 1997, or the dwindling number of small farmers pushed to sell, consolidate, or go under. In 1970 there were 300,000 cotton growers in the United States; today there are about 25,000. Still, doomsday prophecies may be premature. No one who's been around American cotton for any length of time underestimates its resilience. If a civil war and a nasty bug couldn't kill it off, it just might survive the tectonic shifts of global power that characterize the first decade of the new millennium. One thing seems certain: there has already been a high toll levied for that survival. It takes the monetary form of exorbitant subsidies doled out by our federal government to keep American cotton from failing, and, perhaps even more perilously over the long term, it has begun to translate into contempt for our country among millions of poor people around the world whose livelihoods have been stripped bare by our cotton policies. Seeking their loyalty in the war against terrorism, we have fueled their rage. That controversy, among others, has moved cotton out of the commodity listings in the business section onto the front page of the *New York Times* and into the heat of battle when nations gather to negotiate trade agreements—"the innocent cause of all the trouble" once again.

TWELVE

Fields of Conflict

See, we in America believe we can compete with anybody, just so long as the rules are fair, and we intend to keep the rules fair.

—President George W. Bush, Labor Day 2003

On a bright June morning in the Delta region, grower Kenneth B. Hood surveys his family's 10,000-acre cotton plantation at Perthshire Farms in Gunnison, Mississippi—fields that stretch out for fifteen square miles. Hood is one of about 1,700 white Delta cotton farmers and landlords whose federal subsidies, written into a 2002 farm bill supported and signed into law by President George Bush, amount to hundreds of millions of dollars. He stands beside one of his twelve Case tractors, each costing $125,000 and fitted with air-conditioned seats, on-board computers, and global positioning satellite display monitors.

Half a world away in the impoverished sub-Saharan country of Mali, Mody Sangare, twenty-two, walks barefoot behind two oxen to till his fifteen acres of cotton with a single-blade plow, a chore that will take two weeks of dawn-to-dusk labor. He's one of eleven million West African cotton growers working small plots with primitive equipment that was being used three hundred years ago. Still, Hood and Sangare sell their bales in the same world market and compete for the same dollar. Hood's subsidized

American cotton gets dumped below the cost of producing it in years where supply outdistances demand. That brings down the price per pound without jeopardizing the fortunes of American growers because our heavily subsidized cotton farmers are assured of a guaranteed return no matter how wildly world cotton prices fluctuate. By contrast, Sangare gets no Malian government support to cover his expenses or to keep him afloat during hard times. The $2,000 he might net at best for his 2002 crop has to support two dozen family members for a year, and as cotton prices continue to fall, due in part to dumping by Americans, he fears that he is on the brink of financial collapse. He will no longer be able to replenish his cattle stock or support his brother's high school education.

Hood and Sangare gained national attention when they were profiled in a jarring front-page *Wall Street Journal* article headlined "In U.S. Cotton Farmers Thrive; in Africa, They Fight to Survive" that ran shortly after the 2002 U.S. farm bill was signed into law. Putting a human face on esoteric concepts like reciprocal trade agreements and protective tariffs, *Journal* reporters Roger Thurow and Scott Kilman exposed an ugly contradiction between America's alleged commitment to fair, ethical world trade practices in the global community and its widely perceived screw-everyone-else farm policy. The journalists pointed out that we spend $40 million on social programs in Mali while creating a $30 million deficit in that poor West African nation's economy, largely due to domestic cotton subsidies. They revealed that America's 25,000 cotton farmers have an average net worth of $800,000 per household; and that these growers now reap in $3.2 billion in subsidies, about five times as much as grain farmers.

"By widening the wealth gap," Thurow and Kilman wrote, "the subsidies sow a potentially bitter harvest. Citizens of West and Central Africa, where Islam is the major religion, are crowding into the cities of Europe. Those who stay are seeing more clerics from Pakistan and the Middle East visit their mosques and Quarnic schools. In Mali, Western diplomats hear reports of some Malians crossing the Algerian border for religious training abroad." These journalists and many others hold America's cotton subsidies accountable for contributing to the extreme, hopeless poverty that promotes terrorism—and for sabotaging our efforts at improving the conditions that underlie it. In the wake of 9/11, any argument about boorish American behavior in the Third World immediately escalates to a debate on national security.

Thurow and Kilman also take readers into warrens where eighty-six members of a Malian cotton-farmer family huddle together in two-room mud huts with no running water or electricity, where a battered TV set is wired to a car battery. They quote a village elder who still believes all farmers everywhere are part of the same family. "We shouldn't let one group of brothers make all the profits while the others get nothing," he says.

Back in Mississippi, Ken Hood, then chairman of the National Cotton Council, or NCC, expresses a different viewpoint: "Maybe the farmers in Africa should be the ones not raising cotton. The Delta needs cotton farmers, and they can't exist without subsidies." Hood does not add that 75 percent of the $3.2 billion in cotton subsidies is awarded to only 12 percent, or about 3,000, of America's cotton growers—most of whom currently receive well over $1 million a year. The cotton industry's biggest problem, he would later tell a trade publication, is "finding a way to make the public feel good about six- and seven-figure payments to farmers."

Firestorm

Within weeks of the farm bill passage, pundits declared open season on America's heavily subsidized cotton growers. "It's tough for the United States to preach capitalism to developing countries and then go and use government money to undercut their farmers," said Brian Riedl, budget analyst for the conservative Heritage

Foundation. "It's hypocritical." *Business Week* called the bill, which passed the House by an overwhelming vote of 280–141, "a dreadful piece of legislation—bad for most farmers, bad for consumers, and horrendous for taxpayers." The formulas used to calculate handouts in the new program are so convoluted that if you can grasp them you can probably understand your adolescent children, but in the end they all add up to free or low-cost money. In general they work something like this: cotton farmers with a history of planting 500 or more acres receive a direct payment of $40,000; simple enough, except that up to a cap, wives and relatives working the farm can be listed as "entities," each qualifying for the same sum. In past times a "Christmas tree" approach—hanging on new entities, such as illiterate pickers, like ornaments to collect higher payments—became a widely known ploy used by some growers.

Today those abuses have been largely curtailed; still, Ken Hood and his three partner brothers at Perthshire Farms receive $160,000 a year whether or not they plant one seed, all in accordance with the provisions of the bill. In addition, there's the "counter-cyclical payment," which sets a seventy-two-cent price per pound on cotton, and when the world market price falls below that figure, as it did in 2002, it delivers up as much as thirteen cents per pound on 85 percent of a grower's acreage up to $65,000 per eligible person. Then there's the option of a nonrecourse marketing assistance loan up to $360,000. As anyone in the industry will tell you, cotton is on welfare in this country.

A chief architect of that largesse, the National Cotton Council, also came under attack for its manipulation of state and federal politicians to ensure that its members prospered even as Africans and others starved. In a series of blistering editorials and op-ed pieces, the *New York Times* took up the assault. The surplus created by this "absurd form of rural workfare" that keeps wealthy and politically powerful growers in business at taxpayers' expense is indefensible, the paper editorialized. It happens "at a grievous cost to 10 million West African cotton growers." As a nation we have abdicated our moral leadership on trade matters, the paper added.

Rarely in recent years has the *Times* worked itself into such a lather over economic policies. There was no mistaking the passion or the disgust. "If it weren't killing them, people in Burkina Faso might get a good laugh at America's unprofitable cotton-growing fetish. . . . Burkinabe, after all, are known for their sense of humor. And what could be more absurd than the sight of the world's richest nation—a fiery preacher of free-trade and free-market values at that—spending $3 billion or $4 billion a year in taxpayer money to grow cotton worth less than that and selling its mounting surpluses at an ever greater loss? But those American subsidies are killing the Burkinabe farmers, so the inclination to laugh hardens to sorrow and resentment."

Freedom Cloth

A brief historical detour puts the *Times*'s outrage in perspective. Ever since England invaded India in the mid-eighteenth century on its way to colonizing as much of the known world as possible, many conscientious citizens of First World countries, media companies included, have viewed the concept of might-makes-right to be a faulty, inherently immoral premise. In essence the *Times*'s editorial board was echoing the arguments that had been raised for hundreds of years against imperial arrogance; in the process it was also enlisting the spirit of India's pacifist warrior, Mahatma Gandhi, against global exploitation. Coming along in the first decades of the 1900s after a century or more of oppressive British rule, Gandhi devoted himself to self-determination for India with such passionate commitment that he made all other concerns in anyone's daily life seem criminally selfish by comparison. Gandhi believed that cotton and Indian pride were so closely linked as to be inseparable, and in the 1920s he launched the Khadi Movement, a large-scale boycott of British-made cotton fabric. He argued that if peasants were allowed to home spin and weave khadi, a simple white cotton cloth, millions of starving men, women, and children would find work, clothe themselves, and learn to become self-sufficient. Caste does not matter, he said. Neither does formal education. We are all spinners; the thread in our hands does not know or care about our station in life. He urged Indians at every level of society to spin in the name of freedom. Khadi became the national uniform of the Indian Congress, and spinning wheels and weaving looms were symbolically elevated to the pacifist equivalents of muskets and sabers for the independence movement. All Gandhi disciples wore khadi—in uniforms that usually included a high-collared jacket worn with dhoti, or loin cloths, a far cry from the black coat and striped pants donned by Gandhi during his youthful days as a lawyer in South Africa.

Winston Churchill, among others, was none too happy to see Gandhi wearing his homespun-cotton peasant khadi to meetings with the British viceroy in Delhi: "It is alarming and also nauseating to see Mr. Gandhi, a seditious Middle Temple lawyer . . . striding half-naked up the steps of the vice regal palace. . . ," said Sir Winston. He hated the man in the dhoti, and the dhoti on the man.

Gandhi, unfazed, wore his khadi to Buckingham Palace on a visit to take tea with King George V and Queen Mary. Was he wearing anything to protect himself against a bitterly cold British winter? asked a worried friend. "It's quite all right," Gandhi replied. "The King had enough on for both of us."

Returning to India, Gandhi established an ashram at the country's textile center, Ahmedabab, known as "the Manchester of India," where the English were now churning out the mill-made cloth sold to the Indian populace. At his ashram Gandhi and his disciples learned to spin and weave by hand. The work, they learned, required infinite patience, but it rewarded the individual with a sense of deep satisfaction as loose fiber strands slowly but surely transformed into yarn, and then into sheets of cotton cloth. Much like the independence movement itself, Gandhi was fashioning something from nothing and succeeding. Working at his spinning wheel each morning for two or more hours, he developed an affection for it. "His thin slightly nervous hands worked rapidly, the spinning wheel made a soft, warm, comfortable buzz, and his lap was soon filled with fluffy cotton fiber," one biographer noted.

In the 1920s there were an estimated 700,000 villages in India, and Gandhi set out with characteristic brio to visit as many of them as possible and convert them into khadi centers. Urging all Indians to follow his nonviolent lead, he offered a prize to villagers for the most efficient hand-spinning-wheel, called a *charkha*. "The music of the spinning wheel will be as balm to your soul," he said. "Khadi is more than a cloth . . . it is a symbol of national emancipation." By 1934, he had introduced khadi making to 5,000 villages, and by 1940 there were an estimated 15,000 villages producing simple white cotton cloth; it provided a versatile wearable fabric and a tangible inspiration to free India from the control of a government that was determining its destiny at a distance of thousands of miles.

When Gandhi publicly burned his own British-made garments in a bonfire and asked the men and women of each village to do so as well, shirts, hats, coats, and shoes were tossed to the flames. Gandhi himself hesitated before throwing the foreign silk sari of his wife, Kasturbai, into the fire, knowing he was about to catch hell at home, but in the end he did. During World War II shortages led to a sudden upsurge in demand for khadi, and in one nine-month period 15,000 villages produced 16 million yards of cloth, enough to clothe 3.5 million people. The British Raj had seen enough. This "Freedom Cloth" was deemed subversive, an insult to Her Majesty's colonial rule, and subject to confiscation. In many areas, stocks were burned and workers who resisted were jailed. The suppression only increased resistance. When Gandhi was assassinated in 1948, with the help of khadi he had realized his goal: India had been returned to its people.

J'accuse

There may appear to be a sizeable distance between Ahmedabab and Burkina Faso, and between the British Raj and Ken Hood at Perthshire Farms on the Mississippi Delta. At the next level down, however, a similar battle continues to rage between the privileges we assume to be rightfully ours as masters of the universe and our apparent disregard for the price paid by less fortunate people in the Third World. It's that age-old willful insensitivity that set the teeth of the *New York Times* on edge. A few months after its first editorial, the paper warned, "There is nothing that creates more anger and disillusionment in poor countries than the refusal of rich nations to play by fair rules when it comes to agriculture," and it capped off 2003 with a full-bore attack on "The Unkept Promise." "By then the WTO talks in Cancun had collapsed because of America's intractable position on dumping cotton, which provoked a walk-out by West African delegates and a legal challenge by Brazil. A few month later, China, India, South Africa, and a dozen other developing nations joined Brazil in filing to present the first subsidy-related international court case to a WTO trade panel. In it the United States stood accused of breaking its own rules by inflicting hardship on poor farmers. The *Times*, which had already tried, convicted, and built the gallows for American cotton in its editorials, ran a front-page business section feature pointing out that Brazil relied almost exclusively on USDA data for its evidence.

Perhaps the stakes were highest in cotton-producing West Africa, ranked by the United Nations among the poorest and least capable of providing for basic human needs. Mali and its neighbors—Burkina Faso and Benin and Chad—collectively export more raw cotton than any area of the world with the crucial exception of the United States, which exports 40 percent of the total. Indonesia is American field cotton's largest foreign market.

Since many countries that grow the largest amount of cotton do not take the lead in exporting it, and some large producers use little in their own factories, things can get confusing in a hurry. China, the world's largest producer, exports more finished cotton goods but uses almost all of its raw material domestically to provide for its massive population and to feed its export factories. That more than likely makes the shirt on your back a multinational endeavor. Some of its raw cotton might come from Ken Hood's farm, then get blended by a factory in China or Indonesia or India with another variety or two of *G. hirsutum,* possibly from Uzbekistan or India. Spun and woven into unfinished fabric, or gray goods, it may be cut and sewn in the country where it was manufactured, or be sent to a second country for that work, and to still another for final touches like pockets.

West African countries grow but do not manufacture cotton. Still, more than one-third of Mali's population relies on the crop for all of its income. The governments of these countries argue that if able to fairly compete, their barefoot growers would produce a quality product that undercuts American cotton's price by 20 to 30 percent. In Burkina Faso, the *Times* reported, one native explained away our bizarre program by telling his friends that George Bush was a cotton farmer. "He was wrong," said the *Times*. "It is some leading members of Congress responsible for the $180 billion farm bill who are cotton farmers, or who blindly follow the dictates of the so-called King Cotton lobby."

That lobby is the proud creation of the NCC, which finally decided to respond to the *Times* with a letter published in the op-ed pages. "The [latest] editorial follows the pattern of repeating unsubstantiated claims made by other sources, fails to check background material appropriately and resorts to outright fabrication when reality fails to conform to the writer's blindly held views," said the National Cotton Council, just warming up. "This is the same combination that finally chased one reporter from employment with the institution and contributed to the departure of two senior editors."

With that the council bitch-slapped the gray lady. Somehow it concluded that hurling the ugly Jayson Blair affair back at the *Times* would strengthen its case, as if to suggest every other story in that fishwrap rag was written by a lying, plagiarizing sociopath. This issue, however, was about cotton, and if the *Times* had it wrong, the council needed calmly to present its own facts to challenge these assertions point by point, but no—"A review of the circumstances surrounding world agricultural markets would take more than a column in the *New York Times*," it protested. "Attempts to 'sound-bite' policy positions and otherwise obfuscate and misrepresent the facts are at least a disservice, and at worse, wholesale lies."

Hifalutin empty words from Mark D. Lange, signing as president and CEO of the NCC. The first thing Mark might want to do, it seemed to some readers, was to fire himself. In the short space of a rebuttal letter he had managed to spew insults like a gutter-fighter, then turn and run like a coward. What lies, exactly? What truth instead? At best, Lange left behind a hint that there was more to the story than anyone knew; at worst, it appeared that his industry organization didn't have the first clue about how to make its case in a convincing fashion to the public, and also that the NCC was either too haughty or out of touch with reality to think it needed to do so in this forum. You simply can't argue, as he did, that there isn't room in the *Times* to present your position, not in a newspaper with enough real estate to invest 10,000 words to air out its own Blair scandal, or one that reprints entire State of the Union speeches.

With Lange's unintentional help, both the conservative *Wall Street Journal* and liberal *Times* were seizing on cotton as a symbol of the cruel arrogance of power that could well provoke another attack against America. For a crop with such a profound emotional connection to our country's history, from Jamestown to Lowell to Vicksburg, that scorn represented an agonizing fall from grace. Still, important questions remained unanswered: Was there a genuine correlation between world cotton prices and the amount of money that West African farmers actually received? Were there intervening parties, legitimate or not, who controlled their income? Were America's growers purposely dumping or simply trying to jostle for position in the elbow-swinging universe of commodity trading? It became clear to some observers—this author included—that the media's attack on American cotton was long on accusation and short on supporting information. Better to hit the road and investigate on one's own.

It made sense to first seek out Kenneth Hood, the grower who came across in the *Journal* article as the embodiment of self-centered American callousness, yet a man highly respected for his integrity within the cotton industry. After being pilloried by Thurow and Kilman, the odds were long that Hood would welcome more media attention, but when reached at his Perthshire Farms, he didn't flinch or hesitate. "Come on by the plantation; we'll have us a talk," he said, and patiently gave directions.

Golden Handcuffs

The Delta cotton fields of Gunnison, Mississippi, in the early spring of 2003 lay barren under heavy pewter skies. Towering irrigation pipes arched over mile after mile of empty farmland like mammoth, airborne centipedes. The flat, straight road out of Shelby off Highway 61 that led to Perthshire Farms at first passed tar-papered shacks and weather-scarred trailer homes at the edge of town, some converted into ramshackle churches with posted prayer-meeting schedules. Highway 61, immortalized by the blues, runs parallel to the Mississippi River and now hosts the usual sprawl of KFCs and Pizza Huts, but when you turn off to seek directions in a town like Shelby you are thrown back immediately to a different era and find yourself roaming the block-long potholed main streets of towns that are not towns in any recognizable sense of the word.

Most are stripped raw of amenities, lacking the comfort of a bright ice cream parlor or the reassuring spiral outside a local barbershop where regulars gather to swap lies. Boarded-up, deserted storefronts huddle shoulder to shoulder like drunks trying to support each other. Abandoned buildings sometimes display a curling cardboard sign bearing handwritten news of a come-and-gone event: "Hot Town Blues! Live! Friday Night!" and that's about it.

Your thoughts travel back to B.B. King, playing for nickels on the curb in Indianola, less than an hour south. There was no more money in those days, but there was life, people who bought and sold things in stores, sidewalks teeming with workers and bosses, lovers and fighters. Now, long after the last wave of black laborers migrated north with the arrival of the mechanical harvester in World War II, and long after most whites as well left to seek their fortunes elsewhere, towns like Shelby are not much more than a gas pump in front of a half-empty cinderblock convenience store with a concrete floor whose most prominent features are a gleaming Dr. Pepper soda dispenser and a blurry television set on a cardboard box near the cash register, always on and never in focus. A few locals sit or lounge, and say little. This kind of threadbare existence is different, perhaps, from the kind that Mody Sangare endures in Korokoro, Mali, but it wears down the spirit, too, as surely as a rasp.

Perthshire Farms sits in the middle of an endless expanse of dirt that surrounds Shelby and Gunnison. In early spring its cotton fields, still so deep in alluvial soil that they remain among the most fertile in the country, had already been ripped, hipped, and rolled—turned, contoured, and flattened on top—in precise parallel rows that trailed off to the horizon on all sides. When the ground warmed to 65 degrees in a few weeks, farmers would plant their Bt- and Roundup-ready cotton seeds, and not before then. Seeds planted in cooler Delta soil either fail to germinate or produce inferior plants.

Traveling down a long road to the main house and compound,

you arrive at a converted rustic building that once housed Perthshire's commissary and post office, and which now holds a conference room and management offices filled with powerful computers. Hood, sixty-one, greets you with a strong handshake; he is a muscular man with neatly trimmed graying hair and a square face with narrow eyes that seem to look through you as well as at you. His features are weathered and wind-burned but not deeply scored.

Within minutes, Hood, seated diagonally across a corner of the pine conference table, is busy explaining his view of the current situation. "I used to compete with the fellow across the road," he says. "Now I have to compete with producers—that's growers—in China, India, Pakistan, and Third World countries. Our country's farmers have been dealt a severe blow in the past few years. We historically harvested about eighteen and a half million bales annually, sent about eleven million of those to domestic mills, and exported the rest. That changed about six, seven years ago when all our textile mills began to go under and we exported more than we sold here for the first time. But as a protector of the world, our government's let us down. We're being outtraded in tariffs and in quotas. That shirt on your back didn't come from cotton made here. It came from cotton imported with a 9 percent average U.S. tariff. It might have circumvented a quota restriction by having one pocket sewed onto it by another country just to use that country's more favorable quota. There are too many loopholes. But guess what—when we ship out our cotton it arrives at the other countries' ports with an average 63 percent tariff stuck on it by them. How do you compete?"

Hood is more than ready to take up the West African issue, too.

"As for Mali, it's very interesting to say we're subsidized and they're not. Mali, like its neighbors, has a central cotton marketing department run by the state. Prices are dictated to the farmers. The irrigation water, fertilizers, and inputs are given to the farmers in some of these underdeveloped countries I've visited as the cotton council chairman. Also, they don't have to worry about the environmental restrictions imposed on us. Their governments make sure they're protected."

How would he explain the media uproar, then?

"Wrongheaded. Our prices are set by the New York Cotton Exchange; they're a function of the market in comparison to the rest of the world."

That argument, you soon realize, is repeated with only minor variations by the majority of American cotton farmers. John Deere dealer and grower Tommy Pease in Yazoo City, Mississippi, sees state-controlled corruption whenever he conjures up an image of West Africa. "There's so much government graft it doesn't matter what price a small dirt-poor farmer's supposed to get, he won't get it."

Chuck Earnest in Missouri echoes the same thought: "In Africa, state trading monopolies set the price of cotton, and if there are profitable futures, those peasant-producers will never see a penny of the money. Their own governments are suppressing them so much they can never make it."

Although Hood claims that Mody Sangare and millions more like him are protected by their governments and Earnest and most others claim they're being exploited, these are two sides of the same argument, which is that in underdeveloped cotton-producing countries the state maintains total control over a farmer's destiny.

Terry Townsend, executive director of the International Cotton Advisory Committee, or ICAC, in Washington, D.C., challenges that assumption: "These countries, Benin in particular, are slowly but surely transferring power to private cotton trading companies, and that benefits their farmers. But even where the government may still be exerting its influence there's a direct correlation between what these peasants earn and world cotton prices. The U.S. and China both seriously distort that market."

The WTO, finally ruling in June 2004 on Brazil's protest, emphatically agreed with Townsend. American cotton subsidies, it said, were illegal—because they amounted to twice as much as the fair trade rules permit ($1.6 billion). Brazil argued that America's cotton exports would drop 41 percent, and its production 29 percent, without government support. That in turn would boost cotton prices about 13 percent on the world market, at least theoretically increasing the income for Mody Sangare and millions of poor growers like him. "You can be 100 percent sure we are going to appeal," U.S. Trade Representative Robert Zoellick immediately told Congress. Helping to compile those figures for Brazil was agricultural economic analyst Dr. Dan Sumner from UC Davis—the Benedict Arnold of American cotton, according to the NCC. To the council, dissention is treason. The United States is currently appealing. Two months later, in August, the European Union, representing twenty-five countries, agreed to cut their cotton subsidies by 60 percent in the next few years. As for America—"The new framework agreement outlining cuts in the United States' $19 billion annual subsidy package remains full of loopholes and points still to be negotiated," the *Chicago Tribune* reported. Should there be any substantial federal cuts in farm handouts, they won't come anytime soon, nor without raising a firestorm of protest.

If cotton, the innocent cause of all the trouble, happens to be the one crop that pits the world's poorest farmers against some of the richest, America's growers are not passing the hat. Their concerns remain closer to home. Cotton may still be the reigning player in the tenuous Delta economy, providing 16,000 jobs on 3,300 farms, but in recent times rice, soybeans, and catfish farming have threatened to upstage it. Down in Benoit, Charles Coghlan doesn't have to be reminded. His large plantation would have gone belly up ten years back in the absence of state aid, he assures you. Add persistent insect control and GM seeds in this region to the high cost of fighting tough alluvial weeds, and soon enough you've racked up extras that make Delta cotton about the most expensive to produce anywhere in the country at $500 an acre. "Without subsidies I'd be off doing something else," Coghlan says. He received just under $3 million in U.S. government assistance in the past three years. His fellow farmers know they're no longer truly competitive. In a depressed world market they admit they can't afford to grow cotton most years for what they'd get for it unsubsidized, but to a man they lay the blame on forces beyond their control—the sharp practices of unscrupulous foreign rivals. And they express little or no guilt at exporting close to half of their cotton production to China or Indonesia or India at artificially low prices when the market is soft. An unspoken irony is that America's subsidies help produce the cotton glut that drives down the price.

Maybe more to the point, even with federal aid a sizeable majority of small American cotton growers barely manages to scrape by. The *Times* and *Journal* chose to overlook or simply dismissed the fact that many of the 22,000 or so farmers who receive only 25 percent of all federal cotton money do not even own their own acreage. They rent from landlords like Chuck Earnest, whose father-in-law cleared out malarial swampland in the Boot Hill area of Missouri first for its timber, and then later put in drainage ditches, pulled out the stumps, and converted it to about 4,400 acres of cotton. Earnest now rents to ten tenant farmers whose plots range from 56 to 800 acres—a common practice in many parts of the mid-South that harks back to sharecropping.

Earnest decided early on he didn't want to drive a tractor. For years he ran the local cotton gin, also family-owned, until he had to put it up for sale—for lack of income to support it. He shares in the cost of production with his tenants, disperses their harvest to a buyer, and takes a 25-percent cut. "It's rough, very, very rough," Earnest says. "Rough on us all." He estimates his tenants average about forty dollars an acre in federal payments. The newspapers, he adds, have got it all wrong: European subsidies are five times what ours are. "Also, we aren't killing one African peasant on account of them." But Earnest acknowledges a significant difference. "I can borrow a fabulous amount based on my farm, whereas a Burkina Faso farmer has no equity; he can never borrow to improve his land. If his village tries to put in electric transmission lines, somebody will steal the copper off it overnight. And it won't be us."

As for Kenneth Hood, in forty-odd years of farming he's crawled under one too many tractors to fix a leaky universal joint and torn up his fingers pulling burrs off the barbed spindles of a mechanical harvester once too often to view himself as a slacker feeding in the government trough. Most of his wealthy contemporaries have stepped back from daily operations, while Hood still gets his hands dirty. Clearly that visceral connection to his land—he and his three brothers also grow soy and sorghum—fuels his impatience at being cast as remote and indifferent. The U.S. cotton program may be trapping farmers in a set of golden handcuffs, encouraging them to invest more heavily to produce a crop that cannot compete fairly at world prices, but as he walks you around the compound and introduces you to his high-tech equipment, Hood impresses you as a man who will probably never pause long to consider how anything he owns contributes to the plight of a poor African farmer; in his view he too has paid for it in sweat, the international currency of hard labor.

Cultivating Progress

A small fixed-wing aircraft flies more than a mile above Perthshire Farms at regular intervals during cotton growing season and produces precise digital images that analyze the moisture level, nitrogen content, and elevation of the soil in minute detail. They are then downloaded into the farm's office computers, where sophisticated software converts the remote-sensing data into color-coded maps that reveal the current health of each plant in every row of Hood's 10,000 cotton acres. Based on that multispectral data, fertilizer and insecticide and growth-hormone prescriptions are customized to meet each plant's requirements, then digitally embedded in a wallet-sized "pigma-card" that slips into a slot on an on-board computer, or controller, in the cab of each Case tractor on Hood's property. Attached to the tractor is a computerized feeder-sprayer that spans forty feet and can service twelve rows at once. Based on the card's digitized instructions, as it moves down the rows the feeder will continually adjust the amount of chemicals it emits to suit the need of each of thousands of individual plants, spraying some but not others in precise doses.

"We program that twelve-row planter-sprayer to drop cotton seeds in every row at the exact interval we set, usually four inches apart," Hood offers, sitting in a comfortable tractor cab that rises about fifteen feet off the ground near a spindle picker that can consume enough cotton in one day to produce 150 bales—as many as Mody Sangare will harvest, if lucky, in a lifetime. "As for insecticides, well, a bug's like you and me when we go to the supermarket; we choose the shiny red apple, not the brown-spotted one. Insects head right to healthy plants for the same reason, so when we map and monitor an acre, we'll hit only the biggest plants 'cause we know that's where the bugs are congregated. It saves us lots of money, and it doesn't mess up the environment half as much as a blanket spray."

A global positioning satellite guides the tractor, which travels more than a mile down a row of cotton without having to turn around. The driver in his air-conditioned seat has no responsibility other than to look out the window and hold onto the steering wheel. NASA is using Perthshire to conduct remote-sensing research. As one of NASA's program directors observes, "Farmers like Ken Hood can now farm by the foot, not by the acre."

That's where an increasing portion of the American taxpayer's investment in the country's cotton and other farming industries is going or about to go—into "precision agriculture." Cotton, in all of its time on earth, has never been so attentively served by such a dedicated, sophisticated staff of human and computerized courtiers, but to what end? For the moment this technology helps us to flood the world market with a product that most years it does not necessarily want or need in the volume we deliver it.

"Nothing much will change until the government de-couples support of cotton from production," says Gerald Estur, statistician for the ICAC. "De-coupling" for him means paying growers not to plant—or to diversify. The 2002 farm bill actually tries to encourage that by making a direct payment to cotton farmers whether or not they grow a crop—but with little success.

The same approach has been tried before, in the '30s. It didn't work then, and it isn't working now. Cotton gets in the blood of third- or fourth-generation growers like Ken Hood and J. G. Boswell II, the California agricultural titan whose 125 miles of cotton fields in the center of the state spread across an area one and a half times the size of Rhode Island. The Boswell operation, which received more than $10 million in cotton subsidies between 1995 and 2001, was sufficiently powerful to coerce the state to drain the largest lake west of the Mississippi in order to grow and irrigate its crops. These free-range individualists may be rich enough to buy the hotel they're staying in anywhere in the world, probably the whole chain, yet they still feel most at home in a dusty pick-up heading out to check a boll weevil trap.

They will always see themselves as misunderstood and put-upon, too, fairly or not. It might just be impossible to explain to anyone in a suit how much you've got to be willing to roll the dice every spring. You till and plant and hope for the best; sooner or later it's all still up to the rain and sun gods no matter if it's cotton or soy or peanuts or if you got intergalactic satellites sending remote-sensing images from Mars. It's this thing that gets instilled in you, the miracle of what develops from one small seed. You can say your job is running the whole operation with the help of your brothers if you're Ken Hood, and it is. But you know your *real* job is to do whatever it takes to forge firm alliances with the right politicians and appear before committees and press the flesh and muster all your resources so that at the end of the day you go to bed knowing you'll wake up tomorrow morning and still have a farm to run and a field to go back out to, the one place on earth you truly belong.

The United Way

Those same skills honed by Hood extend out to the National Cotton Council. While the umbrella organization for the nation's growers seems unable or unwilling to present a responsive, flexible public image to its numerous critics, it has developed a lobbying strategy since its founding in 1938 that stands as a model of applied persuasion.

The politicians it lobbies quickly learn the meaning of "no wiggle room." Bruised egos have been placated and contentious issues tackled to the ground behind closed doors long before John Maguire, the NCC's Washington lobbyist, comes a'callin. All seven industries wholly or partially dependent on American cotton wrangle it out: the mills, producers, co-ops, merchants, warehouses, seed-crushers, and shippers. To break ranks is to find yourself in a fallow field without a hoe. "We hash the problem out, however long it takes. Then we move forward with a unified voice," says former chairman Hood. When cotton emerges from these closed-door sessions, it is no longer merely a fiber or food product. It has been transformed into a patriotic act. Our national security is at stake, as well as the country's economic future. Cutting subsidy levels to cotton farmers threatens our ability to sustain ourselves in the event of impending worldwide conflict. We cannot and must not rely on foreign providers. If cotton fails, so goes the party line, all of these related industries and the industries that depend on them, by ripple effect, suffer and possibly collapse. A vote against cotton is a vote for apocalyptic chaos.

That message, a stimulating brew of truth and conjecture, finds its way to some of the nation's key legislators as it has for close to eighty years. In that time cotton has become so politically entrenched that it deserves its own seat in the House and Senate. Never mind that wheat and corn and soy occupy vastly more of this country's agricultural landscape, or contribute more tax money to its coffers. No one wrote an epic like *Gone With the Wind* as a paean to the glory and hardship inspired by the rise and fall of textured soy protein or fructose. Somewhere along the way the myth and reality of cotton merged to form a perfect union, much like the red and white stripes on the American flag.

Office-holders from the South and Southwest learn early in their careers to salute when cotton passes by. Three Southern senators—Saxby Chambliss, Blanche Lincoln, and Thad Cochran from Georgia, Arkansas, and Mississippi, respectively—wrote an impassioned letter at the start of the 2003 WTO summit in Cancun to our trade representative, Robert Zoellick, warning him in so many words not to buckle under to international pressure: "In reality, ending the U.S. cotton program would only harm U.S. producers. It would have no long-lasting effect on the four [West African] countries seeking assistance," they counseled him, even if he hadn't sought out their advice. The angry media protests that resulted from that conference might have flattened the ears of any other special interest group—but cotton's $3.2 billion subsidy program survived intact. It also withstood attacks by outraged congressional leaders from Midwestern food-farming states like Iowa's Senator Charles Grassley, who railed against cotton for skipping off with a lion's share of the federal agricultural booty, at least in relative terms.

If, as Tip O'Neill famously remarked, all politics is local, in cotton country all politics is rural, too. Senator Thad Cochran, recipient of the National Cotton Council's Distinguished Service Award, heads up the Senate Agricultural Committee. Former House Agricultural Committee chairman Larry Combest from Texas, the nation's largest cotton-producing state, helped push through the 2002 farm bill. "I stand by cotton," he said. That committee's ranking minority member, Rep. Charles Stenholm, a powerful West Texas rancher who once held an executive position with a cotton group, announced to the *New York Times* that he would not consider dropping subsidies until other countries agreed to drop theirs. That is, never.

In 2001–2002, cotton contributed $216,000 to the campaigns of politicians it endorsed. These days that's pocket change. "You don't need to spend a bundle when you've got the ear of legislators on key committees," an expert in interest-group politics, Burdett Loomis, explains. "They have to know how important you are to their constituents, however. And don't underestimate cotton's mythical, symbolic hold on regional pride. Those kind of intangibles are enormously influential."

Also, as one observer remarked, cotton is "the politically perfect subsidy." She meant that it is so unsexy most taxpayers and legislators would rather pick up the tab than be forced to read the fine print. For decades the industry followed the example of its product, that pay-no-attention-to-me fiber and seed that makes itself useful in everything from fishing nets in Japan to Pepperidge Farm cookies in Cleveland to 600-thread-count Frette bedsheets in Italy to solid rocket fuel at Cape Canaveral. It moved by night under the cover of blandness. The more nobody noticed, the freer it became to exert its will. Suddenly thrust into the harsh glare of an unflattering spotlight at Cancun as a puffed-up symbol of American greed, cotton lost its anonymity and with it the protection that it afforded.

What happens next remains uncertain. As an agricultural product, American cotton now has to contend with an extremely hostile world press and a precarious market. Like many trade organizations populated by the old guard, the National Cotton Council seems ill-equipped to take on the feisty, contentious, in-your-face global community that defines the warp and woof of the new millennium. As with children who have matured into thinking, independent, and assertive adults, these developing nations recoil at condescending attitudes and trust no one until proven worthy. "You have to understand we are fed up with these farm subsidies and hearing for 25 years that things will get better," Aluisio G. de Lima-Campos, an economic adviser at the Brazilian Embassy in Washington, told the *New York Times*. "The only way to deal with it is to turn this into a make-it-or-break-it proposition."

If Brazil's impending legal action against American cotton results in a court case that receives wide broadcast coverage, loyal politicians might well begin to run for cover. Imagine a CNN graphic of a bale of cotton handcuffed to a witness chair, with "U.S.A." stamped on its prison-stripe burlap covering. Then imagine your mug shot as a subsidy-supporting Southern senator also in prison garb superimposed over it. The verdict: guilty by association. Whatever the outcome of this impending battle, and however unjustly all of America's cotton farmers have been lumped together by the *New York Times* and others as a club of rich, conniving freeloaders, the contestants on both sides may soon be turning their attention elsewhere. At the beginning of 2005, under the auspices of the WTO, all trade quotas on apparel and textiles are set to be eliminated in the 143 participating countries, including the United States and China. If enacted, that change will certainly move up the supply chain to directly affect subsidies around the world, and it could lead to a trade war between these two giants that dwarfs all other issues.

The China Syndrome

In recent decades China has become the 10,000-pound mercantile gorilla that steals your seat and sits wherever it wants to eat your lunch. If cotton as a crop and as a fabric mirrors the larger debate about how the United States can adapt to remain a key player in the global marketplace, China provides the ultimate test case. In one year alone, 2002, our trade imbalance with that country exceeded $103 billion. Today China grows 21 million bales of cotton annually, more than any other place on earth. China's mills manufacture by far the most finished cotton product, about twice as much as their closest competitor, India, and four times as much as U.S. mills turn out.

Approximately 42 million cotton farmers in China grow crops on about 3 million acres, primarily in the southern Yangtze River area and in the Yellow River valley, as well as in the remote, arid northwest Xinjiang Province, a vast, desolate expanse naturally irrigated by mountain runoff. The yield of the country's cotton farmers averages out to about half a bale per grower, grown on one-half acre. By contrast, 25,000 American growers produce about eighteen million bales, or 720 bales per farmer, but in reality most American cotton comes from 10,000 acre-and-up plantations.

Xinjiang is home to China's large cotton farms, still small by comparison to ours. With rare exception the cotton grown in China, sticky green-seed upland, is the same species that continues to blanket the American South, and that now accounts for 97 percent of all cotton grown around the world. Surprisingly, while the equipment of the small farmers—mule-drawn plows—is extremely primitive by Western standards, their government-issued seeds now most often carry Bt genes. China is developing the largest plant biotechnology capacity outside of North America.

Zhang Jiannong has been working the same cotton field in Tongxu County, Henan Province, for the last twenty-two years—a half-acre plot leased to him after the break-up of the People's Agricultural Collective in 1982. He pays $27 a year as an "agricultural tax" on his plot, and no property tax for his house. All of the 400 farmers in his village in the Yellow River floodplain grow cotton on plots of roughly the same size in rich alluvial soil. Ancient irrigation canals still water the area. Heavy rains come in the summer.

Now in his mid-forties, Zhang, whose circumstances typify China's family growers, works his field with his wife and teenage son. The family's annual income, derived entirely from cotton, is $420. Zhang, his sun-browned face deeply creased, plows with a mule; his family picks cotton entirely by hand, collecting three rounds of harvested crop to maximize the yield. The work is back-breaking, but less so now than in the past. Before the introduction of genetically modified, pest-resistant Bt cotton, Zhang and his family had no choice but to kill invasive bollworms by hand, one at a time, plant by plant, in the high heat of summer.

Zhang and his family live in a modest, 800-square-foot brick house with concrete floors that Zhang built himself three years ago. He owes nothing on it. There's a coal-fired stove in the kitchen, a common room, and a bedroom, each with a heated brick bed typical of North China. The walls are papered over with old newsprint and posters of the God of Wealth and of Chairman Mao, who has morphed into the new God of Luck for millions of Chinese peasants. Zhang bought a twenty-two-inch Changhong color TV last year, and a videotape player as well. Together they cost him one-fourth of the cash he made selling his cotton crop at market in the Tongxu county seat.

He keeps four pigs and a half-dozen chickens. Like all the farmers in the region, Zhang grows winter wheat for the family's major staple food. He plants corn with his cotton, too—just enough to feed his hogs. Some years he plants peanuts, but this year he's decided to give more of his land over to cotton, his only cash crop. In fact, he saved enough from the sale of last year's harvest to consider leasing land from his neighbor, Old Liu, whose daughter and son-in-law recently lit off for the city to work in a factory rather than face another season of toil in the fields.

While $420 provides enough income for farmers like Zhang, the hard work required to earn it has spurred hundreds of thousands of rural Chinese to seek higher-paying, less-taxing factory labor in China's flourishing industrial cities.

Li Yujiao, a twenty-three-year-old woman, has worked spinning cotton yarn in the Zhengfang Yarn and Textile factory in the capital of Central China's Henan Province for just over two years, along with millions of others like her. She came there from a small cotton-growing village ninety miles to the east, where her parents and brother still farm their half-acre plot. Ms. Li has an eighth-grade education, and decided that, like Old Liu's young daughter and her spouse, she would rather migrate to the city than stay on the farm. She makes about just under $50 a month, about 32¢ an hour including overtime, but her expenses are so minimal that she usually saves over 80 percent of her monthly pay.

The 1,800-employee Zhengfang factory, which is privately owned, has managed to stay in business by modernizing its equipment, and now runs computer-controlled air-shuttle power looms. Numerous competitors, especially the state-run mills, went out of business in 2003 after flooding damaged much of the country's cotton crop. Li Yujiao considers herself fortunate to work here. She knows girls from her village who went to work in mills in Jiangsu and had their pay taken from them by corrupt managers who declared false bankruptcies only to reopen new factories with the cash they had embezzled. The women took part in strikes and work stoppages, but ultimately they were dragged out of the factories by their hair.

Since she spins and doesn't weave, Li Yujiao is at a disadvantage. Spinning is still unskilled labor, done on relatively primitive machines, and she earns only half of what a trained loom operator makes. Still, she's grateful for the new looms, installed in 2002. They've reduced the noise level. Her ears don't ring at night and she no longer needs to wear ear plugs. Zhengfang has a relatively progressive management. The factory is neatly kept and well-lit. In the mornings, after a breakfast of rice porridge, pickled vegetables, steamed bread dumplings, and soy milk, she lines up at 8:00 with the workers from her section, all dressed in the same uniform of blue smock and headscarf, to sing the company song, chant slogans about the importance of safety and quality, and do some simple calisthenics. Lunch is good: the cafeteria serves rice, two vegetables, one meat dish, and soup every day. Dinner is another story—usually, it's an unpleasant rehash of lunch. She does not pay for her meals, but is docked six-tenths of a cent for her meat per meal on her paycheck.

The work day ends at 5:00, but Li Yujiao works overtime

whenever she can. Although she is not paid time-and-a-half, she believes her enthusiasm will help her to advance. Another girl from her dorm has already been selected to begin training on the power looms. Li Yujiao aspires to work in finished garments. To train, she bought her own sewing machine last year and often makes clothes for her roommates in her dormitory on the factory campus. She shares a free room with three other girls, all unmarried and all from the countryside. She gets two days off a week. In two years, by the age of twenty-five, she expects to have saved up enough money to open a seamstress shop in the county town near her home village, settle down, and get married.

All similarities between Li Yujiao's living and working conditions and those of the Lowell girls about two hundred years earlier are probably more than merely coincidental. China is in the throes of its own twenty-first-century industrial revolution. Times may change, but they don't necessarily move forward uniformly. Essentially the United States is competing with itself in a technically updated rendition of its own birth as a manufacturing nation. Many of the same elements prevail—the migration of workers off the farm, the sea changes in society and in family structures that any industrial revolution brings about, and the national pride that develops as well.

Dangerous Delusions

Currently, China's apparel and textiles account for about 20 percent of the U.S. market, but that figure is deceptively low due to past quota restraints. For items where these limits have already been lifted, Chinese imports claim 55 percent of the market. In 2002 China exported more cotton textiles and clothing to the United States than all other countries in the world combined. It dominates the European Union as well. In Japan, Chinese goods now occupy 90 percent of many clothing categories.

"Why can't the Americans stick to making things we can't?"

asks Yang Rong, manager of the privately run Jinhua Asset Underwear Company outside Beijing. "For things like bras," he told *Time*'s Beijing bureau chief Matt Forney, "nobody can compete with China." The simple logic behind that sends shivers up and down the fragile spine of America's textile industry when it contemplates the impending elimination of all quotas. In the past China has kept its currency, the yuan, artificially low to reduce the purchase price of Chinese imports—a strategy that raises hackles across the industrial United States. The Chinese yuan is pegged to the dollar, which has slipped in recent times, legitimately leading to lower Chinese prices; still, legions of critics argue that China has a tradition of unfairly manipulating its currency to subsidize its textile factories and cotton growers with our administration's tacit consent. "We need a government that is as aggressive on behalf of our workers as China is on behalf of its workers," National Spinning Company president James Chestnutt warned at the beginning of 2004. He noted that as many as 650,000 jobs lay in the balance and that 1,300 additional U.S. textile mills face imminent closure.

In order to survive, apparel and soft goods companies, from Gap to Wal-Mart, have for decades been manufacturing offshore. The prevailing wisdom has always been that we provide the brainpower and innovation and countries like China provide the happy, eager, and inexpensive fingers. That may have been a safe assumption in the past, but at the dawn of the twenty-first century it is already proving to be a fatuous, dangerous delusion. Evidence of state-of-the-art native technology is everywhere, as close at hand as a tour through any one of China's numerous highly automated, robotic textile factories. American companies doing business in China now contend with technology-transfer clauses in their joint venture agreements that give Chinese industries a legal right to license and duplicate foreign companies' patented technology after a designated time period.

Although these stipulations became less onerous when China joined the WTO in 2001, they reinforce that country's single-focused ambition: to slap us silly at our own game. State-controlled industries like cotton are now opening to entrepreneurs in a society that functions unlike ours in just about every meaningful way except one: the bottom line. Pragmatism reigns supreme. Until recently the military and paramilitary controlled all aspects of China's cotton industry. The government classified cotton's status as a state secret, along with things like nuclear warheads, and refused to disclose any information on surplus stockpiles and so forth. From a business perspective, that approach impeded China in the world market. Ideology swiftly bowed to efficiency. The industry is now becoming privatized. Factories that fail usually get propped up by the government. There are simply too many people in need of jobs.

"China trumps the world," says Cass Johnson, interim president of the American Textile Manufacturers Institute. He's in a position to know. His organization keeps track of the welfare of the industry; these days that amounts to counting body bags. More than 210,000 domestic textile workers have lost their jobs since 2000, close to 300,000 in total. At least 280 plants have now closed down, primarily in the Piedmont area of North and South Carolina. Among them are the venerable Burlington Industries, Pillowtex, WestPoint Stevens, and Fruit of the Loom, companies that built the industrial South out of the ashes of the Civil War. Cone, the former denim king and pride and joy of Moses and Ceasar, has also filed for bankruptcy. The collapse of these mills is the textile equivalent of GM, Daimler-Chrysler, and Ford all going under within months. Bill Griffin, a Memphis industry consultant who's been in cotton all his adult life, holds out little hope for U.S. textile factories. "Dead dogs," he says, "don't bounce."

Facts and figures illuminate the crisis at hand but not the human toll it exacts. With the closure of the mills, of course, go the heart and soul of the towns they built or fortified with the spending money they made available. Thousands of mill workers and managers and executives in these plants, quite apart from suffering a critical loss of income, are now without a place to go to put their skills to work, to be productive, to accomplish goals and be acknowledged for getting the job done—intangible rewards, but for many of us as vital as oxygen. There are levels of devastation that can be calculated, and some that simply cannot.

An obvious solution, it might seem, would be to safeguard the remaining domestic textile plants by imposing temporary trade barriers—informally known as "surge protectors"—that counter the elimination of quotas in 2005. It may happen. You don't have to be a Marxist to be convinced that in a capitalist society like ours, government is ultimately the handmaiden of business. But viewed narrowly through that same spyglass, there are conflicting forces at play that make across-the-board solutions untenable. Pressuring China to limit exports will drive up the cost of consumer apparel and textile goods; retailers want no part of any changes that scare away customers. It will also have a chilling effect on investor confidence at a time when America's exports to China—led by raw cotton—are growing at a healthy rate, and the goods we import now are increasingly driven by U.S. and foreign investment.

Across the battle lines, what's left of the U.S. textile industry is marshaling its troops to lobby for protectionism. "We are on the verge of a catastrophe," Mark Levinson of Unite, an umbrella group for labor unions representing U.S. textile workers, told the *Southwest Farm Press.* "When textile import quotas expire, as they are scheduled to do in 2005, we will see textile manufacturing plant closings on a massive scale." Unlike growers, manufacturers receive little if any subsidy money from the government.

If politicians traditionally bend to survive the mightiest wind, in this situation it is blowing their way from Target, Gap, and Wal-Mart, and not from the silent spindles of Burlington Industries. Congress and the administration are likely to appease giant retailers first, then make an effort to stem the tsunami of cheap finished cotton goods and other fabrics that will arrive from Asia. Members of both Houses might decide to enforce a pledge signed by China when it became a member of the WTO that enjoins any country from creating the "threat of serious damage" to another member. If proof of serious damage is needed, it's there: one textile worker in America out of every four has lost his or her job in the past two years.

And where does cotton find itself in this battle? Predictably, right in the middle of the fray. It encapsulates the schizophrenic nature of our relationship with friends and foes around the globe. Those vast, voracious cotton mills in China need huge amounts of raw material. In good years, they can rely on millions of bales produced by Chinese family growers, but when crops fail, as they did in 2003, U.S. planters find their cotton eagerly purchased in a seller's market where profits are suddenly available for the taking. To China's mill managers, the USDA's superior grading system ensures consistent quality, a highly prized attribute in a country suspended somewhere between the First and Third World, where tight quality control through government supervision lags far behind productivity. At the same time, China's quickly expanding demand for American raw cotton puts growers squarely at odds with the U.S. mills they also supply. Every shipment of bales to China potentially costs jobs in North Carolina. Jeff Coey, China and Asia regional director for the Cotton Council International, takes a philosophical stance. In an ever-changing global economy that rewards agility, he says, there are no moral absolutes. There are only mixed emotions.

Bold New World

Most analysts lending an ear to the future of American cotton these days hear a funeral dirge, not a rousing trumpet fanfare. Few would argue that sweeping changes are needed, but curiously, little effort appears to be channeled into proposing reforms in the factory and field. Textile unions and organizations deluge us daily with broadsides against unfair trade policies. Growers protest that without continued generous subsidies they face ruin. The anger is palpable and understandable, but in the end it will also prove to be futile if it does not inspire the kind of outrageous, outlandish visionary daring that brought cotton fabric to America in the first place.

To this observer, for one, the colossally audacious spirit of Francis Cabot Lowell, Samuel Slater, and all those heroic individualists in cotton's past could profitably be revitalized. Lowell was a man of means, you might recall, with the hunger to succeed of a pauper. Undeterred by the adamantine resistance of the English cotton mill barons, he took a huge calculated gamble in betting on his own acumen and the practical application of the technology he swiped from England. He was also fearless, driven, resourceful, and quite able to think outside the box since at the time there was no box to think within.

By contrast, America's cotton industry on all fronts today seems to be filled with leaders who are wary of the future, infuriated by having to countenance foreign intrusion, and determined to preserve a dysfunctional way of conducting business. Isn't there something smug about congratulating ourselves on our high standard of living while we stand by and watch hundreds of thousands of industrial jobs being taken away from us? Complacency doesn't work; neither does a campaign of blame. John Wayne would probably urge us to quit bellyaching. He might remind us, too, that we managed to find the strength of purpose to heal from slavery despite a common belief after the Civil War that the nation's cotton economy would soon perish in the absence of forced labor. It survived, and so did our free enterprise system—but only to become enslaved again by a selfish sense of entitlement that leads to impotent anger. Pragmatism works, as China keeps reminding us. With that in mind, what if America's cotton power elite were to accept that we've lost the battle? If we have any hope at all, we have to be willing to radically change the rules of engagement.

That would mean adopting a no-fault approach. It would substitute cooperation for competition. We would give up fighting a war we cannot win and embrace the rest of the cotton world as equal partners, with something valuable of our own to offer. What if, as an example, Ken Hood and his colleagues brought their precision farming skill and equipment to the sprawling farms of Xinjiang in China and to Brazil and India in exchange for a portion of the profits derived from them? What if America opened up its doomed, abandoned textile factories to Asia and even to its workers in combination with ours to manufacture innovative fibers based on nanotechnology? What if we developed a very twenty-first-century point of view, which is that, as a matter of survival for all concerned, a global perspective is critical?

Unlikely scenarios, perhaps. But cotton has always invited invention and experimentation. As this chronicle hopefully illustrates, the immensely difficult process of domesticating the fiber and then putting it to work as the most widely used fabric on earth became a reality only when extraordinarily determined and gifted people refused to accept failure and allowed the potential of cotton to inspire them. In turn, they inspired changes in their societies that continue to influence the way we live and work today. Whatever course its future may take, cotton's creative gift to the American people has always been its capacity to engage us at the highest levels of accomplishment and to transform the impossible into the doable. Maybe we simply need to be reminded that for every problem in its past, we found a solution. If we approach its future with that attitude, we might again surprise ourselves. Cotton, the friendliest fiber on earth, will be there for us when we do.

Afterword

I was lost, but not. Traveling back to my hometown, Lowell, from California for the first time in many decades to research its place in the history of cotton, I took a side trip to the neighborhood where I grew up. On a gray, drizzly day in October I drove aimlessly up and down a few streets in South Lowell that looked vaguely familiar but slightly unsettling, too, like college friends you run into in your later years and try to picture without their lines and jowls even as you exchange pleasantries. The wood frame houses I passed seemed to have sunken down further into their foundations over the intervening years. After circling a few blocks with no luck, I spotted a fire station, pulled up to it, and got out to ask directions from a fireman on the front walk.

"I'm looking for my old neighborhood, somewhere around here. Could you tell me where Pine Street is?"

He eyed me closely for a few seconds as if checking for leaks in a faulty transmission. Then he said, slowly, "You're standing on it."

It was a perfect Lowell moment. I was yet another dunderhead out-of-towner, this one asking directions to a street that was right beneath him. I probably needed help tying my shoelaces. If you're from Lowell, so does everybody who's not from Lowell. It's always something of a miracle to come across anybody from elsewhere who can tell one end of a dog from the other. Clearly I wasn't among them. But I brought that fireman a fine surprise, too—a gem of a tale that he could share with the boys. Because, as he'd explain, it's not every day at the Engine Seven, corner of Stevens and Pine, that a guy comes up and asks—get this!—where's Pine Street? What I shudda done, I shudda started him with an easy one like, Where's your nose? Good! Okay, now where're your feet? Good! Now keep lookin' down. . . .

Eventually I made my way past the house I grew up in—now gutted, on jacks, in the throes of a major makeover—and past the Morey School, a construction made out of massive haphazard blocks of weathered brick, to the cotton mills along the Merrimack. It was in that school in the seventh grade that one of my close friends and I decided to play a joke on a substitute teacher. Since she didn't know either of us, we decided to switch identities for the day. He wrote my name on his Friday spelling quiz, and I signed his name on mine; we handed them in and doubled over in laughter at recess. That afternoon we were called to the principal's office. The sub had nailed us, and looking back, I guess we shouldn't have been too surprised, since my pal happened to have an identical twin sister, Thalia, who sat right beside him and who didn't look at all like me. We flunked our spelling tests and got marked down a grade in deportment. My coconspirator, Paul Tsongas, straightened out quickly after that walk on the wild side, decided a life in forgery was not for him, and went on be elected to Congress and then to the U.S. Senate. While the rest of us fled crumbling Lowell at the first opportunity, Tsongas recognized something about the value of remaining true to his roots that we seemed to miss, and he stayed behind to try to resurrect his decaying hometown—a lost cause by any standard except his own. By the time he died of cancer in 1997 at fifty-five, Tsongas, who ran against Bill Clinton for the Democratic presidential nomination and won in New Hampshire before eventually losing, had somehow been able—through a combination of persistence, intelligence, and arm-twisting—to reinvent Lowell as a national living museum to mark the birthplace of America's Industrial Revolution.

He looked at the same hulking abandoned cotton mills I saw as a boy, imagined them as they once had been, bright and brimming with activity, and managed to bring the force of his persuasive power to bear to realize that vision. But historical renovation alone wouldn't suffice. You might find money to clean up the mills and surrounding foul canals and turn them into an educational recreation like Williamsburg, Virginia, with a working loom room and canal barge tours and a spinning wheel museum and such, but then there was the town itself. Tourists were not likely to flock to a city matted with layers of grime from a century of neglect. In the 1970s and 1980s, Tsongas got the city elders of Lowell and their political machine to rally behind a massive, expensive downtown revitalization program that demolished every building and storefront I remembered from my youth as I passed along Merrimack Street. They replaced them with scrubbed, vibrantly colored urban-village bookstores, restaurants, trendy clothing stores, and more; all were surrounded by flowerbeds and bordered by clean sidewalks interspersed with planter boxes housing robust young saplings. The ethnic Europeans who once worked the mills were now largely replaced by South Asians and Hispanics, some attracted by Lowell's brief, doomed encounter with Wang Computer Laboratories, others drawn to the city's more recent service and educational job opportunities.

Today you cannot walk more than a block or so in any direction in downtown Lowell without running into Tsongas. Turn here and discover the Paul E. Tsongas Arena, "New England's newest state-of-the-art venue for quality entertainment." Turn there and enter the Tsongas Historical Industrial Center, where, according to its literature, "Students 'do history' by weaving, creating a canal system and testing water wheels, working on an assembly line, role-playing immigrants, and becoming inventors."

Not the Lowell we grew up in. Tsongas and I "did history" by raiding my father's collection of old silver dollars in search of the one rare double-stamped 1903 coin that would make us rich. Because this was Lowell, the citadel of dashed hope, we never found it. Now, too, there were "Ranger-led tours that explore old ethnic neighborhoods." That would have given the Irish, Greek, French Canadian, Italian, and Syrian kids I went to school with a deep chuckle. Little did we know our sweaty streets were destined to become theme-park attractions. Never mind. I was here on another mission now in the early years of the new millennium, with Tsongas's curling, mischievous boyhood smile not much more than a faint image in my mind as I entered the courtyard off Kirk Street leading to the National Park Service's Mill Museum.

It was at that moment, all alone on this drizzly morning, that I got the full impact of the revolutionary power of the American cotton industry. The narrow four-story rectangular brick mills that rose up on three sides conveyed a spiritual heft accentuated by their imposing wooden octagonal bell towers—spires reaching up to God. They might have been built with something secular in mind, the conversion of white fibers to green money, but the mortal awe they inspired even now hewed much closer to a cathedral than to a mundane workplace. I could almost hear and see those farm girls two centuries before me arriving at this spot in covered wagons, looking up, gasping, crying, giddy, shaken, breathless, excited, and terrified all at once. This was their new place of worship, and soon the country's as well—empowered by the miracle of industry.

I recalled an early visitor from France comparing the nineteenth-century Lowell girls to nuns. Standing in that courtyard, it suddenly made sense. They surely must have felt the presence of a higher being the moment they gazed up at those intimidating structures that humbled and dwarfed them. Some would flourish, others suffer within their walls, but each of these women would leave Lowell changed by the experience—and soon enough, the course of America's future would evolve as well, when the basic fundamentals of cotton manufacture were transplanted and adapted to a dozen or more unrelated products and the world's mightiest industrial empire took root.

Making my way inside the main exhibit, the Boott Mill, now refitted and reopened with a working weave room and ground-floor display that walks the visitor through the entire cotton conversion process from bale to fat tube of sliver to roving to thread to fabric, I met up with Lowell historian Gray Fitzsimons. Like many historians, Fitzsimons comes across as a gallant warrior fighting vainly to correct common and persistent misconceptions in a world that plays fast and loose with historical accuracy. Speaking with me while lines of school children filed past on tour, he zeroed in on the legend of Francis Cabot Lowell's miraculous feat of re-creating the English power loom entirely from memory. "Not completely true," he assured me. "Although he was searched by customs repeatedly, even in Nova Scotia, he managed to smuggle back at least a few drawings." Were they ever found? "No, but he did."

Fitzsimons, like half a dozen or more staff members I met at the mills during my stay, greeted my interest in cotton as an invitation to share a wealth of special knowledge accumulated over many decades. Invariably, these guides delivered the goods with total authority. Park Ranger Alex Demas wondered if I knew the connection between the invention of the rubber sole and the history of the mills. In 1896, he told me, a calico printer named Humphrey O'Sullivan decided to do something about the vibrations from the looms on the top floor that rattled the entire building, so he brought to work a slab of rubber to stand on. But each new piece kept getting stolen, so one day he decided to draw his shoe outline on the rubber, cut it out, and glue it to the bottom of his leather sole. The idea quickly caught on, Humphrey took on a patent three years later, and, with a capital investment of $25,000, O'Sullivan went on to start his own rubber sole company.

I've always been a sucker for the odd happenstance that

changes the way we lead our lives, or at least the way we stand while going about them. Everybody had a story, and usually a story about that story as well. Upstairs I was greeted by a deafening roar as I entered the reconstructed weave room, where a hundred or more belt-driven looms clattered incessantly as harnesses lifted and lowered their attached metallic heddles, and shuttles flew with blinding speed and an ear-ringing clack. I understood why many went deaf working on these antiquated Draper looms that churned out fabric now as in the past with such a thunderous crack that I felt trapped in the jaws of a giant rodent as it furiously devoured a mouthful of hard nuts. "Can I help you?" I heard someone holler. Weave Shed Operator Gary Hudson emerged, we shouted at each other for a few minutes over the clamor, then he flipped a few switches—these looms now run on electricity, of course, not waterpower—and by doing so he created a pocket of relative serenity around us. Nearby, a loom's hundreds of heddles resembling long needles danced vertically in repeated patterns and fabric materialized as if from air, inch by inch, at the far end.

"You have to understand," Hudson said, "any cloth that comes out of a loom, you have to do the threading and treadling, handweaver terms that go way back." We were alone, we both had time; I was interested in what he knew, and that was all Hudson needed to know. For the next three hours he put me through an intensive tutorial on the intricacies of automated weaving. At one point he leapt from Cartwright's eighteenth-century invention of the power loom to the inventor's remark that his shuttle springs were powerful enough to heave a Congreve rocket, all the way to those rockets' red glare and how Congreve provided Francis Scott Key with a central image for "The Star-Spangled Banner." When the English fired those malfunctioning rockets during the War of 1812, Hudson exclaimed, the safest place to be was the target.

A few decades after that, he continued, English manufacturers introduced the looms that were now surrounding us in this room—Fancy Weaving Machines, by name. "They were fitted with binary programming in a mechanical format," Hudson explained, barely stopping to catch his breath, and I could sense we were stepping gingerly into the advanced course on historical textiles. He showed me something that looked like a metal abacus with wheels on five parallel bars. He described it as an analog computer with pass-or-go instructions. Linked to the loom on a circulating chain, the moving wheeled bars activated or blocked each row of threaded heddles on the harness much like a punch card to create ornate patterns in the cloth—a second generation of the Jacquard technique. The cams that controlled the harnesses could now be programmed in an endless variety of alternating sequences to produce the desired pattern.

The facts were fascinating, at least to me, but there was something more personal that captured my attention. You could feel in Hudson and Demas and others I met during my stay the rustle of ideas and incidents and anecdotes all jostling for position like school kids vying to get to the head of the line. That youthful exuberance ran counter to the sardonic sheen that coats so many conversations in Lowell. There was an undisguised delight in meeting up with someone who shared similar interests, a depth to the bond of exchanged information, and it came back to me like a bright, clear beacon from the past. I'd forgotten that passion of the mind, which was as much a part of the world I grew up in as the collapsed glory of the town itself. Not unique to Lowell, perhaps, but clearly welcome and thriving here. That recognition connected me back to the keen, frisky impatience with unsatisfactory answers and a relish for knowledge that had sent me off into the world as a young man with a healthy curiosity about its secrets and a nose for the bogus. On a journey to learn about one history, I'd stumbled across another, too—my own. I'd come home to cotton.

Glossary

ACID WASH: Used primarily for blue jeans. Process involves soaking pumice stones with chlorine, then using their abrasive power to bleach jeans into sharp contrasts. Also known as moon, fog, marble, ice, and frosted.

AMERICAN PIMA COTTON: A cross between Sea Island and Egyptian cotton. Grown in Arizona. Length averages 1⅜ inches to 1⅝ inches.

BASKET WEAVE: A variation of the plain weave in which two or more warp and filling threads are woven side to side to resemble a plaited basket.

CALICO: Cotton cloth originally imported from Calicut, India; is one of the oldest cottons. Today, it is usually coarse and light in weight. Designs are often geometric in shape, but originally they were elaborate birds, trees, and flowers. Similar to percale.

CHAMBRAY: The lightest of indigo-dyed fabrics, chambray is a plain cotton weave normally used for shirts and womenswear. A lightweight plain-weave cotton fabric having a color.

CHINTZ: Any plain-weave fabric with a shiny lustrous finish, often printed in bright floral designs.

COMBED COTTON: An operation that removes the short fibers and foreign impurities, creating a higher-end fabric whose longer fibers are straightened and laid parallel.

CRETONNE: A printed fabric, originally, and usually of cotton and of heavier weight than chintz.

DRAFTING: Attenuating or drawing out the fibers before twisting them.

ECRU: Undyed denim; the natural shade of the cotton yarn, now considered an alternative basic color for jeans worn year-round.

FASTNESS: A color's resistance to washing, sunlight, and abrasion.

FINISHING: A step in fabric production. In denim, as one example, the cloth may be passed between rolls that remove lint and raise loose fibers. Next, the cloth passes through a gas flame, or "singer," that burns the loose fibers off. It is then run through a vat of liquid finish solution and afterwards through squeeze rolls to remove excess liquid. A series of steam cans dries the fabric while setting the widths.

FUSTIAN: A hard-wearing type of clothing fabric containing a large amount of weft yarn. Used to describe a class of heavily wefted fabrics.

GREIGE FABRIC: Also commonly called gray fabric. It refers to cotton fabric in the unfinished state, often yellowish, after it has been woven but before it is dyed or finished.

HAND: The feel or touch of a fabric, inherent to its contents or achieved through washes and finishes.

HEAVYWEIGHT DENIM: Denim fabric weighing over 14 oz. per square yard.

JACQUARD LOOM: The automated patterns woven into fabric owe their origin to Joseph Jacquard, who in 1801 invented a method for using punched cards to mechanically regulate the sequence of harnesses controlling the warp thread. As they individually opened and closed on a loom according to a prefigured program to let the horizontal weft pass through, they replicated the intended design. While punched cards have given way to high-tech methods, Jacquard's principle remains viable—both in fabric creation and in computer technology, which later adopted the same stop-pass functionality.

LINSEY-WOOLSEY: A fabric made of mixed linen and wool.

MERCERIZED COTTON: In 1851, Englishman John Mercer patented a method for treating cotton and other fibers with caustic soda. Although the soda—mixed with sulfuric acid, initially—caused the fibers to swell, become round, and straighten, no one paid attention for forty years. Then another inventor showed how to make cotton fabric lustrous using Mercer's caustic soda, applied under pressure. The cotton took on a silky look, and a revolution was born.

MUSLIN: Today, a cloth used for sheeting with a thread count fewer than 180 threads per inch. Muslin is more loosely woven and feels coarser than percale.

NOILS: Short fibers that are removed in the combing operation of yarn formation. Noils sometimes are mixed with other fibers to make low-quality yarns or are used for purposes other than yarn making, such as for padding and stuffing.

OPEN-END: An air-driven yarn-spinning technology that produces denim and other cotton fabrics with a more regular, flatter appearance.

PERCALE: A fine, closely woven cotton cloth variously finished for clothing, sheeting, and industrial uses. Medium weight, firm, smooth, with no gloss and thread counts above 180. Percale sheeting is the finest sheeting available, with a soft silklike feel.

PICK: A filling thread or yarn that runs crosswise (horizontally) in woven goods. The pick interlaces with the warp to form a woven cloth.

PIECE-DYED: The process of dyeing pieces of fabric, rather than the yarn.

RING-SPUN: Yarn that is produced by using a "ring" for spinning. The ring, which spins and winds yarn in one continuous motion onto bobbins, produces yarn that has a characteristic natural unevenness. The hand is softer than that of open-end fabric. Quality denim companies favor ring-spinning, which has been around (and around) since 1850.

SANFORIZED COTTON: Named for its inventor, Sanford Cluett, who invented a way to wash cotton clothes without causing them to shrink. His machine, patented in 1933, preshrank the cloth by pulling and pushing it mechanically over an elastic felt blanket.

SATEEN: A manufacturing process that provides an extra soft sheet by smoothing the yarn. Usually made of a higher than average thread count fiber for extra softness and durability.

SELVAGE: The narrow edge of woven fabric that runs parallel to the warp. It is made with stronger yarns in a tighter construction than the body of the fabric to prevent raveling; it is usually stitched with a colored thread. Because of the simpler, narrower looms, the selvage of original denim fabric was visible inside the trousers along the inner leg and became a distinctive signature of the manufacturer—red for Levi's, green for Lee, and yellow for Wrangler. Exposed selvage is a key to recognizing period originals.

THREAD COUNT: The number of threads in one square inch of fabric. Often used as a mark of quality, but can prove unreliable.

WARP: Set of yarn found in every fabric woven on the loom, running lengthwise, interwoven with the weft or filling.

WEFT: In woven fabrics, the set of yarn that runs at right angles to the warp. Also known as the filling or woof.

WEIGHTS: Fabric weight is measured in ounces per square yard. The heaviest, for denim, is over 15 oz.; regular is 13¾ to 14¾ oz.; midweights are from 10 to 13 oz.; and lightweights or shirtings are from 4 to 9 oz.

Sources:

http://www.alrashidmall.com/clothing4.htm
http://www.apparelsearch.com/glossary_p_.htm
http://www.ifi.org/consumer/dictionary.html
http://www.atmi.org/EconTradeData/textterm.asp
http://www.towels4u.com/glossary.html

Notes

ONE: Spun in All Directions

9 ***"the women wearing a slight covering of cotton over the nudities":*** See Fordham University, excerpts from the Journals of Christopher Columbus, http://www.fordham.edu/halsall/source/columbus1.html.

11 ***"Columbus discovered approximately 12,000 pounds of cotton":*** James Scherer, *Cotton as a World Power*, 113.

13 ***anything approaching spinnable lint:*** Cotton's precise botanical history remains uncertain; this account comes from Jonathan F. Wendel, Iowa State University, a leading expert, presented in an e-mail to author, December 12, 2002.

13 ***in Central America and Mexico:*** See Iziko Museum, Capetown, South Africa: http://www.museums.org.za/bio/plants/malvaceae/gossypium.htm. This site offers a succinct overview of the four domesticated species, including their countries of origin, the areas of the world they inhabited, and their social significance.

15 ***a practice that continues to this day:*** James M. Vreeland, "The Revival of Colored Cotton," *Scientific American* vol. 280, issue 4, April 1999, 112 ff, provides an excellent introduction to the origins of colored cotton. Vreeland has dedicated himself to reviving vividly pigmented strains of the fiber in Peru using organic, sustainable farming methods. A full-text version of the article can be found at Vreeland's Web site, http://www.perunaturtex.com/scientif.htm.

15 ***while contributing to the community:*** Elizabeth Wayland Barber, *Women's Work*, 29. She quotes Judith Brown, who says that women's labor contribution in early communities depended upon "the compatibility of this pursuit with the demands of child-care." The work had to be repetitive, not the sort that required total concentration, safe for children to be near, and easily interruptible. Anthropologists report that men did not rear children in any early societies.

17 ***infinite patience and meditative concentration:*** In their incisively written and superbly photographed chronicle of women's crafts around the world, *In Her Hands*, Gianturco (text) and Tuttle (photography) explore the social and artistic ramifications of Guatemalan hand-spinning techniques still used today in the villages of that country and of many others in the nontechnology world to produce fine, strong yarns. Spinning details from Perry Walton, *The Story of Textiles*, 71–72.

19 ***give or take a few hundred years:*** See the Sampler House, a cross-stitch design company: http://thesamplerhouse.com/body_history_lesson.html.

22 ***to help preserve them:*** See the Sampler House, a cross-stitch design company: http://thesamplerhouse.com/body_history_lesson.html.

23 ***"God is marveyllous in his Werkes":*** Sir John Mandeville, *Voiage and Travaile of Sir John Mandeville Kt.* (Haliwell), reprinted from 1725 edition. London, 1883, xxvi. A modern-English full text is available at http://www.worldwideschool.org/library/books/hst/biography/TheTravelsofSirJohnMandeville/Chap0.html.

24 ***"fell prostrate to the earth and died":*** See Dave's Mythical Creatures and Places, http://www.eaudrey.com/myth/lamb_tree.htm, for Claude Duret's account. A full chronicle of the long, robust history of this mysterious animal-plant can be found at the National Agricultural Library Web site, http://www.nal.usda.gov/pgdic/Probe/v2n3/legend.html. These pages include Judith J. Ho's illuminating account and an early illustration of the vegetable lamb from Duret's *History of Remarkable Plants* (Paris: Nicolas Buon, 1605, [xxiv], 341).

TWO: Star Turns

25 ***In one hall you see . . . manufacturers:*** Joyce Burnard, *Chintz and Cotton*, quoted on 51.

26 ***more so than any linen:*** See Fordham University History Sourcebook: http://www.fordham.edu/halsall/mod/1497degama.html.

28 ***remained permanently fixed:*** Mattobelle Gittinger, *Master Dyers to the World*, 19–29. An excellent source for an illustrated description of the ancient dyeing process.

30 ***"Condolences to my husband's mistress":*** Duncan Clarke, *The Art of African Textiles*, 22.

31 ***in the words of one English import company:*** Burnard, *Chintz and Cotton*, 16.

31 ***"This trade (the woollen) is very much hindered":*** Quoted in Edward Baines, *History of the Cotton Manufacture in Great Britain*, 77–78. After 170 years, Edward Baines's history, published in 1835, remains the most thorough examination of the origins of the English cot-

ton industry. Baines had the advantage of knowing people who knew Richard Arkwright, and of knowing Arkwright's son, Richard, who would inherit his father's business and become the wealthiest commoner in England. Like his father, a newspaper owner in Leeds, Baines attacked factory child-labor reforms; an adamant apologist for the exploitative, reviled manufacturers, he insisted that the treatment of children in the Manchester mills was fair and humane. Since Baines refused to acknowledge the obvious, damning truth of the situation, historians ever since have viewed him with disdain, even contempt. His historical study of cotton, still available in libraries and through collectors, does not suffer greatly from his misguided sentiments. It is, in fact, marvelously informative, lively, and filled with rich detail.

31 ***"About half the woolen manufacture":*** Quoted in Burnard, *Chintz and Cotton*, 9.

31 ***"The general fansie of the people":*** Quoted in Bertha Dodge, *Cotton: The Plant That Would Be King*, 72. Originally in *Weekly Standard*, volume 4, 606.

32 ***"A tawdery, pie-spotted, flabby":*** Quoted in Natalie Rothstein, "The Calico Campaign," *East End Papers*, 1964, 6.

33 ***"We will make together a rude war":*** Quoted in M.D.C. Crawford, *The Heritage of Cotton*, 88–89.

35 ***"smeared the body with pure water":*** From Fashion India, an Indian designers' showcase and information site: http://www.fashionindia.net/history_fashion/history_fashion.htm.

35 ***"dance with nothing else upon them":*** "Cotton: A Great Yarn," *The Economist*, December 18, 2003.

36 ***one pound of it could be stretched over two hundred miles:*** Phyllis G. Tortora, *Understanding Textiles*.

THREE: The Barber from Preston

39 ***It is not from the benevolence of the butcher:*** See Pinecrest School Web site, Fort Lauderdale, FL: www.teacherweb.ftl.pinecrest.edu/crawfor/apcg/Unit1smith.htm.

41 ***"His customers that had employed him":*** Baines, *History*, 78.

42 ***"But oh how happy should they be":*** See Early Origins of the Western Family, www.peterwestern.f9.co.uk/westernorigins.htm.

44 ***first passed under King James I in 1624:*** See Wikipedia, the Free Encyclopedia: http://en.wikipedia.org/wiki/Patents#Early_history_of_patents.

45 ***"The age is running mad after innovation":*** T. S. Ashton, *The Industrial Revolution*. Quoted on p.10.

50 ***"she, convinced that he would rather starve":*** Dodge, *Cotton*, 33.

52 ***"am sorry to find matters betwixt you & Mr. Arkwright":*** Baines, *History*, 87.

53 ***"When I rote to you last":*** Quoted, and background on relationship between Arkwright and partners drawn, from R. S. Fitton et al., *The Strutts and the Arkwrights*, 66.

57 ***"fame . . . resounded throughout the land":*** Baines, *History*, 67.

58 ***"I found myself compelled to form a lower estimate":*** Ibid., 70.

59 ***"I have no doubt that Arkwright was an inventive genius":*** Author interview with Arkwright Society members, Pawtucket, RI, October 2002.

59 ***cotton fabric without stiff taxation penalties:*** Information drawn from Arkwright Society educational handout: *Patent*, April 10, 1775, 111. Anyone interested in visiting the Cromford mill and nearby museum is advised to contact the Arkwright Society, Cromford, Derbyshire DE4 3RQ; Tel: 01629 824297; Fax: 01629 823256; Web site: www.arkwrightsociety.co.uk.

59 ***"every servant girl has her cotton gown":*** Anne Buck, "Variations of English Women's Dress in the Eighteenth Century," *Folk Life*, 1971, 22–24.

59 ***"It is impossible to estimate":*** Dodge, *Cotton*, 41, quoting Baines.

60 ***two hundred in operation a decade later:*** Donald C. Wellington, "Sir Richard Awkright: Premier Crook in the Industrial Revolution," *International Journal of Social Economics*, vol. 20, no. 12, December 1993, 37(13).

60 ***"to realize a more perfect principle of spinning":*** Crompton quoted in Walton, *Story of Textiles*, 84–85.

61 ***employing 500,000 workers:*** Ibid., 88.

61 ***"Happening to be in Matlock in the summer of 1784":*** James Scherer, *Cotton as a World Power*, 73.

62 "[A]***nd you will guess my astonishment":*** Ibid., 90.

62 ***"Few persons could tell a story so well":*** Ibid., 91.

63 ***"grave and polite, but full of humor and spirit":*** Much of the history in this section is drawn from a comprehensive overview of cotton's manufacturing origins, first published in 1924. Ibid., 77.

63 ***"From the year 1770 to 1778":*** Ibid., quoting Radcliffe, 64.

63 "[E]***ven old barns, cart-houses, and out-buildings":*** Ibid., 78.

64 ***"For desolate moors and fens":*** Ibid., 79.

64 ***"I had gone for a walk on a fine Sabbath afternoon":*** Jeannette Mirsky et al., *The World of Eli Whitney*, 19. This book contains Whitney's correspondence and offers the best insight into his character as well as his inventions, such as the steam engine that complemented his cotton gin. In later life, Whitney became the first manufacturer to

make interchangeable parts for muskets, gave birth to the machine-tool industry, and died a wealthy man in Connecticut.

64 ***"The trunk of an elephant that can pick up":*** Scherer, *Cotton*, 81.

65 ***"This machine represents, at the present time":*** Baines, *History*, 83, quoting M. Charles Dupin in an address to the Mechanic of Paris.

65 ***"The Indies, so long superior to Europe":*** Dupin quoted in ibid., 83.

66 ***"His concerns in Derbyshire, Lancaster":*** Ibid., 196.

66 ***"The most marked traits in the character of Arkwright":*** Ibid., 193.

68 ***slept six or more to a windowless basement room:*** Anthony Burton, *The Rise and Fall of King Cotton*, 78–79.

68 ***from ankle irons to their hips:*** Scherer, *Cotton*, 99.

FOUR: Revolutionary Fiber

71 ***"I have a set of the most Depraved villains":*** Scherer, *Cotton as a World Power*, 165.

71 ***"Toil, anxiety and disappointment have broken me down":*** John F. Stegeman et al., *Caty*, 171.

72 ***from English dependence:*** Scherer, *Cotton*, 120.

73 ***"I hope it will not be a great while":*** Ibid., 121.

73 ***arrived on our shores:*** Florence Montgomery, *Printed Textiles*, 46.

73 ***the country's homespun ingenuity:*** Frances Little, *Early American Textiles*, 195–196.

74 ***"New England's First Fruits" in 1642:*** Walton, *Story of Textiles*, 125–126.

75 ***"There was not reason to doubt":*** Ibid., 144.

75 ***"the four southernmost States make a great deal of cotton":*** Ibid., 144.

76 ***"Several of these Southern colonies":*** Scherer, *Cotton*, Hamilton to Carroll D. Wright, quoted on 125.

76 ***"In this manufactory, they have the new":*** Ibid., 173.

77 ***"When the horses went too fast":*** Quoted in Walton, *Story of Textiles*, 156.

78 ***"Under my proposals":*** Quoted in ibid., 171.

78 ***"We shall be glad to engage":*** Quoted in ibid.

79 ***"The manufactory at Providence":*** Ibid., 174.

79 ***"You must shut down thy gates":*** Ibid.

79 ***"What Peter the Great did to make Russia dominant":*** Quoted in Scherer, *Cotton*, 127.

81 ***"I find myself in a new natural world":*** Quoted in Constance Green, *Eli Whitney*, 53.

83 ***"One man and a horse":*** Ibid., 46.

83 ***"As I [manufacture cotton] myself":*** Quoted in Scherer, *Cotton*, 161.

84 ***"Favorable answers to these questions":*** Ibid., 161.

84 ***"I had the satisfaction to hear it":*** Quoted in Green, *Eli Whitney*, 161.

85 ***these two transplanted Yankees "who demand, as I am informed":*** Quoted in Scherer, *Cotton*, 164.

86 ***and by 1820, close to 128 million pounds:*** Montgomery, *Printed Textiles*, 46; and The University of Wisconsin–Eau Claire Web site: http://www.uwec.edu/Geography/Ivogeler/w188/planta1.htm. It was this astronomical increase in demand that not only brought wealth to the largest Southern planters but cemented a bond of interdependence with England's mills. That convinced the South its raw cotton was critically important to the British economy, which it was. It also convinced the South that England could not afford to refuse to back its bid for secession. It couldn't, at least in the short run, but it did nevertheless for noneconomic reasons.

86 ***fifty times as much by 1860:*** Dodge, *Cotton*, 52.

87 ***One square inch has the tensile strength of 100,000 pounds:*** Norm Thomas catalog, summer 1992.

87 ***"the skeins of this plant":*** Scherer, *Cotton*, 127.

89 ***"The machine of which Mr. Whitney claims the invention":*** Quoted in Green, *Eli Whitney*, 90.

89 ***"Individuals who were depressed with poverty":*** Quoted in Scherer, *Cotton*, 166.

FIVE: Camelot on the Merrimack

92 ***"I am going home where I shall not be obliged":*** Benita Eisler, ed., *The Lowell Offering*, 161.

94 ***"In the sweet June weather":*** Lucy Larcom, *A New England Girlhood*, 182.

94 ***"Despite the toil we all agree":*** William Moran, *The Belles of New England*, 23.

95 ***mattered more than the workers':*** Ibid., 22.

96 ***brown lung disease, or byssinosis:*** Ibid., 22–23.

98 ***"Years after, this scene dwelt in my memory":*** Harriet Robinson, *Loom and Spindle*, 39.

98 ***"The daughter leaves the farm":*** Quoted in Moran, *Belles*, 22.

99 ***"Certainly we mill girls did not":*** Lucy Larcom, "Among Mill Girls: A Reminiscence," *Atlantic Monthly*, vol. 48, no. 289, November 1881, 610.

99 ***". . . [T]he village speckled over with girls":*** Quoted in Hannah Josephson, *The Golden Threads*, 51.

102 ***"Everywhere heaps of debris":*** Quoted in Burton, *Rise and Fall*, quoting Frederick Engels in *The Conditions of the Working Class in England*, 92.

102 ***"looking for nature's weak side, we have found our own":*** Barbara Freese, *Coal*, 72.

102 ***"You are considerably deformed":*** Quoted in Moran, *Belles*, 49.

103 ***"From this foul drain the greatest stream":*** See Annenberg/CPB Web site: www.leaner.org/biographyofamerica/prog07/transcript/page02.html.

104 ***could have arrived at them:*** Josephson, *Golden Threads*, 21.

105 ***"I well recollect the state of admiration":*** Quoted in Burton, *Rise and Fall*, quoting Nathan Appleton, *Introduction of the Power Loom, and the Origin of Lowell*, 100.

107 ***Waltham sold $412 worth of cotton fabric:*** National Bureau of Economic Research: papers.nber.org/papers/W9182.pdf. To subscribe, visit https://secure.ssrn.com/secure/secureForm.cfm?function=chargePapers&nber_id=w9182.

107 ***a staggering increase:*** Thomas Dublin, *Women at Work*, 18.

107 ***taking advantage of his tariff exemption:*** Josephson, *Golden Threads*, 30.

108 ***in the future, one of them speculated:*** Ibid., 20.

108 ***about $6.6 million in today's dollars:*** See Federal Reserve Bank of Minneapolis Web site for conversion of historical currency: http://minneapolisfed.org/research/data/us/calc/hist1800.cfm$1 *($550/51)= $11*$600,000. See site for explanation of equivalency figures.

109 ***was returning 30 percent yearly profits:*** Josephson, *Golden Threads*, 41.

110 ***to more than 18,000 inhabitants:*** Dublin, *Women at Work*, 21.

110 ***for twenty consecutive years, until 1845:*** Ibid., 20.

110 ***"There was nothing peculiar about the Lowell mill girls":*** See Larcom's article in *Atlantic Monthly*, vol. XLVIII, November 1881, 60.

111 ***"A pile of huge factories":*** Quoted in Josephson, *Golden Threads*, 5.

111 ***"bricks had been in the mold but yesterday":*** Quoted in ibid., 51.

111 ***"Who shall sneer at your calling?":*** Quoted in Robinson, *Loom and Spindle*, 45.

111 ***"more as lads and girls of a great seminary":*** Quoted in David Cohn, *The Life and Times of King Cotton*, 195.

113 ***from the daily lives of his textile operatives:*** Ibid., 145. Appleton wears the thorny crown of remote, insensitive industrial baron uneasily. While he was hardly a man of the people who mingled comfortably

with his workers, he took a hands-on approach to mill ownership and genuine interest in working conditions, unlike many of the Boston Associates.

113 ***fair treatment of labor were moral issues for him:*** Quoted in Moran, *Belles*, 54.

113 ***"A Repository of Original Articles":*** Benita Eisler, *The Lowell Offering*, 34. Reading through these poems, articles, and reflections today, one is struck by the desire of the working-women contributors to satisfy the readers' insatiable curiosity about their lives. They describe their surroundings and activities in rich detail as if reporting back from a newly inhabited planet. A sense of wonder, excitement, and hopefulness permeates these pages. Clearly, a new age is dawning, and these women are for the most part thrilled to be among its pioneering explorers and eager to share their discoveries.

114 ***"I am now going to state three facts":*** Quoted in ibid., 184.

114 ***"It is a pleasure to find that many":*** Quoted in ibid., 188–189.

114 ***"One who sits on my right hand":*** Quoted in ibid., 64–65.

114 ***"They set me to threading shuttles":*** Quoted in Dublin, *Women at Work*, 71.

115 ***"With wages, board, &c., we have nothing to do":*** Quoted in ibid., 124.

115 ***"They say* The Offering *does not":*** Quoted in ibid., 22.

116 ***"The children of New England":*** Quoted in Robinson, *Loom and Spindle*, 59.

116 ***"If in our sketches there is too much light":*** Quoted in Eisler, *Lowell Offering*, 207.

SIX: Looming Conflicts

118 ***If the entire state of South Carolina should disappear:*** You can read a richly detailed account by Kevin Baker at his Web site: http://www.kevinbaker.info/c_cp.html.2. Baker, a novelist and nonfiction writer, roams across a wide range of personal interests with brio and an energetic intelligence—as his compelling chronicle of the Sumner beating demonstrates.

119 ***"We consider the act good in conception":*** Ibid., 4. Time hasn't blunted the venemous hatred that spews forth from these editorials. The war that followed shortly looks to be a foregone conclusion.

120 ***enough to circle the globe more than twice:*** Thomas O'Connor, *Lords of the Loom*, 95.

120 ***"unless requested by my brethren of the Slave-holding States . . .":*** Quoted in ibid., 50–51.

120 ***"We went to bed one night old-fashioned, conservative":*** Quoted

in ibid., 109. Historians still hotly debate the Lawrences' genuine commitment to abolition; many view it as a token gesture, more expedient in the end than heartfelt, given the public outrage at Sumner's beating.

121 ***"ideas of a contrariety of interests":*** *The Papers of Alexander Hamilton*, vol. 1, chapter 7, document 23, (Columbia University Press, 1961), 1778.

122 ***"the horrid sight of the sale of human flesh":*** Quoted in Moran, *Belles*, 54.

122 ***"African slavery . . . is a curse":*** Ibid., 301.

122 ***"I am no opponent to manufactures or manufacturers":*** Quoted in ibid., 67.

125 ***"direct slavery is as much the pivot":*** Quoted in the *Socialist Review*: See http://pubs.socialistreviewindex.org.uk/sr217/bennett.htm.

125 ***"We gain nothing by allowing":*** Quoted in O'Connor, *Lords of the Loom*, 81.

125 ***"I believe they have done mischief":*** Quoted in Moran, *Belles*, 63.

125 ***"There is something better than Expedience":*** Quoted in ibid.

126 ***"The word* liberty *in the mouth of Mr. Webster":*** Quoted in ibid. Considering the prominence of Emerson, and his close social ties to many of Webster's colleagues, this condemnation stands as a declaration of war, and it was a shot heard throughout the North. The public looked to Emerson and his group for moral leadership.

127 ***"men of the North, subdued":*** Quoted in ibid., 62.

127 ***"the light withdrawn which once he wore!":*** Quoted in O'Connor, *Lords of the Loom*, 83. The words of John Greenleaf Whittier, like Ralph Waldo Emerson, were not only read but often quoted as inspiration in times of crisis—both by the media and general public. By the example they set, putting moral probity before friendship, they emboldened the public to take action against slavery's apologists.

128 ***"whose soul has been absorbed in tariffs":*** Quoted in ibid.

128 ***"with a calm belief that he has placed a vexed question":*** Quoted in ibid., 84. Webster committed the politician's fatal error—he lost touch with his constituents and took comfort instead in the illusory support of his rich, powerful backers.

129 ***"the U.S. Marshal and his hired assistants":*** Quoted in ibid., 100.

129 ***"You may rely upon it":*** Quoted in ibid., 101.

129 ***close to 5,000,000 spindles in operation:*** Robert Greenhalgh Albion, *The Rise of New York Port*, 63.

129 ***every ounce of it relied wholly on slave labor:*** Ibid.

129 ***Virginia alone accounted for 425,757:*** Scherer, *Cotton*, 416–417.

130 ***60 percent of the country's exports:*** O'Connor, *Lords of the Loom*, 47. Figures like these tend to last in the mind about as long as a

snowflake on a warm window pane. But this one in particular holds meaning. Cotton put America on the world trading map, but by 1860 the economic welfare of the entire Union had become far too dependent on the same commodity that was rupturing the nation into two warring factions.

132 ***according to an 1852 report to Congress:*** Albion, *Rise of New York Port*, 97.

133 ***"The combined income from interest":*** Ibid., 96.

134 ***"condemned by the public as a heartless extortioner":*** Dodge, *Cotton*, 132–133.

136 ***in the words of a Charleston report:*** Albion, *Rise of New York Port*, 120.

137 ***"a few thousand glittering bayonets are on their way":*** Quoted in ibid., 1.

137 ***"A slave too goes voluntarily to his task but":*** Quoted in ibid., 18.

137 ***"They cum swarmin out of the factories like bees":*** Quoted in Burton, *Rise and Fall*, 105. It took a brave Southern reporter to challenge the popular, fiercely defended belief throughout the South that mill workers in the North were no better off—worse, in fact—than slaves. Whether he changed any minds is not known.

138 ***"had never been an organ":*** Ibid., 201.

139 ***"To my mind it is slavery":*** Quoted in Moran, *Belles*, 35.

139 ***the extension of slavery "while thousands of the fair daughters":*** Gray Fitzsimons, Martha Mayo, et al., *Cotton, Cloth and Conflict*, 7.

139 ***"In the strength of our united influence we will show these driveling cotton lords":*** Quoted in Josephson, *Golden Threads*, 252.

140 "[*I*]***n view of our condition—the evils already come upon us":*** Quoted in ibid., 256.

141 ***"Grass plots have been laid out, trees have been planted":*** Report quoted in Robinson, *Loom and Spindle*, 141.

141 ***the committee decided that although "we think there are abuses":*** Ibid., 135–148: Reprint of the full 1845 Legislative Investigation Into Labor Conditions report.

142 ***"It would be impossible to restrict the hours of labor":*** Ibid., 147.

142 ***"Bad as is the condition for so many women":*** Dublin, *Women at Work*, 125.

143 ***"If they grind and cheat as brethren should not, let us go":*** Quoted in Josephson, *Golden Threads*, 293.

144 ***and, in one instance, an acting United States Treasurer:*** Ibid., xii.

144 ***with the Irish alone comprising 47 percent:*** Dodge, *Cotton*, 2.

145 ***"Would any sane nation make war on cotton?":*** Hammond's oration quoted in Scherer, *Cotton*, 236–241. The senator's fiery declara-

tion of Southern independence, "Cotton is King!" is of course remembered long after its central premise has been forgotten. Hammon is trying to convey the importance of surplus production. At the time, we were still a hand-to-mouth nation. Our production capacity barely met our needs. Hammond felt compelled to point out that the South's cotton exports ($185 million annually) accounted for close to twice those of the North ($95 million), and offered the nation its only hope of gaining wealth and economic security by building up inventories of a valued commodity. Rattling off a barrage of supporting figures and calculations, Hammond concluded that "the South would never go to war." Why? Because in essence it held the rest of the country hostage from a position of power, and war was a disruption. But far worse for the North to pick a fight with anyone, especially the South—because in terms of real wealth it was a beggar who would soon starve without the South's largesse. A skewed argument? Yes. Delivered with thunderous conviction? Absolutely.

145 ***That, said Hammond, "constitutes the very mud-sill of society":*** Ibid., 230.

146 ***"any intention on the part of political agitators":*** Ibid., 230.

SEVEN: Southern Exposure

147 ***and Israel Gillespie was no longer a slave:*** James Cobb, *Most Southern Place on Earth*, 38. Much of the material on the Delta's history is drawn from Cobb's engaging, detailed chronicle of the land and its people. Another book that brings the Delta to vivid life is John M. Barry's *Rising Tide*, a compelling exploration of the devastating Mississippi River flood in 1927 that looks at the engineering of levees—essentially man-made walls of dirt that could rise as high as four stories—as well as the lawless camps they spawned, and the powerful dynasties that rose up along with them.

148 ***"a chaos of vine and canes and bush," in one visitor's opinion:*** Quoted in Cobb, *Most Southern Place*, 5.

149 ***"not often did our ladies give vent to their indignation":*** Quoted in ibid., 36.

149 ***She declared her contempt for "a people who instigate a race . . .":*** Ibid., 37.

150 ***Mamma Chaney, who "had held us all from babyhood":*** University of North Carolina Web site: http://docsouth.unc.edu/thomas/thomas.html, p. 13; electronic reprint of Edward J. Thomas's *Memoirs of a Southerner*, 1912. UNC's Web site, Documenting the American South, archives the Thomas memoir along with one of the great collections of first-person (primary source) regional American history available to the public. As compelling as it is ambitious, the collection contains full-length texts of innumerable slave narratives, as well as white Southerners' recollections. There is no substitute for authenticity, and these materials, which cover two centuries, bring to life as no secondhand accounts can the day-to-day relationships between slave and master, slave and slave, and members of white society. An extraordinary achievement.

150 ***"I remember the great big cotton house":*** Quoted at UNC: http://docsouth.unc.edu/thomas/thomas.html, p. 12.

151 ***"It is . . . of a perfect cream color":*** Quoted in Dodge, *Cotton,* 117.

152 ***"Then, on the old plantation," one typically wrote:*** Quoted in Scherer, *Cotton*, 305.

153 ***"Nine times out of ten I slept in a room with others, in a bed which stank . . .":*** Quoted in Cohn, *Life and Times*, 52.

155 ***"The hands are required to be in the cotton field as soon as it is light":*** UNC: http://docsouth.unc.edu/northup/northup.html: Solomon Northup, *Twelve Years a Slave* (1853), 170–171.

157 ***"There are few sights more pleasant to the eye":*** Ibid., 166.

157 ***"The burden of a deep melancholy weighed heavily":*** Ibid., 258–259.

158 ***"What planter would exchange his cotton field for a silver mine?"*** Story in Michael Wayne, *The Reshaping of the Plantation Society*, 1.

159 ***In 1839 they dropped to three cents:*** Journal of Southern History, *Cotton Planters' Conventions in the Old South*, vol. 19, issue 3, August 1953, 322.

160 ***600 planters in the district owned fifty or more slaves:*** Wayne, *Reshaping*, 7.

161 ***less than 25 percent of the South's white population owned slaves:*** Cohn, *Life and Times*, 76.

161 ***formed influential alliances with the Boston Associates:*** Charles Reagan Wilson et al., eds., *Encyclopedia of Southern Culture*, vol. 1, (New York: Doubleday Anchor Books: 1989), 43. Four volumes and 2,600-plus pages later, you will know anything you ever wanted to about the South, and possibly answer William Faulkner's questions: "Why do they live there? Why do they live?"

162 ***"the farce of the vulgar rich":*** Frederick Law Olmsted, *The Cotton Kingdom*, 411. Traveling through as a New York journalist in 1860, Olmsted created a class first-person historical narrative, steeping his readers in the filth and poverty of small farmers, the pretentions of the rich, and the brutality of a slave society. He keeps his voice calm and his perceptions sharp.

162 ***aristocratic Royalist exiles in Cromwell's England:*** Wilson et al., *Encyclopedia of Southern Culture*, 310.

163 ***"Why, all we have is cotton and slaves and arrogance":*** Margaret Mitchell, *Gone With the Wind* (New York: Scribner, 1936). Full text is at Nalanda Digital Library Web site, http://www.nalanda.nitc.ac.in/resources/english/etextproject/Margaret_Mitchel/gone_wind/part-1 chapter6.html. Butler quote is from chapter VI.

163 ***more than twice the available recruits:*** Visit the Henry Form Museum Web site: http://www.hfmgv.org/education/smartfun/hermitage/house/statesmap.html; also, Wilson et al., *Encyclopedia of Southern Culture*, 43, put figure at 8,039,884 whites in 15 slave states.

163 ***South's fragmented 9,000-plus miles:*** Cyber Essays: www.cyber essays.com/History/141/htm.

164 ***"From the rattle . . . of the child born in the South":*** Cohn, *Life and Times*, 93.

164 ***shortly after the war began:*** Wilson et al., *Encyclopedia of Southern Culture*, 348.

164 ***This crime, this worse than crime, this* blunder":** Quote in Burton, *Rise and Fall*, 195.

165 ***half of England's export trade was in cotton textiles:*** The war between the states may be history, but it isn't dead in pockets of the South, not by any stretch. Visit Shotgun's Home of the American Civil War: www.civilwarhome/kingcotton.htm, 2.

165 ***British merchants "would, if money were to be made by it":*** Dodge, *Cotton*, 124.

165 ***"Can any sane man believe that England and France":*** See Arthur James Lyon Fremantle: www.colfremantle.com/recognition.html, *London Times*, November 29, 1860, and January 4, 1861, editorials. Many sane men did in fact believe England would ignore its nonslavery policy to support the South, as it came a whisker away from doing.

166 ***"We from that hour shall cease to be friends":*** Quoted in John Taylor, *William Henry Seward, Lincoln's Right-Hand Man*, 179. Seward ups the ante for Britain's support and in effect scared Her Royal Highness out of the pot.

166 ***"How we cling to the idea of an alliance with England":*** Quoted in Dodge, *Cotton*, 126.

166 ***"we shall find the same causes that produced it":*** Quoted in ibid., 133.

167 ***The Civil War, he said, "has originated from the efforts":*** Quoted in ibid., 123–124.

167 ***"When you look around you, how dare you talk":*** Quoted in Burton, *Rise and Fall*, 171.

168 ***if needed, to placate the North . . . in producing the South's cotton:*** Harold Woodman, *King Cotton and His Retainers*, 160–170.

169 ***"The rebellion is already arrested":*** Like President George W. Bush on the air carrier flight deck, Seward spoke way too soon. Quoted in Scherer, *Cotton*, 265.

169 ***were being supported by organized charity:*** Ibid., 264.

169 ***half-a-million factory hands were out of work:*** Ibid., 265.

169 ***"No crisis in modern times has been so anxiously watched":*** Ibid., 268.

171 ***At one point Alcorn reports that 400 bales:*** All may not be fair in love and war, but all is fare in war. Quoted in Cobb, *Most Southern Place*, 38.

172 ***alternately trading and fighting with one another:*** Shotgun's Home of the American Civil War: www.civilwarhome.com/kingcotton.htm. Information Now encyclopedia, "The Confederacy," article by Orville Burton and Patricia Bonnin, no page numbers listed.

172 ***Defeat is impossible. Defeat is unthinkable:*** "Empire, Before the Fall," *New York Times*, translation by Ellen McLaughlin, June 11, 2003, Section B1.

EIGHT: Changing Fortunes

173 ***complained angrily about this "extortion, genteel swindling":*** Quoted in Frank E. Smith, *The Yazoo*, 177.

174 ***artificial limbs for Confederate veterans:*** Cohn, *Life and Times*, 152.

174 ***"a magnate of the very first rank":*** Quoted in Eric Foner, *A Short History of Reconstruction*, 100.

175 ***unable to cover back taxes:*** Ibid., 228.

176 ***for lease the following year for six bales of cotton:*** Cobb, *The Most Southern Place*, 54.

176 ***land forfeited for nonpayment of back taxes:*** Ibid., 64.

177 ***all this life for sure and maybe the next:*** See University of Houston History Department Web site: http://vi.uh.edu/pages/mintz/45.htm.

178 ***the furnisher at Frenchman's Bend in William Faulkner's* The Hamlet *was also "the largest landholder":*** Faulkner quoted in Harold Woodman, *King Cotton and His Retainers*, 8–9.

178 ***"Charge up another barrel of flour":*** Smith, *The Yazoo*, 182.

178 ***dry goods and sundry items instead of money:*** Ibid., 176.

178 ***The farmer, he added, "has passed into a state of helpless peonage":*** Quoted in ibid., 178–179.

179 ***"Credit supports agriculture, as cord supports the hanged":*** Quoted wisely in Cohn, *Life and Times*, 164.

179 ***"Hurrah! Hurrah! 'Tis queer I do declare!":*** Quoted in Smith, *The Yazoo*, 180.

179 ***"Time was sure better long time ago than they be now":*** Quoted in Bernice Hurmence, ed., *Before Freedom*, 130.

179 ***"Like all the fool niggers of that time":*** Quoted in Andrew Waters, ed., *Prayin' to Be Set Free*, 38–39. Both of these short collections are part of an extensive series of former-slave narratives collected during the Great Depression. The Federal Writers Program sent field workers to interview ex-slaves under the supervision of the Library of Congress, where these narratives can be read.

182 ***"The lines were so grand and so beautiful":*** Quoted in Steve Dunwell, *The Run of the Mill*, 103.

182 ***perfection of mankind through science and technology:*** The shrewdest observers of this Gilded Age were less impressed. Coming along a few decades later when the whole Western world seemed to be swooning over the horseless carriage and electric light bulb, George Bernard Shaw pointed to where inventive ingenuity truly flourished—not in improving lives but in destroying them. In his play *Man and Superman*, Shaw imagines Satan taunting us. After walking the Earth up and down, he announces that "in the arts of life man invents nothing; but in the arts of death he outdoes Nature herself and produces by chemistry and machinery all the slaughter of plague, pestilence, and famine." We eat and drink and essentially make clothing the way we always have, although more efficiently. As for Man: "I have seen his cotton factories and the like, with machinery that a . . . dog could have invented if it wanted money instead of food." But Man's heart, says Satan, is in his weapons. All other inventions "are toys compared to the Maxim gun, the submarine torpedo boat." Walking the world today, Shaw's Satan would probably note that in the age of the Stealth bomber and heat-seeking Patriot laser-guided missile, cotton continues to be ginned, spun, and woven by mechanical processes first implemented about two hundred years ago. The changes have been refinements rather than new inventions.

182 ***second in cotton production only to Manchester, England:*** Dunwell, *Run of the Mill*, 106.

182 ***"but a town also":*** Ibid., 105, quoting *Niles Weekly Register.*

183 ***in one early decade of the twentieth century:*** Hall et. al., *Like a Family*, 197. By 1929, fifty-one New England mills, with 1.3 million spindles, had shut down and moved south. This is the definitive study of the Southern cotton mill culture, exhaustively researched and filled with first-person narratives.

183 ***"disruptive technology"—the new product or improvement:*** Harvard Business School, Baker Library Web site: http://www.library.hbs.edu/hc/exhibits/distech/exhibit.htm, 1.

184 ***"A public job is more interesting than one-horse farming . . .":*** Quoted in Hall, *Like a Family*, 56.

184 ***children were paid 10 to 12 cents:*** Cohn, *Life and Times*, 216.

185 ***At least 90 percent of Southern working children under fifteen in 1900 were employed by cotton mills:*** Wilson et al., *Encyclopedia of Southern Culture*, vol. 4, 200.

185 ***through the unlighted streets . . . to their squalid homes:*** Cohn, *Life and Times*, 216.

185 ***Another mill visitor, W. J. Cash, wrote that "[b]y 1900 the cotton-mill worker . . .":*** Quoted in ibid., 216–217.

186 ***the misery of a sick, dismal childhood:*** Dunwell, *Run of the Mill*, 119.

186 ***Profits between 30 and 75 percent were common:*** C. Vann Woodward, *Origins of the New South*, 132–133.

186 ***for a new factory and hope to attract outside finance:*** Laurence Gross, *The Course of Industrial Decline*, 95.

187 ***"We have fallen in love with work":*** Henry Grady, cited at American Social History Project/Center for Media and Learning, City University of New York Web site: www.historymatters.gmu.edu/d/5745.

187 ***they represented monthly wages of less than one dollar per person per week:*** Burton, *Rise and Fall*, 195.

188 ***Each community tried to outdo the other:*** Cohn, *Life and Times*, 214.

188 ***the hearts of men in whom the American dream was dying:*** Quoted in ibid., 213.

189 ***"Let croakers against enterprise be silenced":*** Quoted in Jacquelyn Hall et al., *Like a Family*, 61.

189 ***"When I was a little fellow, my daddy was":*** Quoted in ibid., 180.

190 ***They just fight for theirself:*** Quoted in ibid., p. 250.

190 ***" 'Bynum's red mud. If you stick to Bynum, it'll stick to you when it rains' ":*** Quoted in ibid., 252.

190 ***"They said you couldn't talk to the other one on the other side of you":*** Quoted in ibid., 97.

190 ***"I seen a time when I'd walk across the road to keep from meeting my supervisor":*** Quoted in ibid., 94.

192 ***"The telegraph is used freely, and the buyer knows hour to hour":*** Quoted in Woodman, *King Cotton*, 276.

193 ***The sidewalks of Carondelet, home to the city's banks, "were perpetually crowded":*** Quoted in Christopher Benfey, *Degas in New Orleans*, 154.

193 ***"In the American quarter, during certain hours of the day, cotton is the only subject spoken of":*** UNC Web site: http://docsouth.unc.edu/nc/king/menu.html, Edward King, "The Great South," 50.

195 ***more than 400,000 bales valued at $16 million:*** Cohn, *Life and Times*, 185.

195 ***when to Cain it was Strict Low Middling:*** Ibid., 187.

196 ***"The classer takes a tuft of cotton, . . . grasps both ends":*** Quoted in ibid., 190.

196 ***with a small fraction measured above and below that in length:*** Ibid., 188–189.

196 ***"Four days of rain was the damnedest excuse to get drunk that a man ever had":*** Quoted in John Branson, "The Last Days of Cotton Row," *Memphis Flyer*, July 13, 2001.

198 ***machines were driven entirely by electricity—self-generated in this case:*** Cohn, *Life and Times*, 217.

200 ***"Lasting work clothes are an economic necessity," Ceasar announced:*** Uncredited copywriter, *A Century of Excellence*, Cone Mills, self-published, 16.

201 ***80 percent of the nation's spindles, and 82 percent of its looms had moved south:*** Cohn, *Life and Times*, 218.

NINE: Two-Horse Power

205 ***a tailor up in Virginia City, Nevada:*** Commonly thought to be Reno, but Jacob's grandson Ben Davis identifies it as Virginia City. See http://www.bendavis.com/home/royal.html.

205 ***In his letter to Strauss, Davis claimed:*** Quoted in Graham Marsh et al., *Denim*, 8.

208 ***"In short," he wrote in his diary, "the whole seemed perfect":*** Quoted in Alice Harris, *The Blue Jean*, 10.

208 ***"A garment that squeezes the testicles":*** Quoted in ibid., 95.

215 ***"I want to thank you many many times":*** Neslin and woman are quoted in *This Is a Pair of Levi's Jeans* (San Francisco: Levi Strauss & Co., 1995), unnumbered pages.

217 ***maintaining quality was another:*** Marsh, *Denim*, 64.

219 ***"They ain't tactics, honey,":*** Quoted in Alice Harris, *The Blue Jean*, unnumbered pages.

219 ***"About 90 percent of American youths wear jeans . . .":*** New Internationalist Magazine Web site: www.newint.org/issue302/blue.htm/ *The Blue Jeans Story.*

228 ***There was only one notable weakness:*** Information drawn from author in-person interview with former Levi's design director for jeans, Stefano Aldighieri, April 2002.

228 ***Some high-end designer jeans:*** Marsh, *Denim*, 124.

228 ***generally considered to be the world's best denim fabric maker:*** Andrew Olah: See www.olah.com. Andrew Olah's "Denim Survival

Guide" is an entertaining and informative introduction to the history and lore of denim.

229 ***"the single best item of apparel ever designed":*** Quoted in Harris, *The Blue Jean*, unnumbered pages.

229 ***"put it on the asses of America":*** Quoted in ibid., unnumbered pages.

230 ***$180 million from $25 million the previous year:*** Marsh, *Denim*, 105.

230 ***more than one billion square yards of denim a year were being manufactured in the United States:*** Figures supplied by Gary Raines, Cotton Inc.

231 ***beaded jeans sold out at $3,500 each:*** Austin Bunn, "Not Fade Away," *Fashion of the Times, New York Times*, December 1, 2002.

TEN: Boll Weevil Blues

238 ***becomes the resonating box on the guitar:*** L. Robert Gordon, interviewed by Ben Manilla for Ben Manilla Productions, June 20, 2002.

239 ***"I kept loneliness close to my heart":*** B.B. King, *Blues All Around Me*, 34. The material on King is drawn largely but not exclusively from King's engrossing, forthright autobiography.

239 ***"It gave you a double feeling":*** Ibid., 36.

239 ***"In the Mississippi Delta of my childhood":*** Ibid., 57.

240 ***"The mule will shit, piss, and fart in my face":*** Ibid., 59.

240 ***"Cotton, you see, is fragile":*** Ibid., 63.

242 ***"Where dozens and dozens of mules had been used before":*** Ibid., 72.

243 ***"I got the keys to the highway":*** Alan Lomax, *Blues in the Mississippi Night* (Cambridge: Rounder Records, 2003). This is a CD-remaster of classic archival blues material that includes interviews with Big Bill Broonzy, Memphis Slim, and Sonny Boy Williamson as well as song tracks.

244 ***two-room shotgun or dogtrot cabin:*** Shotgun and dogtrot: two common Southern plantation shacks—primitive, unheated, and rarely waterproof, both synonymous with rural poverty. The shotgun was so-named because the corner front and rear doors were aligned at either end of a corridor opening to the rooms. A shotgun, it was said, could be fired from the front to the rear and exit without hitting anyone inside. The dogtrot two-room shack got its name from the unenclosed "hallway" separating its two rooms, all under one roof—plenty of space for a dog, horse, or any animal to trot through. Rough logs and planks of raw pine or cedar usually served as the building material, along with any available scraps.

245 ***Once the weevil lays its eggs, he continued:*** Quoted in Theodore Rosengarten, *All God's Dangers*, 222–223.

245 ***"You can't thoroughly understand the nature of a boll weevil":*** Quoted in ibid., 224.

245 ***"And you watch him, just watch him":*** Quoted in ibid.

246 ***the only natural monopoly of a worldwide necessity:*** Scherer, *Cotton*, 338.

246 ***In 1916 Southern mills alone consumed about 2,300,000 bales:*** William Dickerson et al., eds., *Boll Weevil Eradication*, 12–13.

246 ***"Everything, every creature in God's world, understands how to protect itself":*** Rosengarten, *All God's Dangers*, 226.

247 ***More than 2,100 gins in Louisiana and Mississippi:*** William Dickerson, *Boll Weevil Eradication*, 12–13.

247 ***after 1914 fanned out to educate growers:*** Pete Daniel, quoted in Carolyn Merchant, *Major Problems in Environmental History*, 242.

248 ***"the advantages of the Delta as a home for Negroes":*** Robert Palmer, *Deep Blues*, 140.

249 ***often for the slightest infractions:*** Rosengarten, *All God's Dangers*, 113–114.

249 ***"Cotton," he said, "is a drug, and for three hundred years":*** Richard Wright, *Twelve Million Black Voices*, 59.

250 ***"We look out at the wide green fields":*** Quoted in ibid., 92.

250 ***all but emptying out many rural communities:*** Ibid., 98, and Palmer, *Deep Blues*, 139–140. Palmer quotes David L. Cohn in *Where I Was Born and Raised* on the number of blacks leaving every night, and on 12,000 blacks leaving Mississippi within a ninety-day span in the 1920s.

250 ***step into the gutter when a white woman passed:*** Wright describes in vivid detail the visceral dread of blacks toward whites and the dreamlike quality of their journey north on trains. Blacks might doze off in a cabin alongside white people for the first time in their lives, and wake up with a start, expecting to still be on a cotton farm, cursed out by the white riding boss for falling asleep. Wright's book, hard to find but worth the effort, combines a pulsating text with arresting photography to capture the feeling of black life in Northern cities during these years of early migration.

251 ***"Look at me. Look at me":*** Lizzie Miles, "Cotton Belt Blues," 1923.

252 ***"Only wars have bailed the industry out":*** Quoted in Cohn, *Life and Times*, 221.

252 ***Between 1925 and 1939, one-third of America's cotton mills shut down:*** Ibid.

252 ***"If you were a black man on Beale Street":*** Rufus Thomas, interviewed in PBS series *Martin Scorsese's "The Blues."* Thomas appears in the segment entitled "The Blues Comes Home to Memphis," first broadcast 2003.

253 ***before a packed auditorium of cheering whites:*** Wherever the blues went, blue jeans of course followed. The irony was that no genuine black Delta bluesman would ever be caught dead onstage in denim. That was what you wore to sharecrop, when you were obliged to chop and pick from can to can't. No way would you be seen in your work clothes before the public. When B.B. King first went to Memphis, he later told the slide guitarist Roy Rogers that he felt like he had arrived in Paris. Delta blues musicians had never walked down the paved street of a town with more than a handful of rickety threadbare storefronts on it. King was overwhelmed by the size and grandeur of the city, and in that context, wearing jeans onstage would be an insult to his audience and to himself.

253 ***600,000 square miles of infested fields:*** Mississippi State University: http://www.bollweevil.ext.msstate.edu/webpage_history.htm.

254 ***"It was clear," said North Carolina's agricultural commissioner":*** Dickerson, *Boll Weevil Eradication*, 384.

254 ***If only 10 percent of those generations survived:*** History Channel and Killerplants, co-sponsors: www.killerplants.com/renfields-garden/20020220.asp.

254 ***"The control man has secured over nature":*** From Quote Garden Web site, created by Terri Guillemets: www.quotegarden.com/environment.html.

255 ***arsenic, the first insecticide, came into widespread use:*** An arsenic compound, Paris green, had been first applied in limited use as early as 1872. For a full explanation, see http://entweb.clemson.edu/pesticid.history.htm/htm.

255 ***"I carried poison to the cotton fields maybe four or five rows":*** Rosengarten, *All God's Dangers*, 224.

255 ***"Old weevil, he can't stand that*** [***arsenate***]***":*** Ibid., 224–225.

256 ***more than 500 insects have built up resistance to insecticides:*** Author phone interview with Bruce Tabashnik, entomologist, University of Arizona, October 2003.

257 ***even if they didn't have positive proof:*** Author phone interview with Marshall Grant, September 2003.

257 ***from 1.5 million to 190,000 acres:*** Author phone interview with Grant and James Tumlinson III, July 2003.

258 ***detention camps, and killed at least one man:*** Georgia State University Web site: http://wwwlib.gsu.edu/spcoll/Labor/work_n_progress/34Strikebackground.htm.

258 ***"It seems to me . . . that a tenant can feel, toward the crop":*** Quoted in Pete Daniel, *Breaking the Land*, 167–168.

259 ***"It would take a blind sentimentalist to mourn their passing":*** Ibid., quoting C. Vann Woodward, 244.

260 ***typhus during the previous decade:*** Dan Fagin, Marianne Lavelle and Center for Public integrity, *Toxic Deception*, 1.

261 ***"DDT is good for me-e-e-e-!":*** Quoted in ibid.

261 ***chemicals the same way some people believe in God:*** Author phone interview with Pete Daniel, January 2003 and in person, April 2003, Peabody Hotel, Memphis.

261 ***"In the decades after World War II . . .":*** Daniel interview.

262 ***only one fledgling was born in an already stressed population of 552 nesting pairs:*** Pacific Biodiversity Institute: http://www.pacificbio.org/ESIN/Birds/BrownPelican/pelican_overview.htm.

262 ***"The 'control of nature' is a phrase conceived in arrogance":*** Rachel Carson, *Silent Spring*, 262–263.

263 ***"More applications or greater quantities of the insecticide":*** Quoted in ibid., 275.

263 ***41 percent of all pesticides in agricultural use:*** USDA Agricultural Research Service: www.ars.usda.gov/is/AR/archive/feb03/boll0203.htm for USDA report.

267 ***He called it "grandlure":*** This account combines information drawn from Dickerson et al., *Boll Weevil Eradication*, 102–112, and a phone interview with James Tumlinson III, July 7, 2003.

268 ***to coordinate timing with the crop-duster:*** Grant interview, 2003.

269 **New York Times *editorial, "Good Boll Weevil News":*** Excerpted from "Good Boll Weevil News," *New York Times*, July 9, 2003.

269 ***"Nature," he says, "has a way of jumping up when you least expect it":*** Drawn from Tumlinson interviews of July and September 2003 and from Dickerson et al., *Boll Weevil Eradication*, 102–108.

ELEVEN: The Shirt on Your Back

273 ***and close to 80 percent by 2003:*** Author phone interview with Monsanto communications director Karen Marshall, December 22, 2003. Monsanto 2003 estimate of 77 percent based on latest USDA statistics.

274 ***"I'm worried more about the big picture, really":*** Quoted in Daniel Charles, *Lords of the Harvest*, xvi.

275 ***"It hardly matters that it's from biotech crops":*** Karen Marshall interview.

276 ***creating a phenomenon commonly known as the pesticide treadmill:*** USDA Agricultural Research Service: www.ars.usda.gov/is/AR/archive/feb03/boll0203.htm and Wilson et al., *Encyclopedia of the South*, vol. 1, 69.

278 ***Monsanto's breakthrough made front-page news:*** See Charles's in-depth account in *Lords of the Harvest*, from which much of this material is drawn, as well as from author interview with Charles. He traces the origins of gene-altering and hones in on the current status of biotech, managing to convey difficult concepts simply while bringing the key players to life.

279 ***"we shouldn't be in this buisness":*** Quoted in Charles, *Lords of the Harvest*, 60.

282 ***"Bt didn't just clobber the tobacco budworm":*** Quoted in ibid., 174.

284 ***Benbrook argues that it "would be analogous to the loss of antibiotics":*** Charles Benbrook, IRM analysis paper, August 20, 2001, delivered to Union of Concerned Scientists, 1.

284 ***"Within the next five years, insects are bound to adapt":*** Tabashnik, B. E., Y. Carrière, T. J. Dennehy, S. Morin, M. S. Sisterson, R. T. Roush, A. M. Shelton, and J. Z. Zhao, "Insect Resistance to Transgenic Bt Crops: Lessons From the Laboratory and Field." In *Journal of Economic Entomology*, no. 96: 2003, 1031–1038.

286 ***"Bollgard was supposed to be sold only by authorized dealers":*** Glenn Stone e-mail to author, December 2, 2003.

288 ***bankrupted cotton growers . . . who committed suicide by swallowing the ineffectual pesticide:*** Jonathan Karp, "Deadly Crop," *Wall Street Journal*, February 18, 1998. He is also quoted in http://www.aworldconnected.org/article.php.305.html, "Bollworm Suicides and the Government That Got Away."

289 ***"It is clear that the Bt cotton has failed on all counts":*** Kisanwatch Public Information Web site monitoring Indian agriculture: http://www.kisanwatch.org/eng/special_reports/feb2002/spr_bt_commercial%20farming.htm.

291 ***Yves Chouinard, after considerable research, concluded that "the most damaging fiber":*** Quoted by Harvest Cooperative, 1996 *Harvest Times* article: http://www.harvestcoop.com/pages/updates/times/1996/sept1996.html#story5996.

291 ***"We've only got one planet":*** Author phone interview with Jill Vlahos, November 17, 2003. For all the lip-service given to sustainability and responsible environmental practices, Patagonia is the only company of any size fully invested in 100 percent organic cotton across the board.

292 ***"only organic cotton can touch its newborn skin":*** Author phone interview with Bill Grieber, November 18, 2003.

292 ***"When planes still sweep down and aerial spray a field":*** Eco By Design: http://www.ecobydesign.com/1/cotton.htm. Excerpt from Paul Hawken, *The Ecology of Commerce*, reprinted on home page.

293 ***one and a quarter pound into every set of queen-sized sheets:*** Or-

ganic Trade Association: www.ota.com. This Massachusetts organization speaks to and for a diverse, often contentious group of individuals, and does so with authority and a refreshing lack of alarmist hyperbole. Its well-organized site also acts as a portal to a wide range of sustainable- and organic-product businesses.

293 ***cyanazine, finally phased out because of its proven link to breast cancer:*** Among the potential carcinogenic pesticides are acephate, dichloropropene, diuron, fluometuron, pendimethalin, tribufos, and trifluralin.

293 ***"The pesticides used on cotton, whether in the U.S. or overseas":*** Pesticide Action Network: http://www.panna.org.resources/pestis/PESTIS.1996.46.html. Sandra Marquardt, *Pick Your Cotton.*

293 ***10 percent of all herbicides and defoliants:*** Organic Trade Association: http://www.ota.com/facts_cotton_environment.htm; also, the Organic Cotton Site: http://www.sustainablecotton.org/NEWS008/news008c.html.

294 ***Union Carbide paid $793 to each family of the deceased:*** Various sources, including International Campaign for Justice in Bhopal: http://www.bhopal.net.

294 ***"because of the potential for genetic contamination and its continued reliance on artificial chemical inputs":*** Sue Mayer, "Genetically Modified Cotton: Implications for Small-Scale Farmers," PAN UK Publications, August, 2002, 5.

296 ***"If cotton were a crop that we ate instead of one that we wore":*** Unique Baby Boutique: http://www.uniquebabyboutique.com/Shop/products/c107. Phillips County, Arkansas, Extensive Service agent Jerry Williams quoted on home page, from article in *New Yorker*, July 8, 1991.

297 ***"I got into organic on account of it was something different":*** Author interview with Pete Coraggia, November 13, 2003.

297 ***"I'm preachy and I know it," LaRhea explains:*** Author interview with LaRhea Pepper, March 2002.

298 ***especially his own kin:*** Sustainable Agricultural Research and Education: http://www.sare.org/newfarmer/pepper.htm. Peppers interview.

298 ***a strong connection between those chemicals and his dad's early death:*** Helen Cordes, "Moral Fiber," *Organic Style Magazine*, June 2002, 78, and several author interviews with LaRhea Pepper, February and March 2002; e-mails November 2003.

298 ***use chemical growth regulators, desiccants, and boll-openers:*** One typically extensive list of agri-chemicals for these purposes can be found at the Texas A&M University Web site: http://216.239.57.104/search?q=cache:6eA8pJWppa0J:soilcrop.tamu.edu/publications/pubs/harvestaid2002.pdf+boll-openers&hl=en&ie=UTF-8.

299 ***"Here's what we got for sale and here's why you need to offer it to your customers":*** Author interview with LaRhea Pepper, March 2002.

TWELVE: Fields of Conflict

306 ***"In U.S. Cotton Farmers Thrive; in Africa, They Fight to Survive":*** Roger Thurow and Scott Kilman, article in *Wall Street Journal*, June 26, 2002. Follow-up author phone interviews with both writers.

307 ***Malians crossing the Algerian border for religious training abroad:*** Ibid., 13.

307 ***finding a way to make the public feel good about six- and seven-figure payments to farmers:*** Delta Farm Press: http://deltafarmpress.com/ar/farming_hood_defends_defense/, August 2, 2002. Hood expresses the popular view among large cotton farmers, which is that bad press is the result of a dismal public relations effort.

307 ***"It's tough for the United States to preach capitalism to developing countries . . .":*** Brian Reidl interviewd by author's research assistant, September 2003.

308 ***"bad for most farmers, bad for consumers, and horrendous for taxpayers":*** Business Week Web site: http://www.businessweek.com/bwdaily.dnflash/may2002/nf2002057_0341.htm.

309 ***"If it weren't killing them, people in Burkina Faso":*** "The Long Reach of King Cotton," *New York Times*, August 5, 2003, Section A, 14.

310 ***"It is alarming and also nauseating to see Mr. Gandhi":*** Quoted in Burnard, *Chintz and Cotton*, 62.

310 ***"It's quite all right," Gandhi replied:*** Quoted in ibid.

311 ***"His thin slightly nervous hands worked rapidly":*** Ibid., 63.

311 ***but in the end he did:*** Ibid., 64.

312 ***a full-bore attack on "The Unkept Promise":*** Refers to an editorial, "The Unkept Promise," from the *New York Times*, December 30, 2003, Section A, 20.

312 ***almost exclusively on USDA data for its evidence:*** Elizabeth Becker, "Looming Battle Over Cotton Subsidies," *New York Times*, January 24, 2004, World Business Section, 1.

313 ***"The [latest] editorial follows the pattern of repeating unsubstantiated claims":*** NCC letter, *New York Times*, August 8, 2003.

314 ***"Attempts to 'sound-bite' policy positions":*** Ibid. The National Cotton Council was also invited by the author on several occasions to make its case for subsidies to American growers in this book. Each time it refused. "We don't think *Big Cotton* is the proper venue," I was told by the trade association's communications director, Marjorie Walker. "Why not?" I asked. "We're putting it on our Web site." (www.cotton.org.) "Yes," I said, "but you're making your response available only to members, not the public." "That's our choice. This will die over," Walker replied by way of an answer. "We've been through it before." She got off the line. "They don't trust anybody, they're incredibly hunkered down," a newspaper journalist who's been on the cotton beat for decades explained. "The Cotton Council wouldn't even let me interview anybody in the organization for a recent article because I was also talking to someone whose views they didn't condone." And so it goes. Nothing is bound to change anytime soon, not until cotton subsidies to American growers come under much closer congressional scrutiny, and are seriously threatened—a probable turn of events within the next decade as the balance shifts even more toward urban representation in Congress. By then, having alienated most of the national media, the NCC will most likely garner little if any support for its cause barring a major revamping of its defensive, insular posture.

319 ***"The new framework agreement . . .":*** Laurie Goering, "Farm Subsidies: Africa Pays the Price for U.S. Ag," August 8, 2004, http://www.chicagotribune.com/news/opinion/perspective/chi0408080368aug08,1,2205244.story.

319 ***$3 million in U.S. government assistance in the past three years:*** Data compiled by research assistant Jason McClure.

322 ***"Farmers like Ken Hood can now farm by the foot, not by the acre":*** SUN Interactive: www.sustainableusa.or/bestpractices/best_detail cfm?Best ID=70. Quote from Mike Seal.

325 ***"In reality, ending the U.S. cotton program would only harm U.S. producers":*** U.S. Senate Committee on Agriculture, letter to Robert Zoellick, July 8, 2003.

326 ***"You don't need to spend a bundle when you've got the ear of legislators":*** Author interview with Burdett Loomis, August 3, 2003.

326 ***"You have to understand we are fed up with these farm subsidies":*** Elizabeth Becker, "Looming Battle Over Subsidies," *New York Times*, July 24, 2004, World Business Section, 1.

328 ***arid northwest Xinjiang Province:*** Officially, the Uighur Autonomous Region.

328 ***the largest plant biotechnology capacity outside of North America:*** The data on China is relevant through 2004, and was gathered in first-person interviews and from news and governmental sources by Beijing research assistant Kaiser Kuo.

331 ***"Why can't the Americans stick to making things we can't?":*** Background interview for possible article, conducted by *Time* Beijing bureau chief Matt Forney, November 2003.

332 ***1,300 additional U.S. textile mills face imminent closure:*** American Textile Manufacturers Institute: www.atmi.org/Newsroom/Releases/yrendo asp; and /Newsroom/releases/pr200312.asp.

333 ***More than 210,000 domestic textile workers have lost their jobs since 2000, close to 300,000 in total:*** Compilation of ATMI stats, www.atmi.org.

334 ***increasingly driven by U.S. and foreign investment:*** "Aiming at Chinese Imports Again," *New York Times*, November 20, 2003, Section A, 30.

334 ***"When textile import quotas expire, as they are scheduled to do in 2005":*** Quoted in Forrest Laws, "Textile Action Safeguard First Step," *Western Farm Press*, December 6, 2003. Full text at http://westernfarmpress.com/mag/farming.

Bibliography

Abbott, Richard H. *Cotton and Capital: Boston Businessmen and Anti-Slavery Reform, 1854–1868*. Amherst: University of Massachusetts Press, 1991.

Agee, James, and Walker Evans. *Let Us Now Praise Famous Men*. Boston: Houghton Mifflin Co., 1939.

Agins, Teri. *The End of Fashion*. New York: William Morrow & Co., 1999.

Albion, Robert Greenhalgh. *The Rise of New York Port, 1815–1860*. Newton Abbot, England: David & Charles, 1970.

Anawalt, P. R. *Indian Clothing before Cortes: Meso-American Costumes from the Codices,* 2nd ed. Civilization of the American Indian Series, no. 15. Norman: University of Oklahoma Press, 1990.

Arax, Mark, and Rick Wartzman. *The King of California: J. G. Boswell and the Making of a Secret American Empire*. New York: PublicAffairs, 2003.

Ashton, T. S. *The Industrial Revolution, 1760–1830*. Oxford: Oxford University Press, 1948.

Baines, Edward. *History of the Cotton Manufacture in Great Britain*. London: H. Fisher, R. Fisher, and P. Jackson, 1835. Reprinted by Frank Cass & Co., London, 1966.

Barber, Elizabeth Wayland. *Women's Work: The First 20,000 Years*. New York: W. W. Norton, 1994.

Barry, John M. *Rising Tide: The Great Mississippi Flood of 1927 and How It Changed America*. New York: Touchstone Books, Simon & Schuster, 1997.

Benfey, Christopher. *Degas in New Orleans*. New York: Alfred Knopf, 1997.

Bennett, Wendell C., and Junius B. Bird, *Andean Culture History*. New York: American Museum of Natural History Handbook, Series no.15, 1960.

Berg, Maxine. *The Age of Manufactures: 1700–1820*. London: Fontana Press, 1985.

Berlo, J. C., and R. E. Senuk. *Maya Textiles of Highland Guatemala*. St. Louis: University of Missouri Press, 1982.

Bernard, Virginia, et al., eds. *America Firsthand, Volume One*. New York: Brandywine Press, 1980.

Brandfon, Robert L. *Cotton Kingdom of the New South*. Cambridge, Mass.: Harvard University Press, 1967.

Brown, Judith M. *Gandhi: Prisoner of Hope*. New Haven: Yale University Press, 1989.

Browne, William. *Cultivating Congress*. Lawrence: University of Kansas Press, 1995.

Buck, Anne. *Dress in Eighteenth-Century England*. New York: Holmes & Meier Publishers, Inc., 1983.

Buck, Norman Sydney. *The Development of the Organisation of Anglo-American Trade, 1800–1850*. New Haven: Yale University Press, 1925.

Burnard, Joyce. *Chintz and Cotton: India's Textile Gift to the World*. Kenthurst, NSW, Australia: Kangaroo Press, 1994.

Burton, Anthony. *The Rise and Fall of King Cotton*. London: BBC and Andre Deutsch Ltd, 1984.

Carson, Rachel. *Silent Spring*. Boston: Houghton Mifflin, 1962.

Chandler, Alfred, Jr. *The Visible Hand: The Managerial Revolution in American Business*. Cambridge, Mass.: Harvard University Press, 1977.

Charles, Daniel. *Lords of the Harvest: Biotech, Big Money and the Future of Food*. Cambridge, Mass.: Perseus Publishing, 2002.

Clarke, Duncan. *The Art of African Textiles*. San Diego: Thunder Bay Press, 1997.

Cobb, James C. *The Most Southern Place on Earth: The Mississippi Delta and the Roots of Regional Identity*. New York: Oxford University Press, 1992.

Cohn, David L. *The Life and Times of King Cotton*. New York: Oxford University Press, 1956.

Columbus, Christopher. *Christopher Columbus: His Account in Spanish and English*. Santa Barbara: Bellerophon Books, 1992.

Cooper, Ilay, John Gillow, and Barry Dawson. *Arts and Crafts of India*. London: Thames and Hudson, 1996.

Crawford, M.D.C. *The Heritage of Cotton*. New York: G. P. Putnam's Sons, 1924.

Daniel, Pete. *Breaking the Land: The Transformation of Cotton, Tobacco and Rice Cultures Since 1880*. Urbana: University of Illinois Press, 1985.

———. *Lost Revolutions*. Chapel Hill: University of North Carolina Press, 2000.

Davies, Lucy, and Mo Fini. *Arts and Crafts of South America*. San Francisco: Chronicle Books, 1954.

Dickerson, William A. et al., eds. *Boll Weevil Eradication in the United States Through 1999*. Memphis: Cotton Foundation, 2000.

Dickson, Harris. *The Story of King Cotton*. New York: Funk & Wagnalls Company, 1937.

Dodge, Bertha. *Cotton: The Plant That Would Be King.* Austin: University of Texas Press, 1984.

Dublin, Thomas. *Transforming Women's Work: New England Lives in the Industrial Revolution.* Ithaca: Cornell University Press, 1994.

———. *Women at Work.* New York: Columbia University Press, 1979.

Dunwell, Steve. *The Run of the Mill.* Boston: David R. Godine, 1978.

Edwards, David Honeyboy. *The World Don't Owe Me Nothing.* Chicago: Chicago Review Press, 1997.

Eisler, Benita, ed. *The Lowell Offering.* Philadelphia: J. B. Lippincott, 1977.

Evans, David. *Big Road Blues: Tradition and Creativity in the Folk Blues.* New York: Da Capo Press, 1987.

Fagin, Dan, Marianne Lavelle, and Center for Public Integrity. *Toxic Deception.* Monroe, ME: Common Courage Press, 1999.

Ferris, William. *Blues From the Delta.* New York: Da Capo Press, 1978.

Fitton, R. S. *The Arkwrights, Spinners of Fortune.* Manchester, England: Manchester University Press, 1989.

Fitton, R. S., and A. K Wadsworth. *The Strutts and the Arkwrights, 1758–1830: A Study of the Early Factory System.* Manchester, England: Manchester University Press, 1958.

Fitzsimons, Gray, Sheila Kirschbaum, and Martha Mayo. "Cotton, Cloth and Conflict: The Meaning of Slavery in a Northern Textile City." Teacher Guide, 26 pages. Lowell, Mass.: Tsongas Industrial History Center, 1994.

Fogel, Robert William, and Stanley L. Engerman. *Time on the Cross.* New York: W. W. Norton & Company, 1989.

Foner, Eric. *A Short History of Reconstruction.* New York: Harper & Row, 1990.

Foner, Philip. *Business and Slavery: The New York Merchants and the Irrepressible Conflict.* New York: Russell and Russell, 1968.

Ford, Lacy J., Jr. *Origins of Southern Radicalism: The South Carolina Upcountry 1800–1860.* Oxford: Oxford University Press, 1988.

Frader, Judy. *Threads of Identity.* Seattle: University of Washington Press, 1995.

Freese, Barbara. *Coal: A Human History.* New York: Penguin Books, 2004.

Genovese, Eugene. *The Political Economy of Slavery.* New York: Vintage Books, Alfred A. Knopf, and Random House, 1967.

Gervers, Veronica, ed. *Studies in Textile History.* Toronto: Royal Ontario Museum, 1977.

Gianturco, Paola, and Toby Tuttle. *In Her Hands: Craftswomen Changing the World.* New York: Monacelli Press, 2000.

Gittinger, Mattobelle. *Master Dyers to the World.* Washington, D.C.: Textile Museum, 1982.

Gray, Lewis C. *History of Agriculture in the Southern United States to 1860: Volume Two.* Gloucester, Mass.: Peter Smith, 1958.

Green, Constance McL. *Eli Whitney and the Birth of American Technology.* Boston: Little, Brown and Company, 1956.

Gross, Laurence F. *The Course of Industrial Decline: The Boott Cotton Mills of Lowell, Massachusetts, 1835–1955*. Baltimore: Johns Hopkins University Press, 1993.

Guy, John. *Woven Cargoes: Indian Textiles in the East*. London: Thames and Hudson, 1998.

Hall, Jacquelyn Dowd, Robert Korstad, et al. *Like a Family: The Making of a Southern Cotton Mill World*. Chapel Hill: University of North Carolina Press, 1987.

Harris, Alice. *The Blue Jean*. New York: Powerhouse Books, 2002.

Hawken, Paul. *The Ecology of Commerce*. New York: HarperCollins, 1994.

Hills, Richard L. *Richard Arkwright and Cotton Spinning*. East Sussex, England: Wayland Publishers, 1973.

Hobhouse, Henry. *Seeds of Change: Five Plants That Transformed the World*. New York: Harper & Row, 1985.

Hurmence, Bernice, ed. *Before Freedom: When I Just Can Remember*. Winston-Salem: John F. Blair, 1989.

Irwin, John. *Origins of Chintz*. Toronto: Royal Ontario Museum, 1970.

Jacobs, Donald M., ed. *Courage and Conscience*. Bloomington: Indiana University Press, 1993.

Jacobson, Timothy Curtis, and George David Smith. *Cotton's Renaissance: A Study in Market Innovation*. Cambridge: Cambridge University Press, 2001.

Johnson, William. *Soul By Soul: Life Inside the Antebellum Slave Market*. Cambridge, Mass.: Harvard University Press, 1999.

Josephson, Hannah. *The Golden Threads: New England's Mill Girls and Magnates*. New York: Duell, Sloan and Pearce, 1949.

King, B.B., with David Ritz. *Blues All Around Me*. New York: Avon Books, 1996.

King, Mary Elizabeth. *Ancient Peruvian Textiles*. Museum catalogue brochure. Washington, D.C.: Textile Museum, 1965.

Kulik, Gary, Roger Parks, and Theodore Z. Penn. *The New England Mill Village, 1790–1860*. North Andover, Mass.: MIT Press, 1982.

Larcom, Lucy. *A New England Girlhood*. Boston: Northeastern Press, 1889 (first printing). Reprint, Boston: Houghton Mifflin, 1986.

Lichtenstein, Jack. *Field to Fabric: The Story of American Cotton Growers*. Lubbock: Texas Tech University Press, 1990.

Little, Frances. *Early American Textiles*. New York: Century Company, 1931.

Macauley, David. *Mill*. Boston: Houghton Mifflin, 1983.

Mandeville, Sir John. *Voiage and Travaile of Sir John Mandeville*. London: Haliwell, reprinted from 1725 edition, 1883.

Marcus, Robert D., and Davis Burner, eds. *American Firsthand, Volume Two: Readings from Reconstruction to the Present*. New York: Bedford/St. Martin's, 1997.

Marsh, Graham, Paul Trynka, and June Marsh. *Denim: From Cowboys to Catwalks*. London: Aurum Press, 2002.

McCoy, Drew R. *The Elusive Republic.* Chapel Hill: University of North Carolina Press, 1980.

Merchant, Carolyn, ed. *Major Problems in American Environmental Policy.* Lexington: D.C. Heath & Company, 1993.

Merrill, Gilbert, Alfred Macormac, and Herbert Mauerberger. *American Cotton Handbook.* New York: Textile Book Publishers, 1949.

Miller, Mary, and Karl Taube. *Ancient Mexico and the Maya.* London: Thames and Hudson, 1993.

Mirsky, Jeannette, and Allan Nevins. *The World of Eli Whitney.* New York: Macmillan Company, 1952.

Mitchell, Broadus. *The Rise of Cotton Mills in the South.* Gloucester, Mass.: Peter Smith, 1966.

Montgomery, Florence. *Printed Textiles: English and American Cotton Linens, 1700–1850.* New York: Viking Press, 1980.

Moran, William. *The Belles of New England.* New York: St. Martin's Press, 2002.

Morley, Stanley. *The Ancient Maya.* Stanford: Stanford University Press, 1956.

Murphey, Veronica, and Rosemary Crill. *Tie-Dyed Textiles of India: Traditions and Trade.* New York: Rizzoli International Publications, 1991.

Myers, Dorothy, and Sue Stolton, eds. *Organic Cotton: From Field to Final Product.* London: Intermediate Technology Publications, 1999.

O'Connor, Thomas H. *Civil War Boston.* Boston: Northeastern University Press, 1997.

———. *Lords of the Loom.* New York: Charles Scribner's Sons, 1968.

Olmsted, Frederick Law. *The Cotton Kingdom.* New York: Random House, 1984 (first printed in 1861).

Palmer, Robert. *Deep Blues.* New York: Penguin Books, 1982.

Possehl, Gregory L. *Indus Civilization: A Contemporary Perspective.* New Delhi: Vistaar Publications, 2002.

Powell, Lawrence N. *New Masters.* New Haven: Yale University Press, 1980.

Radhakrishna, Sabita. Woven Art, *The Hindu*, June 20, 1999, at www.hinduonnet.com/folio/fo9906/99060060.htm.

Ratnagar, Shereen. *Understanding Harappa Civilization in the Greater Indus Valley.* New Delhi: Tulika Books, 2001.

Robinson, Harriet. *Loom and Spindle.* Kailua, Hawaii: Press Pacifica, 1976.

Rosengarten, Theodore. *All God's Dangers: The Life of Nate Shaw.* New York: Alfred A. Knopf, 1974.

Rossner, Judith. *Emmeline.* New York: Simon & Schuster, 1980.

Rothstein, Natalie. *The Calico Campaign.* London: East End Papers, 1964.

Saffell, Cameron, et al. "When Did King Cotton Move His Throne?" *Agricultural History,* King Cotton Issue, vol. 74, no. 2. Spring 2000.

Sandburg, Gosta. *Indigo Textiles: Technique and History.* London: A & C Black, 1989.

Sansing, David, Sim C. Calhoun, and Carolyn Vance Smith. *Natchez: An Illustrated History.* Natchez: Plantation Publishing Company, 1992.

Scherer, James A. B. *Cotton as a World Power: A Study in the Economic Interpretation of History.* New York: Frederick A. Stokes, 1916.

Schevill, M. B., J. C. Berlo, and E. B. Dwyer, eds. *Textile Traditions of Mesoamerica and the Andes.* Austin: University of Texas Press, 1991.

Simpson, B. B., and Molly C. Ogorzaly. *Economic Botany: Plants in Our World.* New York: McGraw-Hill, 2001.

Smith, Frank E. *The Yazoo.* New York: Rinehart & Company, 1954.

Snyder, Robert E. *Cotton Crisis.* Chapel Hill: University of North Carolina Press, 1992.

Stegeman, John F., and Janet A. Stegeman. *Caty: The Biography of Catherine Littlefield Greene.* Athens: University of Georgia Press, 1977.

Taylor, John M. *William Henry Seward, Lincoln's Right-Hand Man.* New York: HarperCollins, 1991.

Tortora, Phyllis G. *Understanding Textiles.* New York: Macmillan, 1978.

Turner, Alice Curtis. *Story of Cotton.* Philadelphia: Penn Publishing Company, 1918.

Van Den Bosch, Robert. *The Pesticide Conspiracy.* New York: Doubleday & Co., 1978.

Walton, Perry. *The Story of Textiles.* New York: Tudor Publishing Company, 1925.

Waters, Andrew, ed. *Prayin' to Be Set Free.* Winston-Salem: John F. Blair, 2002.

Wayne, Michael. *The Reshaping of Plantation Society.* Baton Rouge: Louisiana State University Press, 1983.

Whitten, David O. *Eli Whitney's Cotton Gin, 1793–1993.* Washington, D.C.: Agricultural History Society, 1994.

Wilson, Charles Reagan, and William Ferris. *Encyclopedia of Southern Culture, Volumes 1–4.* New York: Anchor Books, Doubleday Dell Publishing Group, 1989.

Woodman, Harold D. *King Cotton and His Retainers: Financing and Marketing the Cotton Crop of the South, 1800–1925.* Columbia: University of South Carolina Press, 1990.

Woodward, C. Vann. *Origins of the New South, 1877–1913.* Baton Rouge: Louisiana State University Press, 1951.

Wright, Gavin. *Old South, New South: Revolutions in the Southern Economy Since the Civil War.* New York: Basic Books, Inc., 1986.

———. *The Political Economy of the Cotton South.* New York: W. W. Norton, 1978.

Wright, Richard. *Twelve Million Black Voices.* New York: Thunder's Mouth Press, 1988.

Acknowledgments

Tracing the journey of cotton over six thousand years of human history was a little like rowing a boat across an extremely large body of open water—terrific exercise, but with my back to the bow I wasn't always sure where I was headed. Fortunately, I had a crew of accomplished and knowledgeable friends of the project on board to lend their navigational skills, expertise, encouragement, and in many instances their inspiration as well. I am deeply indebted to these folks for the generosity of spirit they showed in helping to guide me past treacherous shoals and safely into port. One of the pleasures of writing is to be reminded time and again that people you've never met are more often than not willing to contribute at a moment's notice, without reserve.

Charlotte Sheedy, literary agent extraordinaire, suggested a social history when I had something else entirely, and entirely misdirected, in mind; my thanks to her also for leading me to Sterling Lord, dean of New York agents, a committed, invaluable champion of this book from its earliest stages, perspective and astute in ways that truly mattered, and a superb dinner companion on two coasts.

In Lowell, the first person I spoke with, Martha Mayo, director of the Center for Lowell History, offered her flinty, no-nonsense, and much-appreciated opinions on the importance of historical accuracy. There, too, Gray Fitzsimons led me to Alex Demas, and Gary Hudson: all took time to help make the city's past come alive. At the Victoria and Albert Museum in London, Rosemary Crill tutored me on Indian muslin, chintz, and English cotton's Asian connection. For help with the origins of New World cotton I turned to Rosemary Joyce and to Paola Gianturco, coauthor of a magnificent book on women's crafts, *In Her Hands*. Nancy Spies spent many hours compiling helpful information on the lore and social evolution of cotton, with great enthusiasm.

In the South, agricultural journalist Vicki Myers graciously steered my research over many months and made important, shrewd, and beneficial suggestions; Peter Coclanis, who until recently headed up the history department at the University of North Carolina, is a person you hope will befriend you in this sort of pursuit. Standing at the bar of a Memphis hotel, he slipped me enough savvy leads and information in one hour to set me off on several months' fruitful exploration. In New Orleans, Larry Powell, Bill Menary, and Susan Tucker made crucial connections for me between that city and the crop that helped to establish it as a port and trade center. Splendidly outrageous Bethany Bultman kindly directed me upriver to Carolyn Vance Smith and her husband, Marion, in Natchez, who provided information on its past and an opportunity to step back in time, arranging a visit with George Marshall at a plantation house largely unchanged since the Civil War. Also in Memphis, Cotton Nelson at the National Cotton Council (NCC) opened numerous doors, fielded endless questions for more than a year, and was never too busy to help; my special thanks to him. Frank Carter at NCC and Gary Raines at Cotton, Inc., were also ready with information when it mattered. Marilyn Sadler at *Memphis* magazine helped bring to life that city's role in the cotton saga. Bill Griffin at the Cotton Institute, who instructs industry representatives from around the world, generously filled in gaps in my knowledge, and reminded me that the South has a lock on colorful language. About the slim chance of America's textile mills rebounding anytime soon, he remarked, "Dead dogs don't bounce," a line I liked so much that, in the best tradition of cotton, I stole.

In the Delta, Ken Hood answered all my questions with forthright honesty, even those that might cast American cotton growers in an unflattering light, never running for cover: a Southern gentleman to the core. Chuck Earnest took time to fill me in on cotton farming at a family level. For background on the blues, radio producer Ben Manilla generously opened up his House of Blues radio archives to me. LaRhea Pepper, Lynda Grose, Will Allen, Kate Duesterberg, Chris Panos, and especially Chris Hancock all added greatly to my scant initial familiarity with the organic cotton movement, as did Sandra Marquardt at the Organic Trade Association.

In California's San Joaquin Valley, cotton baron Dick Shannon, ninety-three, spent an afternoon walking me through his operation, which includes an enclosed gin about the size of a football field. In Visalia, Clarence Ritchie and David Hyde also offered insights into the history and present status of California acala cotton farming and the crucial role of water usage, while Gary Artis graciously tutored me on the mind-paralyzing details of farm-bill subsidy formulas.

Research assistants Amos Kenigsberg, Jason McClure, and Jules Greenberg proved to be a trusted posse of fact-finders and information hunters; they also branched out to connect with individuals whose personal stories add human dimension to much of the material. In China, energetic and resourceful Kaiser Kuo took hold of the subject with both hands, embraced it like an old friend, and delivered intimate portraits that illuminate cotton's power to transform lives in that country. To him I owe a special debt of thanks.

In Washington, D.C., Terry Townsend, Carlos Valderrama, and Gerald Estur at the International Cotton Advisory Committee gave me a crash course on American cotton farm subsidies and their role in determining the destiny of farmers in developing nations. Jim Borneman at *Textile World* did much the same in limning the Byzantine relationship between domestic manufacturing and our government's shifting foreign policy.

A number of individuals closer to home provided friendship and excellent organic pinot noir at the end of a long writing day. When Michael Durphy and Thom Elkjer were not running my bones around the tennis court, they were lobbing tough questions and shrewd observations about this project that kept me on my toes. Tony Cook, a man of many talents, including wicked impersonations, read my manuscript with a seasoned financial journalist's understanding of economics, and, like a family doctor of old, was happy to make virtual house calls at any time. Ciji Ware provided creature comforts, infinite encouragement, and sage advice. An old friend, Edgar-award-winning mystery writer Julie Smith, and her husband, Lee Pryor, made available their New Orleans apartment to me and allowed me to experience the extreme pleasure of their *bon temps roulé* company.

At Viking, my editor Wendy Wolf—from a cotton family—and I hit it off from the first moment we spoke. Her wise suggestions untwisted snarls and straightened the final manuscript. Working with Wendy was a pleasure in all ways.

As for my immediate family, if they never hear the word *cotton* again for a few harvests, it probably won't be too soon. That said, they held up without complaint, added their sage advice, and offered the kind unstinting support every parent needs, even those like me who don't usually know it. To Josh, Omri, and Yael, who raised me while I was helping to raise them and are now out in the world, my thanks for keeping us all dancing to the music of life. To Jacques Dahan, a toast to the notion that nothing matters more than being open to change. And above all, for a quarter-century of laughter from the heart, unwavering loyalty, and for a thousand rollicking delights, to my wife, Bonnie. How lucky I am to be married to a soul mate, best friend, lover, and trusted midnight counselor all in the same person. Bonnie has brought all of those qualities, plus her impressive editorial gifts and profound touch of class, to this project and, not incidentally, to my life as well.

Index